Signals and Systems:
Continuous and Discrete

RODGER E. ZIEMER

Department of Electrical Engineering
University of Missouri-Rolla

WILLIAM H. TRANTER

Department of Electrical Engineering
University of Missouri-Rolla

D. RONALD FANNIN

Department of Electrical Engineering
University of Missouri-Rolla

SIGNALS AND SYSTEMS

Continuous and Discrete

Macmillan Publishing Co., Inc.

New York

Collier Macmillan Publishers

London

The following figures and table appear in
Principles of Communication by R. E. Ziemer
and W. H. Tranter, copyright 1976 by
Houghton Mifflin Co., Boston; they are
reprinted by permission: fig. 1-7, p. 11; fig.
1-8, p. 12; fig. 1-12, p. 19; fig. 3-5, p. 102;
fig. 3-6, p. 110; table 3-3, p. 125; fig. 3-14,
p. 134; fig. 3-18, p. 140; and fig. 3-19, p. 143

Macmillan Publishing Co., Inc.
866 Third Avenue, New York, New York 10022

Collier Macmillan Canada, Inc.

Library of Congress Cataloging in Publication Data

Ziemer, Rodger E.
  Signals and systems.
  Includes bibliographies and index.
  1. System analysis.   2. Discrete-time systems.
3. Digital filters (Mathematics) I. Tranter,
William H.   II. Fannin, D. Ronald,   III. Title.
QA402.Z53   003   82-6616
ISBN 0-02-431650-4   AACR2

Printing: 2 3 4 5 6 7 8   Year: 3 4 5 6 7 8 9 0

ISBN   0-02-431650-4

PREFACE

This textbook provides an introduction to the tools and mathematical techniques necessary for understanding and analyzing both continuous-time and discrete-time linear systems. We have attempted to give an insight into the application of these tools and techniques for solving practical engineering problems. Our philosophy has been to adopt a systems approach throughout the book for the introduction of continuous-time signal and system analysis, rather than use the framework of traditional circuit theory. We believe that the systems viewpoint provides a more natural approach to introducing this material in addition to broadening the horizons of the student. Furthermore, the topics of discrete-time signal and system analysis are most naturally introduced from a systems viewpoint, which lends overall consistency to the development. We have, of course, relied heavily on the students' circuit theory background to provide illustrative examples.

The organization of the book is straightforward. The first six chapters deal with continuous-time linear systems in both the time domain and the frequency domain. The principal tool developed for time-domain analysis is the convolution integral. Frequency-domain techniques include the Fourier and the Laplace transforms. An introduction to state variable techniques is also included. The remainder of the book deals with discrete-time systems including z-transform analysis techniques, digital filter analysis and synthesis, and the discrete Fourier transform and fast Fourier transform (FFT) algorithms.

This organization allows the book to be covered in two three-semester-hour courses, with the first course being devoted to continuous-time signals and systems and the second course being devoted to discrete-time signals and systems. Alternatively, the material can be used as a basis for three quarter-length courses. With this format, the first course would cover time- and frequency-domain analysis of continuous-time systems. The second course would cover state variables, sampling, and an introduction to the z-transform and discrete-time systems. The third course would deal with the analysis and synthesis of digital filters and provide an introduction to the discrete Fourier transform and its applications.

The assumed background of the student is mathematics through differential equations and the usual introductory circuit theory course or courses. Knowledge of the basic concepts of matrix algebra would be helpful but is not essential. Appendix A is included to bring together the pertinent matrix relations that are used in Chapters 5 and 6. We feel that in most electrical engineering curricula the material presented in this book is best taught at the junior level.

We begin the book by introducing the basic concepts of signal and system models and system classifications. The idea of spectral representations of periodic signals is first introduced in Chapter 1 because we feel that it is important for the student to think in terms of both the time and the frequency domains from the outset.

The convolution integral and its use in fixed, linear system analysis by means of the principle of superposition are treated in Chapter 2. The evaluation of the convolution integral is treated in detailed examples to provide reinforcement of the concepts. Calculation of the impulse response and its relation to the step and ramp responses of a system are discussed. Chapter 2 also contains optional sections* and examples regarding writing the governing equations for lumped, fixed, linear systems and the solution of linear, constant coefficient differential equations. These are intended as review and may be omitted without loss of continuity.

The Fourier series and Fourier transform are introduced in Chapter 3. We have emphasized the elementary approach of approximating a periodic function by means of a trigonometric series and obtaining the expansion coefficients by using the orthogonality of sines and cosines. We do this because this is the first time most of our students have been introduced to Fourier series. The alternative generalized orthogonal function approach is included as a nonrequired reading section at the end of this chapter for those who prefer it. The concept of the transfer function in terms of sinusoidal steady-state response of a system is discussed in relation to signal distortion. The Fourier transform is introduced next, with its applications to spectral analysis and systems analysis in the frequency domain. The concept of an ideal filter, as motivated by the idea of distortionless transmission, is also introduced at this point. The Gibbs phenomenon, window functions, and convergence properties of the Fourier coefficients are treated in optional closing sections.

The Laplace transform and its properties are introduced in Chapter 4. Again, we have tried to keep the treatment as simple as possible because this is assumed to be a first exposure to the material for a majority of students, although a

---

*Optional sections are denoted by an asterisk.

summary of complex variable theory is provided in Appendix B so that additional rigor may be used at the instructor's option. The derivation of Laplace transforms from elementary pairs is illustrated by example, as is the technique of inverse Laplace transformation using partial fraction expansion. Optional sections on the evaluation of inverse Laplace transforms by means of the complex inversion integral and an introduction to the two-sided Laplace transform are also provided.

The application of the Laplace transform to network analysis is treated in detail in Chapter 5. The technique of writing Laplace transformed network equations by inspection is covered and used to review the ideas of impedance and admittance matrices, which the student will have learned in earlier circuits courses for resistive networks. The transfer function is treated in detail, and the Routh test for determining stability is presented. The chapter closes with a treatment of Bode plots and block diagram algebra for fixed, linear systems.

In Chapter 6, the concepts of a state variable and the formulation of the state variable approach to system analysis are developed. The state equations are solved using both time-domain and Laplace transform techniques, and the important properties of the solution are examined. Finally, as an example, we show how the state-variable method can be applied to the analysis of circuits.

The final three chapters provide coverage of the topics of discrete-time signal and system analysis. Chapter 7 begins with a study of sampling and the representation of discrete-time systems. The sampling operation is covered in considerable detail. This is accomplished in the context of formulating a model for an analog-to-digital (A/D) converter so that the operation of quantizing can be given some physical basis. A brief analysis of the effect of quantizing sample values in the A/D conversion process is included as an introduction to quantizing errors. As a bonus, the student is given a basis upon which to select an appropriate wordlength of an A/D converter. The z-transform, difference equations, and discrete-time transfer functions are developed with sufficient rigor to allow for competent problem solving but without the complications of contour integration.

Chapter 8 allows the student to use his knowledge of discrete-time analysis techniques to solve an important class of interesting problems. The idea of system *synthesis,* as opposed to system *analysis,* is introduced. Discrete-time integration is covered in considerable detail for several reasons. First, the idea of integration will be a familiar one. Thus the student can appreciate the different information gained by a frequency-domain analysis as opposed to a time-domain analysis. In addition, the integrator is a basic building block for many analog systems. Finally, the relationship between trapezoidal integration and the bilinear z-transform is of sufficient importance to warrant a discussion of trapezoidal integration. The synthesis techniques for digital filters covered in this chapter are the standard ones. These are synthesis by time-domain invariance, the bilinear z-transform synthesis, and synthesis through Fourier series expansion. Through the application of these techniques, the student is able to gain confidence in the previously developed theory. Since several synthesis techniques depend on knowledge of analog filter prototypes, Appendix B, which discusses several different prototypes, is included.

The discrete Fourier transform (DFT) and its realization through the use of fast Fourier transform (FFT) algorithms is the subject of Chapter 9. Both decimation-in-time and decimation-in-frequency algorithms are discussed. Several examples are provided to give the student practice in performing the FFT operations. We believe that this approach best leads to a good understanding of the FFT algorithms and their function. Basic properties of the DFT are summarized and a comparison of the number of operations required for the FFT as compared to the DFT is made. Several applications of the DFT are summarized and the use of windows in suppressing leakage is discussed. This chapter closes with a discussion and illustration of FFT algorithms with arbitrary radixes and the chirp-z transform.

A complete solutions manual, which contains solutions to all problems, is available from the publisher as an aid to the instructor. Answers to selected problems are provided in Appendix E as an aid to the student.

The authors wish to express their thanks to the many people who have contributed, both knowingly and unknowingly, to the development of this textbook. First, thanks go to our long-suffering students, who have been forced to study from our notes, often while they were still in various stages of development. Their many comments and criticisms have been invaluable and are gratefully appreciated. Many of our colleagues in the Electrical Engineering Department at the University of Missouri–Rolla taught courses that used the book in note form and provided many suggestions for improvement. In this regard, we thank Professors Gordon E. Carlson, Kenneth H. Carpenter, Ralph S. Carson, David R. Cunningham, Thomas J. Herrick, Frank J. Kern, Earl F. Richards, John A. Stuller, and Thomas P. Van Doren. Professors Carlson, Carson, and Stuller critically reviewed much of the manuscript and provided valuable suggestions for improvement. Additionally, we would like to thank the reviewers at other institutions who provided valuable criticism, especially K. Ross Johnson, Michigan Technological University, Bruce Johansen, Ohio Northern University, Saleem Kassam, University of Pennsylvania, and Neal Gallagher, Purdue University. However, any shortcomings of the final result are solely the responsibility of the authors. We also are indebted to John Liebetreu, Graduate Teaching Fellow in the EE Department at UMR, who produced an extremely neat master for the solutions manual. A most sincere thanks goes to our secretaries whose great care and expert typing skills allowed us to generate the final manuscript with a minimum of headaches. The National Engineering Consortium is also due thanks since it was through their series of seminars that much of the material in Chapters 7 and 8 was originally taught.

Last, but not least, we thank our wives and families for putting up with a project whose end at times seemed nonexistent.

R.E.Z.
W.H.T.
D.R.F.

CONTENTS

APPENDICES

Signals and Systems:
Continuous and Discrete

CHAPTER 1

Signal and System Modeling Concepts

1-1

INTRODUCTION

This book deals with *systems* and the interaction of *signals* in such systems. A system, in its most general form, is defined as a combination and interconnection of several components to perform a desired task.† Such a task might be the control of liquid level in a tank or the transmission of a message from New York to Los Angeles. A liquid-level controller might make use of a human operator who closes a valve once the liquid reaches the desired level. An equally unsophisticated solution to the message delivery problem might make use of a horse and rider. Obviously, more complex solutions are possible (and probably better). Note, however, that our definition is sufficiently general to include them all.

We will be concerned primarily with *linear* systems. Such a restriction is reasonable because many systems of engineering interest are closely approximated by linear systems and very powerful techniques exist for analyzing them. We consider several methods for analyzing linear systems in this book. Although each of the methods to be considered is general, not all of them are equally

---

†The Institute of Electrical and Electronics Engineers Dictionary defines a system as "an integrated whole even though composed of diverse, interacting structures or subjunctions."

convenient for any particular case. Therefore, we will attempt to point out the usefulness of each.

A *signal* may be considered to be a function of time which represents a physical variable of interest associated with a system. In electrical systems, signals usually represent currents and voltages, whereas in mechanical systems, they might represent forces and velocities.† In the liquid-level control problem mentioned above, one of the signals of interest represents the level of liquid in the tank.

Just as there are several methods of systems analysis, there are several different ways of representing and analyzing signals. They are not all equally convenient in any particular situation. As we study methods of signal representation and analysis we will attempt to point out useful applications of the techniques.

So far, the discussion has been rather general. In order to be more specific and to fix more clearly the ideas we have introduced, we will expand on the liquid-level control and the message delivery problems already mentioned.

1-2

TWO EXAMPLES

Liquid-Level Controller

Shown in Figure 1-1 is a simple electromechanical control system for controlling the liquid level in a tank. Such a system is often used for sump pumps. Two floats are suspended on a wire that is attached to a spring-loaded switch. The weight of the floats is chosen such that the toggle action switch closes if the liquid level is above the higher float, and stays closed until the liquid level goes below the lower float, whereupon the switch opens.

To analyze this system, the first thing an engineer would do is to replace the actual system with a *model*. Such a model is an attempt at representing only the essential details of the actual system: mathematically, or pictorially, or both. To illustrate the concept of a model, we assume that a model is desired which describes the tank's liquid level for all time.

For example, letting the liquid level in the tank be x centimeters from the bottom float and the volumetric flow rate out of the tank be f liters per second when the pump is on, a set of equations that would be a possible model for the system of Figure 1-1 when the pump is on is

$$f(x) = \begin{cases} 10 \text{ liters/s}, & 0 \leq x \leq 50 \text{ cm} \\ 0, & x \leq 0 \end{cases}$$

where the top float is assumed to be at $x = 50$ cm and the lower one is assumed to be at $x = 0$ cm. Futhermore, if the tank is 100 cm in diameter, the volume

---

†More generally, a signal can be a function of more than one independent variable, such as the pressure on the surface of an airfoil, which is a function of three spatial variables *and* time.

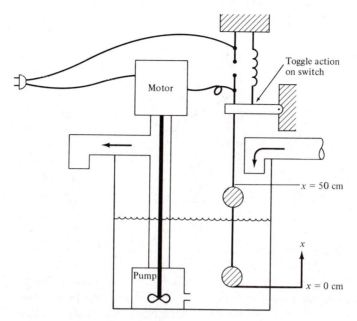

FIGURE 1-1.   Liquid-level controller.

at height x cm is

$$V = \pi(50)^2 x$$

$$= 2500\pi x \text{ cm}^3$$

$$= 2.5\pi x \text{ liters}$$

Thus the time required for the liquid level to decrease 1 cm is

$$\Gamma = (2.5\pi \text{ liters/cm})/(10 \text{ liters/s})$$

$$= 0.25\pi \text{ s/cm}$$

or the rate at which the tank will empty is

$$r = \frac{1}{\Gamma} = \frac{4}{\pi} \text{ cm/s}$$

The total time required for the liquid to reach the bottom float if it began at the top float is

$$T_f = (0.25\pi \text{ s/cm}) (50 \text{ cm})$$

$$= 12.5\pi \text{ s}$$

Clearly, when the pump is on, the change in x with time must be linear, since the volumetric flow rate f and the tank diameter are constants. A plot showing liquid level versus time, assuming that the tank started to empty at $t = 0$, is shown in Figure 1-2.

In order to proceed further with this problem, we might hypothesize a mechanism by which the tank fills so that we could mathematically describe the liquid

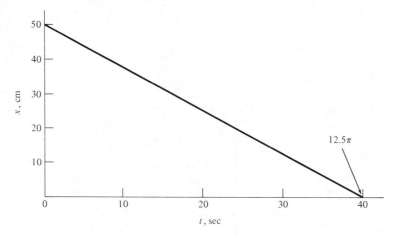

FIGURE 1-2. Liquid level versus time.

level for all time. We could also include information about the pump motor starting and stopping dynamics, or the dynamics of the opening and closing of the switch. We will consider some of these refinements in this and later chapters. All such refinements to the original model would be attempts at representing more closely the real-world situation. Regardless of how refined the model becomes, however, it is still an idealization of the actual set of circumstances. Since it is the model that is analyzed, it therefore follows that the graph showing liquid level versus time is also an idealization.

EXAMPLE 1-1

Suppose that water flows into the tank continuously at a constant rate of $f_{in} = 5$ liters/s. Modify the analysis above to obtain expressions for x that are valid for all time.

Solution: If $t = 0$ is taken as an instant in time at which the tank is full, then $\Gamma = (2.5\pi$ liters/cm$)/[(10 - 5)$ liters/s$)] = 0.5\pi$ s/cm is the time required for the tank to empty 1 cm. Thus $T_f = (0.5\pi$ s/cm$) (50$ cm$) = 25\pi$ s for the tank to empty. The tanks fills at the same rate, so that x is a triangular function of time that repeats every $2(25\pi) = 50\pi$ s. You should sketch the waveform for x and dimension it fully as in Figure 1-2. ∎

To reemphasize, the starting point of any systems analysis problem is a *model* which, no matter how refined, is *always an idealization of a real-world (physical) system*. Hence the result of any systems analysis is an idealization of the true state of affairs. Nevertheless, if the model is sufficiently accurate, the results obtained will portray the operation of the actual system sufficiently accurately to be of use.

Communications Link

Figure 1-3 is a pictoral representation of a two-way communications link as might exist, for example, between New York and Los Angeles. It might consist

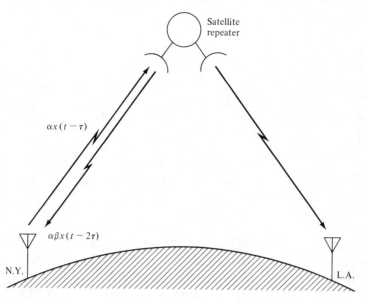

FIGURE 1-3. Satellite communications link.

entirely of earth-based links such as wire lines and microwave links. Alternatively, a relay satellite could be employed. Regardless of the particular mechanization used, the systems analyst must decide on the most important aspects of the physical system for the application and attempt to represent them by a model.

Let us suppose that the link employs a synchronous-orbit satellite repeater, stationed 35,400 km above the earth's equator. Thus the satellite is stationary with respect to the earth's surface. Let us also suppose that the analyst is concerned primarily about echoes in the system due to reflections of the electromagnetic-wave carriers at the satellite and at the receiving ground station. Assume that the transmitter and receiver are equidistant at a slant distance $d = 40,000$ km from the satellite, and that the delay (up or down) is

$$\tau = \frac{d}{c} = \frac{(40,000 \text{ km})(1000 \text{ m/km})}{3 \times 10^8 \text{ m/s}}$$

$$= 0.13 \text{ s}$$

where $c = 3 \times 10^8$ m/s is the velocity of light. The slant distance d accounts for the satellite being located above the earth's equator midway between the ground stations, which are assumed to be at a latitude of 40° north. The signal received at the satellite is

$$s_{\text{sat}}(t) = \alpha x(t - \tau), \qquad \tau \leq t \leq T + \tau$$

where $x(t)$ is the transmitted signal and $[0, T]$ is the interval of time over which the transmission takes place. The parameter α is the attenuation, which accounts for various system characteristics as well as the spreading of the electromagnetic wave as it propagates from the transmitter antenna. A portion of $s_{\text{sat}}(t)$ is reflected and returned to the New York ground station terminal to be received by the receiver at that end; we represent it as

$$s_{\text{refl}}(t) = \alpha\beta x(t - 2\tau), \qquad 2\tau \leq t \leq T + 2\tau$$

where β is another attenuation factor. The remainder of $s_{\text{sat}}(t)$ is relayed to the Los Angeles ground station after processing by the satellite repeater. A portion of this signal may be reflected, but we will assume its effect to be negligible compared to that of $\alpha\beta x(t - 2\tau)$ when received at the New York station. Thus a speaker at New York will hear

$$s(t) \simeq x(t) + \alpha\beta x(t - 2\tau), \qquad 0 \leq t \leq T + 2\tau$$

That is, he will hear his undelayed speech as well as an attenuated version of his speech delayed by approximately

$$\Delta T = 2\tau = 0.26 \text{ s}$$

Psychologically, the effect of this can be very disconcerting to a speaker if the delayed signal is sufficiently strong; it is virtually impossible for him to avoid stuttering. Thus a systems analyst would proceed by relating α and β to more fundamental systems parameters and designing for an acceptably small value of $\alpha\beta$.

1-3

SIGNAL MODELS

Examples of Deterministic Signals

In this book we are concerned with a broad class of signals referred to as *deterministic*. Deterministic signals can be modeled as completely specified functions of time. For example, the signal

$$x(t) = \frac{At^2}{B + t^2}, \qquad -\infty < t < \infty \tag{1-1}$$

where A and B are constants, shown in Figure 1-4a, is a deterministic signal. An example of a deterministic signal that is not a continuous function of time is the unit pulse, denoted as $\Pi(t)$ and defined as

$$\Pi(t) = \begin{cases} 1, & |t| \leq \frac{1}{2} \\ 0, & \text{otherwise} \end{cases} \tag{1-2}$$

It is shown in Figure 1-4b.

A second class of signals, which will not be discussed in this book, are random signals. They are signals taking on random values at any given time instant and must be modeled probabilistically.

Continuous-Time Versus Discrete-Time Signals

The signals illustrated in Figure 1-4 are examples of *continuous-time signals*. It is important to note that "continuous time" does not imply that a signal is a

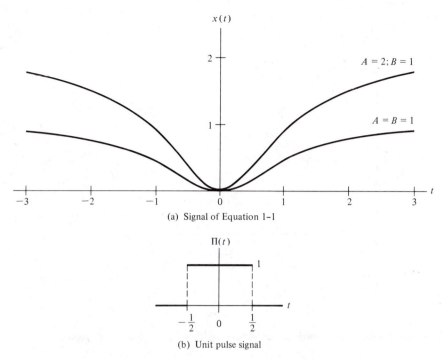

(a) Signal of Equation 1–1

(b) Unit pulse signal

FIGURE 1-4. Graphical representation of two deterministic continuous-time signals.

mathematically continuous function, but rather that it is a function of a *continuous-time variable*.

In some systems the signals are represented only at discrete values of the independent variable (i.e., time). Between these discrete-time instants the value of the signal may be zero, undefined, or of no interest. An example of such a *discrete-time* or *sample-data signal* is shown in Figure 1-5a. Often, the intervals between signal values are the same, but they need not be. We deal with discrete-time signals in the last half of this book.

A distinction between discrete time and quantized signals is necessary. A *quantized signal* is one whose values may assume only a countable† number of values, or levels, but the changes from level to level may occur at any time. Figure 1-5b shows an example of a quantized signal. A real-world situation that can be modeled as a quantized signal is the opening and closing of a switch.

Periodic and Aperiodic Signals

A signal $x(t)$ is *periodic* if and only if

$$x(t + T_0) = x(t), \qquad -\infty < t < \infty \qquad (1\text{-}3)$$

where the constant T_0 is the period. The smallest value of T_0 such that (1-3) is

---

†A *countable* set is a set of objects whose members can be put into one-to-one correspondence with the positive integers. For example, the sets of all integers and of all rational numbers are countable, but the set of all real numbers is not.

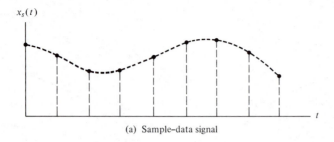

(a) Sample–data signal

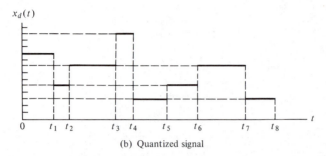

(b) Quantized signal

FIGURE 1-5. Sample-data, or discrete-time, and quantized signals.

satisfied is referred to as the *fundamental period,* and is hereafter simply referred to as the *period.* Any deterministic signal not satisfying (1-3) is called *aperiodic.*

A familiar example of a periodic signal is a sinusoidal signal which may be expressed as

$$x(t) = A \sin (2\pi f_0 t + \theta), \qquad -\infty < t < \infty$$

where A, f_0, and θ are constants referred to as the amplitude, frequency in hertz†, and phase, respectively. For simplicity, $\omega_0 = 2\pi f_0$ is sometimes used, where ω_0 is the frequency in rad/s. The period of this signal is $T_0 = 2\pi/\omega_0 = 1/f_0$, which can be verified by direct substitution and use of trigonometric identities as follows. We wish to verify that

$$x\left(t + \frac{2\pi}{\omega_0}\right) = x(t), \qquad \text{all } t$$

But

$$x\left(t + \frac{2\pi}{\omega_0}\right) = A \sin \left[\omega_0\left(t + \frac{2\pi}{\omega_0}\right) + \theta \right]$$

$$= A \sin (\omega_0 t + 2\pi + \theta)$$

$$= A [\sin (\omega_0 t + \theta) \cos 2\pi + \cos (\omega_0 t + \theta) \sin 2\pi]$$

Since $\cos 2\pi = 1$ and $\sin 2\pi = 0$, the required relationship for periodicity is verified. Note that $T_0 = 2\pi/\omega_0$ is the smallest value of T_0 such that $x(t) =$

†One hertz is one cycle per second.

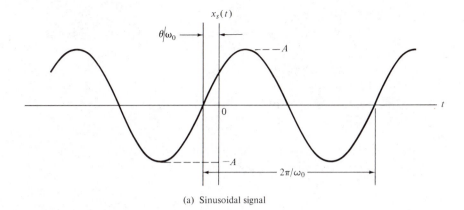

(a) Sinusoidal signal

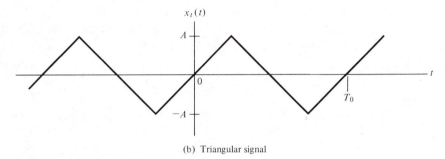

(b) Triangular signal

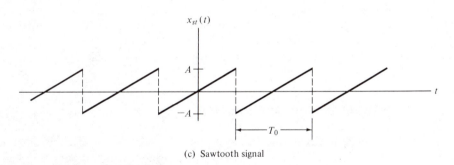

(c) Sawtooth signal

FIGURE 1-6. Various types of periodic signals.

$x(t + T_0)$. Several types of periodic signals are illustrated in Figure 1-6 together with the sinusoidal signal.

==The sum of two or more sinusoids may or may not be periodic, depending on the relationships between their respective periods or frequencies. If the ratio of their periods can be expressed as a rational number, or their frequencies are commensurable,† their sum will be a periodic signal.==

†Two frequencies, f_1 and f_2, are commensurable if they have a common measure. That is, there is a number f_0 contained in each an integral number of times. Thus, if f_0 is the smallest such number,

$$f_1 = n_1 f_0 \quad \text{and} \quad f_2 = n_2 f_0$$

where n_1 and n_2 are integers; f_0 is called the *fundamental frequency*. The periods, T_1 and T_2, corresponding to f_1 and f_2, are therefore related by $T_1/T_2 = n_2/n_1$.

EXAMPLE 1-2

Which of the following signals are periodic? Justify your answers.

(a) $x_1(t) = \sin 10\pi t$
(b) $x_2(t) = \sin 20\pi t$
(c) $x_3(t) = \sin 31t$
(d) $x_4(t) = x_1(t) + x_2(t)$
(e) $x_5(t) = x_1(t) + x_3(t)$

Solution: Only (e) is not periodic. To show that waveforms $x_1(t)$, $x_2(t)$, and $x_3(t)$ are periodic, we note that

$$\sin (\omega_0 t + \theta) = \sin \omega_0 \left(t + \frac{\theta}{\omega_0} \right) = \sin \omega_0 t$$

if θ is an integer multiple of 2π. Taking the smallest such value of θ, we see that the (fundamental) period of $\sin (\omega_0 t + \theta)$ is

$$T_0 = \frac{2\pi}{\omega_0}$$

Thus the periods of $x_1(t)$, $x_2(t)$, and $x_3(t)$ are $\frac{1}{5}$, $\frac{1}{10}$, and $2\pi/31$ s, respectively.

The sum of two sinusoids is periodic if the ratio of their respective periods can be expressed as a rational number. Thus $x_4(t)$ is periodic. Indeed, we see that the second term goes through two cycles for each cycle of the first term. The period of $x_4(t)$ is therefore $\frac{1}{5}$ s.

After a little thought, it is apparent that there is no T_0 for which $x_5(t) = x_5(t + T_0)$, because the ratio of the periods of the two separate terms is not a rational number (i.e., their frequencies are incommensurable). ■

Phasor Signals and Spectra

Although physical systems always interact with real signals, it is often mathematically convenient to represent real signals in terms of complex quantities. An example of this is the use of phasors. The student should recall from circuits courses that the phasor quantity

$$\overline{X} = Ae^{j\theta} = A\angle\theta \tag{1-4}$$

is a shorthand notation for the real, sinusoidal signal

$$x(t) = \text{Re} (\overline{X}e^{j\omega_0 t}) = A \cos (\omega_0 t + \theta), \qquad -\infty < t < \infty \tag{1-5}$$

We refer to the complex signal

$$\tilde{x}(t) = Ae^{j(\omega_0 t + \theta)}, \qquad -\infty < t < \infty \tag{1-6}$$

as the rotating phasor signal. It is characterized by the three parameters A, the amplitude; θ, the phase; and $\omega_0 > 0$, the radian frequency. Using Euler's theorem, which is $\exp (ju) = \cos u + j \sin u$, we may readily show that $\tilde{x}(t)$ is periodic with period $T_0 = 2\pi/\omega_0$.

In addition to taking its real part, as in (1-5), to relate it to the real, sinusoidal signal $A \cos (\omega_0 t + \theta)$, we may relate $\tilde{x}(t)$ to its sinusoidal counterpart by writing

$$x(t) = \tfrac{1}{2}\tilde{x}(t) + \tfrac{1}{2}\tilde{x}^*(t)$$

$$= \frac{A}{2} e^{j(\omega_0 t + \theta)} + \frac{A}{2} e^{-j(\omega_0 + \theta)}, \qquad -\infty < t < \infty \qquad (1\text{-}7)$$

which is a representation in terms of conjugate, oppositely rotating phasors. These two procedures, represented mathematically by (1-5) and (1-7), are illustrated schematically in Figure 1-7. These expressions for $x(t) = A \cos (\omega_0 t + \theta)$ are referred to as *time-domain* representations.

Note that (1-7) can be thought of as the sum of positive-frequency and negative-frequency rotating phasors. This mathematical abstraction results from the fact that it is necessary to add complex-conjugate quantities to obtain a real quantity. It is emphasized that negative frequencies do not physically exist, but are merely convenient mathematical abstractions that result in nicely symmetric equations and figures such as (1-7) and Figure 1-7b.

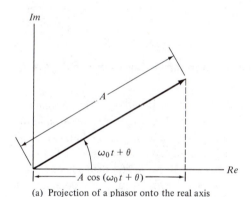

(a) Projection of a phasor onto the real axis

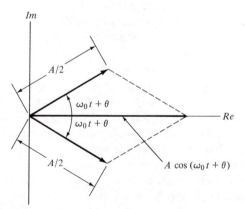

(b) Addition of complex conjugate phasors

FIGURE 1-7.   Two ways of relating a phasor signal to a sinusoidal signal.

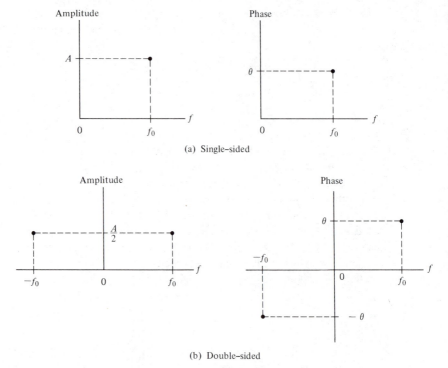

(a) Single–sided

(b) Double–sided

FIGURE 1-8. Amplitude and phase spectra for the signal A cos ($\omega_0 t$ + θ).

An alternative representation for $x(t)$ is provided in the *frequency domain*. Since $\tilde{x}(t) = A \exp [j(\omega_0 t + \theta)]$ is completely specified by A and θ for a given value of f_0, this alternative frequency-domain representation can take the form of two plots, one showing the amplitude A as a function of frequency f, and the other showing θ as a function of f. Because $\tilde{x}(t)$ depends only on the single frequency f_0, the resulting plots each consist of a single point or ''line'' at f = f_0, as illustrated in Figure 1-8a. Had we considered a signal that is the sum of two phasors, each plot would have had two points or lines present. The plot of amplitude versus frequency is referred to as the *single-sided amplitude spectrum*. The modifier ''single-sided'' is used because these spectral plots have points or lines only for positive frequencies.

If a spectral plot corresponding to (1-7) is made, spectral lines will be present at f = f_0 *and* at f = $-f_0$, since $x(t)$ is obtained as the sum of oppositely rotating phasors. Spectral plots for the single sinusoidal signal under consideration here are shown in Figure 1-8b. Such plots are referred to as *double-sided amplitude and phase spectra*. Had a sum of sinusoidal components been present, the spectral plots would have consisted of multiple lines.

It is important to emphasize two points in regard to double-sided spectra. First, the lines at negative frequencies are present precisely because it is necessary to add complex-conjugate phasors to obtain the real sinusoidal signal A cos ($\omega_0 t$ + θ). Second, we note that the amplitude spectrum has *even* symmetry about the origin, and the phase spectrum has *odd* symmetry. This symmetry is again a consequence of $x(t)$ being a real signal.

Figure 1-8a and b serve as equivalent spectral representations for the signal $A \cos (\omega_0 t + \theta)$ if we are careful to identify the spectra plotted as single-sided or double-sided, as the case may be. We will find later that the Fourier series and Fourier transform lead to spectral representations for more complex signals.

EXAMPLE 1-3

We wish to sketch the single-sided and double-sided amplitude and phase spectra of the signal

$$x(t) = 4 \sin \left(20\pi t - \frac{\pi}{6} \right), \qquad -\infty < t < \infty$$

To sketch the single-sided spectra, we write $x(t)$ as the real part of a rotating phasor and plot the amplitude and phase of this phasor as a function of frequency for $t = 0$. Noting that $\cos (u - \pi/2) = \sin u$, we find that

$$x(t) = 4 \cos \left(20\pi t - \frac{\pi}{6} - \frac{\pi}{2} \right)$$

$$= 4 \cos \left(20\pi t - \frac{2\pi}{3} \right)$$

$$= \text{Re} \left\{ 4 \exp \left[j \left(20\pi t - \frac{2\pi}{3} \right) \right] \right\}$$

which results in the amplitude and phase spectral plots shown in Figure 1-9a. To plot the double-sided amplitude and phase spectra, we write $x(t)$ as the sum of complex-conjugate rotating phasors. Recalling that $2 \cos u = \exp (ju) + \exp (-ju)$, we obtain

$$x(t) = 2 \exp \left[j \left(20\pi t - \frac{2\pi}{3} \right) \right] + 2 \exp \left[-j \left(20\pi t - \frac{2\pi}{3} \right) \right]$$

from which the double-sided amplitude and phase spectral plots of Figure 1-9b result.

Singularity Functions

An important subclass of aperiodic signals is the *singularity functions,* which we now discuss. We begin by introducing the unit step and unit ramp functions and using them to represent more complicated signals, such as in the example liquid level controller.

Consider the *unit step function,* $u_{-1}(t)$, defined as

$$u(t) = u_{-1}(t) = \begin{cases} 0, & t < 0 \\ 1, & t > 0 \end{cases} \tag{1-8}$$

The value of $u(t)$ at $t = 0$ will not be specified at this time except to say that it is finite. Other singularity functions are defined in terms of $u_{-1}(t)$ by the

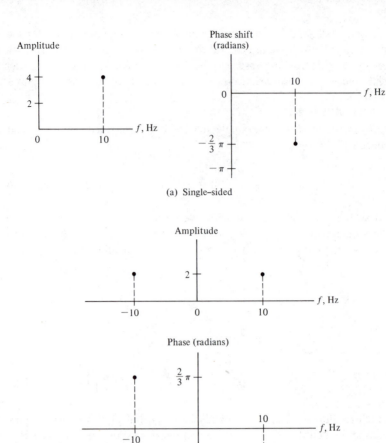

(a) Single-sided

(b) Double-sided

FIGURE 1-9. Amplitude and phase spectra for Example 1-3.

relations

$$u_{i-1}(t) = \int_{-\infty}^{t} u_i(\lambda) \, d\lambda, \qquad i = \ldots, -2, -1, 0, 1, 2, \ldots \quad (1\text{-}9a)$$

or

$$u_{i+1}(t) = \frac{du_i(t)}{dt} \qquad (1\text{-}9b)$$

Figure 1-10a shows the unit step. We may graphically integrate it to obtain $u_{-2}(t)$, shown in Figure 1-10b. This singularity function is referred to as the *unit ramp function*, $r(t)$, and can be expressed algebraically as†

†The subscripts are needed on (1-9a) and (1-9b) to identify different members of the class of singularity functions. Since the step and ramp are very commonly used members, they are often denoted by the special symbols $u(t)$ and $r(t)$, respectively.

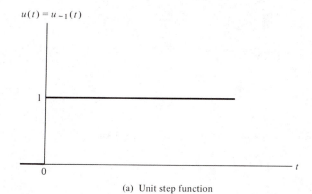

(a) Unit step function

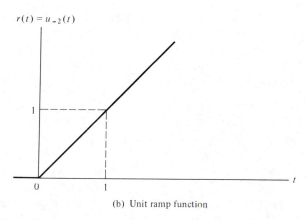

(b) Unit ramp function

FIGURE 1-10.   Examples of two singularity functions.

$$r(t) = u_{-2}(t) = \begin{cases} t, & t \geq 0 \\ 0, & t < 0 \end{cases} \tag{1-10}$$

We may shift any signal on the time axis simply by replacing t by $t - t_0$. If $t_0 > 0$, the signal is shifted to the right; if $t_0 < 0$, it is shifted to the left. For example, replacing t by $t - \frac{1}{2}$ in the definition of the unit step, (1-8), we obtain

$$u(t - \tfrac{1}{2}) = \begin{cases} 0, & t - \frac{1}{2} < 0 \\ 1, & t - \frac{1}{2} > 0 \end{cases} \tag{1-11}$$

or

$$u(t - \tfrac{1}{2}) = \begin{cases} 0, & t < \frac{1}{2} \\ 1, & t > \frac{1}{2} \end{cases}$$

Similarly, replacing t by $t + \frac{1}{2}$ results in a unit step that "turns on" at $t = -\frac{1}{2}$. Using this technique, we can represent other signals in terms of singularity functions. For example, the *unit pulse function,* defined by (1-2), can be represented as

$$\Pi(t) = u(t + \tfrac{1}{2}) - u(t - \tfrac{1}{2}) \tag{1-12}$$

For the definition of $\Pi(t)$ given by (1-2), it would be consistent to define $u(0) = 1$.

We note, also, that replacing t by $-t$ turns a function around.† Doing this in (1-10), we obtain

$$r(-t) = \begin{cases} -t, & t \le 0 \\ 0, & t > 0 \end{cases} \tag{1-13}$$

which is a ramp that starts at $t = 0$ and *increases* linearly as t *decreases*. This observation allows us to represent $x(t)$ in Fig. 1-2 as

$$x(t) = \frac{50}{12.5\pi} \{r[-(t - 12.5\pi)] - r(-t)\} \tag{1-14}$$

The factor $(50/12.5\pi)$ is required to give the ramp the proper slope, while the second ramp, $r(-t)$, is required to flatten out the first ramp for $t < 0$.

EXAMPLE 1-4
Write the liquid level, $x(t)$, of Figure 1-2 in terms of positive-time ramps.

Solution: We start with a constant level of $x = 50$ at $t = 0$ and decrease linearly with slope $= -(50/12.5\pi)$ until $t = 12.5\pi$, at which time $x = 0$, and thereafter. Thus

$$x(t) = 50 - \frac{50}{12.5\pi} [r(t) - r(t - 12.5\pi)] \tag{1-15}$$

Equations (1-14) and (1-15) illustrate that there may be several representations of a signal in terms of singularity functions. ∎

EXAMPLE 1-5
Express the signals shown in Figure 1-11 in terms of singularity functions.

Solution: One possible representation for $x_a(t)$ is

$$x_a(t) = u(t) - r(t - 1) + 2r(t - 2) - r(t - 3) + u(t - 4) - 2u(t - 5)$$

The step function $u(t)$ is unity for $t > 0$. At $t = 1$, the ramp function $-r(t - 1)$ goes downward with slope -1. The sum of ramps $-r(t - 1) + 2r(t - 2)$ go upward with slope $+1$ starting at $t = 2$. The ramp $-r(t - 3)$ then flattens out this upward ramp starting at $t = 3$, and the step $u(t - 4)$ increases the plateau of unity height to a value of 2. The final step, $2u(t - 5)$, "shuts off" the preceding sum of singularity functions so that $x(t) = 0$ for $t > 5$.

For $x_b(t)$ we use a product representation. A possible representation is

$$x_b(t) = 2 u(t)u(2 - t) + u(t - 3)u(5 - t)$$

The product of unit step functions $2u(t)u(2 - t)$ forms the first pulse and the

---

†The negation of the independent variable is often referred to as *folding*, since the resulting reversal of the signal in time can be viewed as folding a paper on which the signal has been plotted along its ordinate and viewing it from behind the paper.

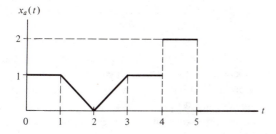

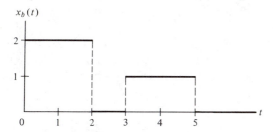

FIGURE 1-11. Signal to be expressed in terms of singularity functions.

product $u(t - 3)u(5 - t)$ forms the second pulse. Note that $u(2 - t)$ and $u(5 - t)$ are step functions that are unity from $t = -\infty$ to $t = 2$ and $t = 5$, respectively, and zero thereafter.

The student should attempt to find other possible representations for $x_a(t)$ and $x_b(t)$. ∎

We turn next to the *unit impulse function* or *delta function*,† $\delta(t)$, which has the properties

$$\delta(t) = 0, \qquad t \neq 0 \qquad\qquad (1\text{-}16a)$$

and

$$\int_{-\infty}^{\infty} \delta(t)\ dt = 1 \qquad\qquad (1\text{-}16b)$$

Equation (1-16b) states that the area of a unit impulse function is unity and (1-16a) indicates that this unity area is obtained in an infinitesimal interval on the t-axis.

The motivation for defining a function with these properties stems from the need to represent phenomena that happen in time intervals short compared with the resolution capability of any measuring apparatus used, but which produce an almost instantaneous change in a measured quantity. Examples are the nearly instantaneous increase in voltage across a capacitor placed across the terminals of a battery, or the nearly instantaneous change in velocity of a billiard ball

---

†The two names stem from the use of this function in engineering and physics; the term "delta function" is associated with Dirac, a physicist, who introduced the notation $\delta(t)$ into quantum mechanics. We will refer to $\delta(t)$ simply as a unit impulse function.

struck by a rapidly moving cue ball. In the first example, the voltage across the capacitor is the result of an extremely large current flowing for a very short time interval, whereas the billiard ball's velocity changes almost instantaneously due to the transfer of momentum from the cue ball at the moment of impact. The current flowing into the capacitor and the force transmitted by the cue ball are examples of quantities that are usefully modeled by impulse functions. In neither case are we able to measure, nor are we particularly interested in, what happens at exactly the moment the action takes place. Rather we are able to observe only the conditions beforehand and afterward.

Other examples of physical quantities that are modeled by mathematical entities having infinitely small dimensions and infinite "weight" are point masses and point charges.

Since it is impossible for any conventional function to have the properties (1-16), we attempt to picture the unit impulse function by considering the limit of a conventional function as some parameter approaches zero.† For example, consider the signal

$$\delta_\epsilon(t) = \frac{1}{2\epsilon} \Pi \left(\frac{t}{2\epsilon} \right) = \begin{cases} \frac{1}{2\epsilon}, & |t| < \epsilon \\ 0, & |t| > \epsilon \end{cases}$$

which is shown in Figure 1-12. We see that, no matter how small ϵ is, $\delta_\epsilon(t)$ always has unit area. Furthermore, this area is obtained as the integration is carried out from $t = -\epsilon$ to $t = \epsilon$. As $\epsilon \to 0$ this area is obtained in an infinitesimally small width, thus implying that the height of $\delta_\epsilon(t) \to \infty$ such that *unity area lies between the function and the t-axis*. This limit is shown as a heavy arrow in the figure.

Many other signals exist which provide heuristic visualizations for $\delta(t)$ in the limit as some parameter approaches zero. Other examples are the signals

$$\delta_\epsilon(t) = \epsilon \left(\frac{1}{\pi t} \sin \frac{\pi t}{\epsilon} \right)^2$$

which is sketched in Figure 1-13a, and

$$\delta_\epsilon(t) = \begin{cases} \dfrac{1 - |t|/\epsilon}{\epsilon}, & |t| < \epsilon \\ 0, & \text{otherwise} \end{cases}$$

shown in Figure 1-13b. Although the three families of functions illustrated in Figures 1-12 and 1-13, are symmetric about $t = 0$, an equally good example of an approximation for a unit impulse function in terms of the properties (1-16) is $\epsilon^{-1} \exp(-t/\epsilon)u(t)$, which is asymmetric about $t = 0$. These examples illustrate that it is not the actual behavior of $\delta(t)$ at $t = 0$ that is important but rather the integral of $\delta(t)$.

---

†The present discussion is aimed at providing an intuitive understanding of the unit impulse function. We will give a more formal mathematical development shortly.

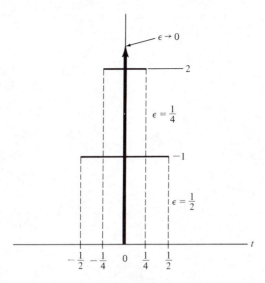

FIGURE 1-12. Square pulse approximations for the unit impulse function.

Note that by integrating the square pulse approximation of Figure 1-12 between $-\infty$ and t, we obtain

$$\int_{-\infty}^{t} \delta_\epsilon(\lambda)\, d\lambda = \begin{cases} 0, & t < -\epsilon \\ 1, & t > \epsilon \end{cases}$$

In the limit as $\epsilon \to 0$, the right side becomes a unit step function. It will be shown later that $u_0(t)$ of (1-9) and $\delta(t)$ have the same properties as expressed by (1-16).

EXERCISE

Verify that all the families of approximating functions $\delta_\epsilon(t)$ discussed above satisfy (1-16) in the limit as $\epsilon \to 0$.

The considerations above illustrate that the unit impulse function is not really a function in the normal sense. Indeed, it was not until the 1950s that Laurent Schwartz's pioneering work with the theory of distributions provided a firm mathematical basis for the unit impulse function. A more formal mathematical definition of the unit impulse function proceeds by defining it in terms of the *functional*

$$\int_{-\infty}^{\infty} x(t)\delta(t)\, dt = x(0) \tag{1-17}$$

where $x(t)$ is continuous at $t = 0$. That is, $\delta(t)$ is a functional as defined by (1-17), which assigns the value $x(0)$ to any function $x(t)$ continuous at the origin.†

---

†For a further discussion, see A. Papoulis, *Signal Analysis* (New York: McGraw-Hill, 1977), or R. N. Bracewell, *The Fourier Transform and Its Applications*, 2nd ed. (New York: McGraw-Hill, 1978).

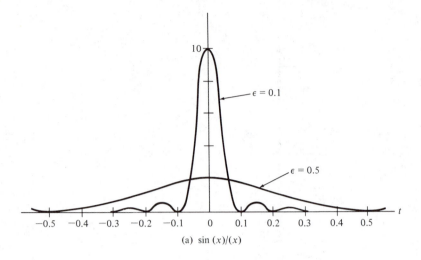

(a) sin (x)/(x)

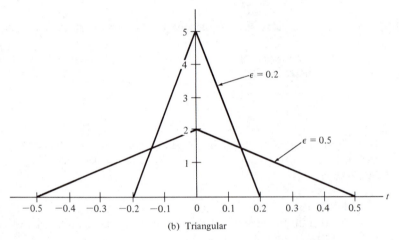

(b) Triangular

FIGURE 1-13. Two approximations for the unit impulse.

Several useful properties of $\delta(t)$ can be proved by insisting that (1-17) obey the formal properties of integrals. For example, by considering $\delta(at)$, where a is a constant, we may show that

$$\delta(at) = \frac{1}{|a|}\delta(t) \qquad (1\text{-}18)$$

by changing variables of integration and considering the cases $a > 0$ and $a < 0$ separately. If we let $a = -1$ in (1-18), we have

$$\delta(-t) = \delta(t) \qquad (1\text{-}19)$$

which agrees with the definition of an even function. Even though $\delta(t)$ is not a function in the conventional sense, we may make use of (1-19) in manipulations involving unit impulse functions as long as we remember that it was a consequence of the defining relation (1-17) and the formal properties of integrals.

A second useful property of the delta function is the *sifting property*, which is

$$\int_{-\infty}^{\infty} x(t)\delta(t - t_0) \, dt = x(t_0) \qquad (1\text{-}20)$$

where $x(t)$ is assumed to be continuous at $t = t_0$. This property can be proved from (1-17) by letting $\lambda = t - t_0$ in the integral (1-20) to obtain

$$\int_{-\infty}^{\infty} x(\lambda + t_0)\delta(\lambda) \, d\lambda = x(t_0)$$

and defining $y(\lambda) = x(\lambda + t_0)$, which is continuous at $\lambda = 0$ because $x(t)$ is continuous at $t = t_0$.

An alternative form of (1-20) which will prove useful in Chapters 2 and 3 is

$$\int_{-\infty}^{\infty} x(\lambda)\delta(t - \lambda) \, d\lambda = x(t) \qquad (1\text{-}21)$$

which is obtained by simple renaming of variables in (1-20) and by using the "even" property of the delta function expressed by (1-19). The form of the integral in (1-21), known as a *convolution*, is dealt with in more detail in Chapter 2.

Integrals with finite limits can be considered as special cases of (1-20) by defining $x(t)$ to be zero outside a certain interval, say $t_1 < t < t_2$. Thus (1-20) becomes

$$\int_{t_1}^{t_2} x(t)\delta(t - t_0) \, dt = \begin{cases} x(t_0), & t_1 < t_0 < t_2 \\ 0, & \text{otherwise} \end{cases}$$

A fourth property of $\delta(t)$ is that

$$x(t)\delta(t - t_0) = x(t_0)\delta(t - t_0) \qquad (1\text{-}22)$$

if $x(t)$ is continuous at $t = t_0$, which follows intuitively from $\delta(t - t_0) = 0$ everywhere except at $t = t_0$.

We may find integrals similar to (1-20) involving derivatives of $x(t)$. For example, assuming $x(t)$ to have a first derivative that is continuous at $t = t_0$ and using the chain rule for differentiation, we obtain

$$\frac{d}{dt} [x(t)\delta(t - t_0)] = \dot{x}(t)\delta(t - t_0) + x(t)\dot{\delta}(t - t_0)$$

where the overdot denotes differentiation with respect to time.† Assuming $\dot{x}(t)$ to be continuous at $t = t_0$ and using (1-22), we have

$$\frac{d}{dt} [x(t)\delta(t - t_0)] = \dot{x}(t_0)\delta(t - t_0) + x(t)\dot{\delta}(t - t_0)$$

†The derivative of the unit impulse function is sometimes referred to as the *unit doublet*, or simply doublet. It can be visualized by graphically differentiating the triangular approximation for the unit impulse shown in Figure 1-13b. Again, we emphasize that $\dot{\delta}(t)$ is to be interpreted in the context of an integral similar to (1-17).

Integrating from $t = t_1$ to $t = t_2$, where $t_1 < t_0 < t_2$, we obtain

$$\int_{t_1}^{t_2} \frac{d}{dt} [x(t)\delta(t - t_0)] \, dt = \int_{t_1}^{t_2} \dot{x}(t_0)\delta(t - t_0) \, dt$$

$$+ \int_{t_1}^{t_2} x(t)\dot{\delta}(t - t_0) \, dt, \qquad t_1 < t_0 < t_2$$

or

$$0 = \dot{x}(t_0) + \int_{t_1}^{t_2} x(t)\dot{\delta}(t - t_0) \, dt, \qquad t_1 < t_0 < t_2 \qquad (1\text{-}23)$$

where we have used

$$\int_{t_1}^{t_2} \frac{d}{dt} [x(t)\delta(t - t_0)] \, dt = x(t)\delta(t - t_0) \Big|_{t_1}^{t_2} = 0, \qquad t_1 < t_0 < t_2$$

and

$$\int_{t_1}^{t_2} \dot{x}(t)\delta(t - t_0) \, dt = \dot{x}(t_0) \int_{t_1}^{t_2} \delta(t - t_0) \, dt = \dot{x}(t_0), \qquad t_1 < t_0 < t_2$$

which follow because $\delta(t - t_0) = 0$ everywhere except at $t = t_0$ and $\dot{x}(t)$ is continuous at $t = t_0$, so that $\dot{x}(t_0)$ is defined. Rearranging (1-23), we obtain

$$\int_{t_1}^{t_2} x(t)\dot{\delta}(t - t_0) \, dt = -\dot{x}(t_0), \qquad t_1 < t_0 < t_2 \qquad (1\text{-}24)$$

In general, if the nth derivative of $x(t)$ exists and is continuous at $t = t_0$, it can be shown that

$$\int_{t_1}^{t_2} x(t)\delta^{(n)}(t - t_0) \, dt = (-1)^n x^{(n)}(t_0), \qquad t_1 < t_0 < t_2 \qquad (1\text{-}25)$$

where

$$x^{(n)}(t_0) \triangleq \frac{d^n x(t)}{dt^n} \bigg|_{t = t_0} \qquad \text{and} \qquad \delta^{(n)}(t - t_0) \triangleq \frac{d^n}{dt^n} [\delta(t - t_0)]$$

A unit impulse function and its derivatives may be treated as generalized functions in the sense that if

$$f(t) = a_1\delta(t) + a_2\dot{\delta}(t) + \cdots + a_n\delta^{(n)}(t)$$

and

$$g(t) = b_1\delta(t) + b_2\dot{\delta}(t) + \cdots + b_n\delta^{(n)}(t)$$

then

$$f(t) + g(t) = (a_1 + b_1)\delta(t) + (a_2 + b_2)\dot{\delta}(t) + \cdots \qquad (1\text{-}26)$$

Equation (1-26) applies also if $a_1, b_1, a_2, b_2, a_3, b_3, \ldots$ are functions that are continuous at $t = 0$.

EXERCISE

Evaluate the following integrals:

(a) $\int_{-\infty}^{\infty} e^{-\alpha t^2} \delta(t - 10) \, dt$

(b) $\int_{0}^{\infty} e^{-\alpha t^2} \delta(t + 10) \, dt$

(c) $\int_{-\infty}^{\infty} e^{-\alpha t^2} \ddot{\delta}(t - 10) \, dt$

(d) $\int_{-\infty}^{\infty} [5\delta(t) + e^{-(t-1)}\dot{\delta}(t) + \cos 5\pi t \, \delta(t) + e^{-t^2}\ddot{\delta}(t)] \, dt$

Answers: (a) $e^{-100\alpha}$; (b) 0; (c) $2\alpha(200\alpha - 1)e^{-100\alpha}$; (d) $6 + e$. ■

We now return to the assertion made earlier that

$$u(t) = u_{-1}(t) = \int_{-\infty}^{t} u_0(\lambda) \, d\lambda = \int_{-\infty}^{t} \delta(\lambda) \, d\lambda$$

or, equivalently, that

$$\delta(t) = \frac{du(t)}{dt}$$

which is (1-9b) with $i = -1$. Letting $x(t)$ be continuous at $t = 0$, we consider

$$\int_{-\infty}^{\infty} x(t) \frac{du(t)}{dt} \, dt = x(t)u(t) \Big|_{-\infty}^{\infty} - \int_{-\infty}^{\infty} u(t) \frac{dx(t)}{dt} \, dt$$

$$= x(\infty) - \int_{0}^{\infty} \frac{dx(t)}{dt} \, dt$$

$$= x(\infty) - x(t) \Big|_{0}^{\infty}$$

$$= x(\infty) - x(\infty) + x(0)$$

$$= x(0)$$

which was obtained through integration by parts. That is, $du(t)/dt$ has the same property as $\delta(t)$ as expressed by (1-17).

Energy and Power Signals

Quite often the particular representation used for a signal depends on the type of signal involved. It is therefore convenient to introduce a method for signal classification at this point. It is useful to classify signals as those having finite energy and those having finite average power. There are signals that have neither finite average power nor finite energy.

To introduce these signal classes, suppose that $e(t)$ is the voltage across a resistance R producing a current $i(t)$. The instantaneous power per ohm is

$$p(t) = \frac{e(t)i(t)}{R} = i^2(t)$$

Integrating over the interval $|t| \leq T$, the total energy and the average power on a per ohm basis are defined as the limits

$$E = \lim_{T \to \infty} \int_{-T}^{T} i^2(t)\, dt \qquad \text{joules}$$

and

$$P = \lim_{T \to \infty} \frac{1}{2T} \int_{-T}^{T} i^2(t)\, dt \qquad \text{watts}$$

respectively.

For an arbitrary signal $x(t)$, which may, in general, be complex, the total energy normalized to unit resistance is defined as

$$E \triangleq \lim_{T \to \infty} \int_{-T}^{T} |x(t)|^2\, dt \qquad \text{joules} \qquad (1\text{-}27)$$

and the average power normalized to unit resistance is defined as

$$P \triangleq \lim_{T \to \infty} \frac{1}{2T} \int_{-T}^{T} |x(t)|^2\, dt \qquad \text{watts} \qquad (1\text{-}28)$$

Based on the definitions (1-27) and (1-28), the following classes of signals are defined:

1. $x(t)$ is an energy signal if and only if $0 < E < \infty$, so that $P = 0$.
2. $x(t)$ is a power signal if and only if $0 < P < \infty$, thus implying that $E = \infty$.
3. Signals that satisfy neither property are therefore neither energy nor power signals.†

EXAMPLE 1-6
Consider the signal

$$x_1(t) = Ae^{-\alpha t}u(t), \qquad \alpha > 0,$$

where A and α are constants. Using (1-27), it can be verified that $x_1(t)$ has energy

$$E = \frac{A^2}{2\alpha}$$

Letting $\alpha \to 0$, we obtain the signal

$$x_2(t) = Au(t)$$

---

†It is easy to contrive examples of signals with infinite energy and zero average power that are nonzero over a finite range of t, but we classify such signals as neither energy nor power. An example is $x(t) = t^{-1/4}$, $t \geq 1$ and zero otherwise.

which has infinite energy but finite power as can be verified by applying (1-28):

$$P = \frac{A^2}{2}$$

Thus $x_2(t)$ is a power signal. ∎

EXAMPLE 1-7

Consider the periodic, sinusoidal signal

$$x(t) = A \cos(\omega_0 t + \theta)$$

The normalized average power of this signal is

$$P = \lim_{T \to \infty} \frac{1}{2T} \int_{-T}^{T} A^2 \cos^2(\omega_0 t + \theta) \, dt$$

$$= \lim_{T \to \infty} \frac{A^2}{2T} \int_{-T}^{T} [\tfrac{1}{2} + \tfrac{1}{2} \cos 2(\omega_0 t + \theta)] \, dt = \frac{A^2}{2}$$

which follows because

$$\lim_{T \to \infty} \frac{A^2}{4T} \int_{-T}^{T} \cos 2(\omega_0 t + \theta) \, dt = 0$$

∎

Average Power of a Periodic Signal

We note that there is no need to carry out the limiting operation to find P for a periodic signal, since an average carried out over a single period gives the same result as (1-28); that is, for a periodic signal, $x_P(t)$,

$$P = \frac{1}{T_0} \int_{t_0}^{t_0 + T_0} |x_P(t)|^2 \, dt \qquad (1\text{-}29)$$

where T_0 is the period. The proof of (1-29) is left to the problems.

EXAMPLE 1-8

Using (1-29), the power of a rotating phasor signal of the form (1-6) is

$$P = \frac{\omega_0}{2\pi} \int_{t_0}^{t_0 + 2\pi/\omega_0} |A e^{j(\omega_0 t + \theta)}|^2 \, dt = A^2$$

Using Euler's theorem, we may write

$$A e^{j(\omega_0 t + \theta)} = A \cos(\omega_0 t + \theta) + jA \sin(\omega_0 t + \theta)$$

The power of the real and imaginary components is $A^2/2$. Thus we may associate half the power in a phasor with the real component and half with the imaginary component. ∎

SYSTEM MODELING CONCEPTS

Some Terminology

Having discussed some appropriate signal models, we now wish to look at ways to represent the effects of systems on signals; that is, we need to be able to construct appropriate *system models* that adequately represent the interaction of signals and systems and the relationship of *causes* and *effects* for that system. Usually, we refer to certain causes of interest as *inputs* and certain effects of interest as *outputs*. For example, in the liquid-level controller example, the input of interest could be the electric current into the motor armature and the output could be the water level in the tank.

An obvious choice for input and output quantities may not always be readily apparent, easily isolated, or physically distinct. For instance, in the communications link example, the input, $x(t)$, and output, $s(t)$, appear at the same physical location: the transmitter. The relationship governing their interaction was very simple: that is,

$$s(t) = x(t) + \alpha\beta x(t - 2\tau), \qquad 0 \le t \le T + 2\tau$$

Usually, we will not have such simple input–output relationships.

It is convenient to visualize a system schematically by means of a box, as shown in Figure 1-14. (Sometimes symbols other than x and y may be used.) On the left-hand side of the box we represent the inputs (excitations, causes, stimuli) as a series of arrows labeled $x_1(t)$, $x_2(t)$, . . ., $x_m(t)$. The outputs (responses, effects), $y_1(t)$, $y_2(t)$, . . ., $y_p(t)$, are represented as arrows emanating from the right-hand side of the box. In general, the inputs and outputs vary with time, which is represented by the independent variable t; sometimes it will not be shown explicitly in order to simplify notation. The number of inputs and outputs need not be equal. Quite often, however, we will be concerned with situations where there is a single input and a single output. Such systems are referred to as *two-port* or *single-input, single-output systems* since they have one input port and one output port.

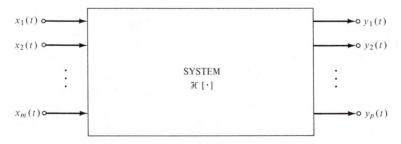

FIGURE 1-14.   Block diagram representation of a system.

Representations for Systems

Usually, we are interested in obtaining explicit relationships between the input and output variables of a system, and we discuss systematic procedures for doing so in Chapter 2. For now, however, we will express the dependence of the system output on the input symbolically as

$$y(t) = \mathcal{H}[x(t)] \tag{1-30}$$

Eliminating the explicit use of t, this can be written more compactly as

$$y = \mathcal{H}x$$

which is read

$$y(t) \text{ is the response of } \mathcal{H} \text{ to } x(t)$$

where initial conditions are included if pertinent. The symbol $\mathcal{H}$, which is known as an *operator,* serves the dual role of identifying the system and specifying the operation to be performed on $x(t)$ to produce $y(t)$. (For a multiple-input, multiple-output system, $x(t)$ and $y(t)$ are vectors and $\mathcal{H}$ is a matrix.)

Several examples of system input–output relationships are given below to illustrate more clearly some of the basic concepts.

1. *Instantaneous (Nondynamic) Relationships.* Many systems are adequately modeled for many situations by relationships such as

$$y(t) = Ax(t) + B \tag{1-31a}$$

or

$$y(t) = Ax(t) + Bx^3(t) \tag{1-31b}$$

where A and B are constants. The former equation could represent an amplifier of gain A with a dc bias of B volts at its output, and the latter equation would be a possible model for an amplifier that introduces *nonlinear distortion, $Bx^3(t)$,* into the amplified output, $Ax(t)$. The concept of nonlinearity for a system will be explicitly defined later.

2. *Linear, Constant-Coefficient, Ordinary Differential Equations.* The student is assumed to be familiar with obtaining the relationships between currents and voltages in an electrical circuit composed of passive elements (resistors, capacitors, and inductors) in terms of ordinary integrodifferential equations through the application of Kirchhoff's voltage and current laws, although systematic procedures for doing so are reviewed at the end of Chapter 2. If all integrals are removed through repeated differentiation, the general form for such equations may be written as

$$a_n \frac{d^n y(t)}{dt^n} + a_{n-1} \frac{d^{n-1} y(t)}{dt^{n-1}} + \cdots + a_0 y(t)$$

$$= b_m \frac{d^m x(t)}{dt^m} + b_{m-1} \frac{d^{m-1} x(t)}{dt^{m-1}} + \cdots + b_0 x(t) \tag{1-32}$$

where $a_n, a_{n-1}, \ldots, a_0, b_m, b_{m-1}, \ldots, b_0$ are constants that depend on the structure of the system and where nonzero initial conditions may be specified.

For systems whose input–output relationships can be represented in this manner, the *order* of the system is defined as the order of the highest derivative of the *dependent variable*, or system output, present.†

Three variations of (1-32) sometimes result from the modeling and analysis of systems. First, time-varying components may be present in the system (e.g., a capacitor whose capacitance varies with time). In such cases one or more of the *coefficients* of (1-32) will be *functions of time*. These are examples of *time-varying systems,* although the class of time-varying systems is broader than that described by ordinary differential equations with nonconstant coefficients.

A second situation that results in a governing differential equation of different form than (1-32) is where one or more nonlinear components are present in the system. For example, consider a voltage source $v_s(t)$ in series with an inductor and a resistor whose resistance depends on the current through it: for example, $R = R_0 + \alpha i(t)$, where R_0 and α are constants. Then application of Kirchhoff's voltage law results in

$$v_s(t) = [R_0 + \alpha i(t)]i(t) + L\frac{di}{dt}$$

and the relationship between the input, $v_s(t)$, and the output, $i(t)$, is no longer linear. This is an example of a *nonlinear* system. Precise definitions for time-varying and nonlinear systems are given later.

The third departure from (1-32) which may result is that of a system described by a *partial differential equation.* Such systems are said to be *distributed,* in contrast to *lumped* systems, for which the cause–effect properties of the system can be ascribed to elements modeled as infinitesimally small in the spatial dimension. This is permissible if their dimensions are small compared with the wavelength of the variable quantities within the system (voltages, currents, etc.), where the wavelength is given by

$$\lambda = \frac{c}{f} \quad \text{meters}$$

where $c = 3 \times 10^8$ m/s is the speed of electromagnetic radiation and f is the frequency of the oscillating waveform in hertz. A light-bulb, for example, has dimensions that are small compared with $\lambda = 3 \times 10^8/60 = 5 \times 10^6$ m, which is the wavelength of the 60-Hz power-line voltage.

3. *Integral Relationships.* In Chapter 2 we discuss an integral representation of a system input–output relation of the form

$$y(t) = \int_{-\infty}^{\infty} h(\lambda)x(t - \lambda)\, d\lambda = \int_{-\infty}^{\infty} x(\lambda)h(t - \lambda)\, d\lambda \qquad (1\text{-}33)$$

which is known as a *superposition integral.* The function $h(\lambda)$ is called the *impulse response* of the system and is its response to a unit impulse applied at $\lambda = 0$.‡ All systems that can be represented by ordinary, constant-coefficient,

---

†The order of a multiple-input, multiple-output system can be determined through state-variable techniques, which are discussed in Chapter 6.

‡To show this, let $x(t) = \delta(t)$, so that $x(t - \lambda) = \delta(t - \lambda)$. Then $y(t) = h(t)$ by the sifting property of the unit impulse function, and $h(t)$ is seen to be the system's response to $\delta(t)$.

linear differential equations can also be represented by an equation of the form (1-33). However, not all systems that can be represented by a relationship of the form (1-33) can be represented by ordinary, constant-coefficient linear differential equations.

For a time-varying system, the generalization of (1-33) is

$$y(t) = \int_{-\infty}^{\infty} h(t, \lambda)x(\lambda) \, d\lambda \tag{1-34}$$

where $h(t, \lambda)$ is the response of the system at time t to an impulse applied at time λ.

EXAMPLE 1-9

Consider the RC circuit shown in Figure 1-15 with input $x(t)$ and output $y(t)$ as shown. We wish to obtain (a) the differential equation relating $y(t)$ to $x(t)$, and (b) convert this differential equation to the integral form (1-33). Assume that $x(t)$ is applied at $t = t_0$ and $y(t_0) = y_0$.

Solution: (a) The differential equation relating $y(t)$ and $x(t)$ is found by writing Kirchhoff's voltage law (KVL) around the loop indicated by $i(t)$. This results in the equation

$$x(t) = Ri(t) + y(t)$$

where $y(t)$, the voltage drop across the capacitor, is related to $i(t)$, the current through it, by

$$i(t) = C \frac{dy(t)}{dt}$$

Eliminating $i(t)$ in the KVL equation, we obtain

$$RC \frac{dy(t)}{dt} + y(t) = x(t)$$

(b) To write the input–output relationship as an explicit equation for $y(t)$ in terms of $x(t)$, we first obtain the solution to the homogeneous differential equation

$$RC \frac{dy_h(t)}{dt} + y_h(t) = 0$$

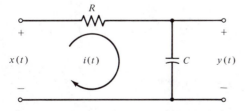

FIGURE 1-15. *RC circuit to illustrate the derivation of input–output relationships for systems.*

Assuming a solution of the form

$$y_h(t) = Ae^{pt}$$

and substituting into the homogeneous equation, we find that $p = -1/RC$, so that

$$y_h(t) = A \exp\left(\frac{-t}{RC}\right)$$

The total solution consists of $y_h(t)$ plus a particular solution for a specific $x(t)$. Assume that the input $x(t)$ is applied beginning at $t = t_0$ but is otherwise arbitrary. Also assume that the value of $y(t)$ at $t = t_0$ is y_0. To find the total solution we use the technique of *variation of parameters,* which consists of assuming a solution of the form of $y_h(t)$ but with the undetermined coefficient A replaced by a function of time to be found. Thus we assume that

$$y(t) = A(t) \exp\left(\frac{-t}{RC}\right)$$

Differentiating by means of the chain rule, we obtain

$$\frac{dy(t)}{dt} = \left[\frac{dA(t)}{dt} - \frac{A(t)}{RC}\right] \exp\left(\frac{-t}{RC}\right)$$

Substituting the assumed solution and its derivative into the nonhomogeneous differential equation, we have

$$RC \exp\left(\frac{-t}{RC}\right) \frac{dA(t)}{dt} = x(t)$$

where the term $-A(t) \exp(-t/RC)$ has been canceled by $y(t)$ in the differential equation. Solving for $dA(t)/dt$ and integrating, we get an explicit result for the unknown "varying parameter." It is given by

$$A(t) = \frac{1}{RC} \int_{t_0}^{t} x(\lambda) \exp\left(\frac{\lambda}{RC}\right) d\lambda + A(t_0)$$

where λ is a dummy variable of integration and $A(t_0)$ is the initial value of $A(t)$ at the time t_0 that the input $x(t)$ is applied. Since $y(t) = A(t) \exp(-t/RC)$, it follows that $A(t_0) = y_0 \exp(t_0/RC)$. Using this result together with the expression for $A(t)$, we obtain

$$y(t) = \left[\int_{t_0}^{t} \frac{x(\lambda) \exp(\lambda/RC)}{RC} d\lambda + y_0 \exp\left(\frac{t_0}{RC}\right)\right] \exp\left(\frac{-t}{RC}\right)$$

$$= y_0 \exp\left(-\frac{t - t_0}{RC}\right) + \int_{t_0}^{t} x(\lambda) \frac{\exp[-(t - \lambda)/RC]}{RC} d\lambda$$

If we assume that the input $x(t)$ is applied at $t = -\infty$ and that $y_0 \triangleq y(-\infty) = 0$, then

$$y(t) = \int_{-\infty}^{t} x(t) \frac{\exp\left[-(t - \lambda)/RC\right]}{RC} \, d\lambda$$

If we define

$$h(t) = \frac{\exp\left(-t/RC\right)}{RC} \, u(t)$$

so that

$$h(t - \lambda) = \frac{\exp\left[-(t - \lambda)/RC\right]}{RC} \, u(t - \lambda)$$

then the solution for $y(t)$, as determined by the variation-of-parameters approach, is exactly of the form given by (1-33). ∎

Properties of Systems

We now give precise definitions of several system properties, some of which were mentioned in the previous discussion.

Continuous-Time and Discrete-Time Systems: If the signals processed by a system are continuous-time signals, the system itself is referred to as a *continuous time* system. If, on the other hand, the system processes signals that exist only at discrete times, it is called a *discrete-time* system. The signals in a system may or may not be quantized to a finite number of levels. If they are, the system is referred to as *quantized*. A quantized system may be continuous-time or discrete-time, although usually a quantized system is also a discrete-time system and is then referred to as a *digital system*.

Examples of continuous-time systems are electric networks composed of resistors, capacitors, and inductors that are driven by continuous-time sources. An example of a quantized, discrete-time system is the optical-wand sensing system of a department store cash register. An example of a device that can be used to process discrete-time unquantized signals is a charged-coupled delay line. This device is described further in Chapter 9, where some of its applications will be discussed.

Fixed and Time-Varying Systems: A system is *time-invariant*, or *fixed*, if its input–output relationship does not change with time. Otherwise, it is said to be *time-varying*. In terms of the symbolic notation of (1-30), a system is fixed if and only if

$$\mathcal{H}[x(t - \tau)] = y(t - \tau) \tag{1-35}$$

for *any* $x(t)$ and *any* τ. The system must be at rest prior to the application of $x(t)$. For example, the system with input–output relationship

$$y(t) = x(t) + Ax(t - T)$$

is fixed if A and T are constants and time-varying if either or both are functions of time. Thus the satellite communication system is fixed if $\alpha\beta$ and τ are constants. Similarly, a system described by a differential equation of the form

(1-32) is fixed if the coefficients are constant, and time-varying if one or more coefficients are functions of time.

Causal and Noncausal Systems: A system is *causal* or *nonanticipatory* if its response to an input does not depend on future values of that input.† A precise definition is that a continuous-time system is causal if and only if the condition

$$x_1(t) = x_2(t) \qquad \text{for } t \le t_0$$

implies the condition

$$\mathcal{H}[x_1(t)] = \mathcal{H}[x_2(t)] \qquad \text{for } t \le t_0 \tag{1-36}$$

for any t_0, $x_1(t)$, and $x_2(t)$. Stated another way, if the *difference* between two system inputs is zero for $t \le t_0$, the *difference* between the respective outputs must be zero for $t \le t_0$ if the system is causal. Thus the definition applies to systems for which the response to zero input is not zero, such as a system containing independent sources, as well as systems with nonzero inputs.

EXAMPLE 1-10

Consider a system described by (1-31a). Let $x_1(t) = u(t)$ and $x_2(t) = r(t)$. For $t \le t_0 < 0$, both inputs are equal. The outputs for $t \le t_0$ for both inputs are $y_1(t) = y_2(t) = B$. In general, it is seen from (1-31a) that if $x_1(t) = x_2(t)$, $t \le t_0$, then $y_1(t) = y_2(t) = B$ for $t \le t_0$ because $Ax_1(t) = Ax_2(t)$. It is easy to see that this holds for any pair of inputs for which it is true that $x_1(t) = x_2(t)$, $t \le t_0$. Thus this system is *causal*. ∎

Dynamic and Instantaneous Systems: A system for which the output is a function of the input *at the present time only* is said to be *instantaneous* (or *memoryless, or zero-memory*). A *dynamic* system is one whose output depends on past or future values of the input in addition to the present time. If the system is also causal, the output of a dynamic system depends only on present and past values of the input. Mathematically, the input–output relationship for an instantaneous system must be of the form

$$y(t) = f[x(t), t] \tag{1-37}$$

where $f(\cdot)$ is a function, possibly time dependent, which depends on $x(t)$ only at the present time, t.

The systems described by (1-31a) and (1-31b) are instantaneous. Systems described by differential equations are dynamic (nonzero memory). For example, an inductor is a system with memory. This is easy to see if the input is taken as the voltage across its terminals and the output as the current through it so that

$$i(t) = \frac{1}{L} \int_{-\infty}^{t} v(\lambda) \, d\lambda$$

It is clear that $i(t)$ depends on past values of $v(t)$.

†Another way of informally defining a causal system is that such a system does not anticipate its input.

Linear and Nonlinear Systems: From your circuits courses, you should already be familiar with the concept of linearity. It allows the analysis of circuits with multiple sources by means of *superposition,* or addition of currents or voltages calculated when each source is applied separately. A precise mathematical definition of a linear system is that *superposition holds,* where superposition for a system with any two inputs $x_1(t)$ and $x_2(t)$ is defined as

$$\mathcal{H}[\alpha_1 x_1(t) + \alpha_2 x_2(t)] = \alpha_1 \mathcal{H}[x_1(t)] + \alpha_2 \mathcal{H}[x_2(t)]$$

$$= \alpha_1 y_1(t) + \alpha_2 y_2(t) \tag{1-38}$$

In (1-38), $y_1(t)$ is the response of the system when input $x_1(t)$ is applied alone, and $y_2(t)$ is the response of the system when input $x_2(t)$ is applied alone; α_1 and α_2 are arbitrary constants.

EXAMPLE 1-11

Show that the system described by the differential equation

$$\frac{dy(t)}{dt} + ty(t) = x(t)$$

is linear.

Solution: The response to $x_1(t)$, $y_1(t)$, satisfies

$$\frac{dy_1}{dt} + ty_1 = x_1$$

and the response to $x_2(t)$, $y_2(t)$, satisfies

$$\frac{dy_2}{dt} + ty_2 = x_2$$

where the time dependence of x_1, x_2, y_1, and y_2 is omitted for simplicity. Addition of these equations multiplied, respectively. by α_1 and α_2 results in

$$\alpha_1 \frac{dy_1}{dt} + \alpha_2 \frac{dy_2}{dt} + \alpha_1 ty_1 + \alpha_2 ty_2 = \alpha_1 x_1 + \alpha_2 x_2$$

or

$$\frac{d}{dt}(\alpha_1 y_1 + \alpha_2 y_2) + t(\alpha_1 y_1 + \alpha_2 y_2) = \alpha_1 x_1 + \alpha_2 x_2$$

That is, the response to the input $\alpha_1 x_1 + \alpha_2 x_2$ is $\alpha_1 y_1 + \alpha_2 y_2$. Thus superposition holds and the system is linear. ∎

EXAMPLE 1-12

The system described by the differential equation

$$\frac{dy(t)}{dt} + 10y(t) + 5 = x(t)$$

is *nonlinear.* This is demonstrated by attempting to apply the superposition principle. Letting $y_1(t)$ be the response to the input $x_1(t)$ and $y_2(t)$ be the response to the input $x_2(t)$, we write

$$\frac{dy_1}{dt} + 10y_1 + 5 = x_1$$

and

$$\frac{dy_2}{dt} + 10y_2 + 5 = x_2$$

where the time dependence is again dropped for simplicity. Multiplying the first equation by an arbitrary constant α_1, the second equation by the arbitrary constant α_2, adding, and regrouping terms, we obtain

$$\frac{d}{dt}(\alpha_1 y_1 + \alpha_2 y_2) + 10(\alpha_1 y_1 + \alpha_2 y_2) + 5(\alpha_1 + \alpha_2) = \alpha_1 x_1 + \alpha_2 x_2$$

We *cannot* put this equation into the same form as the original differential equation for an *arbitrary choice* of α_1 and α_2. Thus the system is *nonlinear*. ∎

1-5

SUMMARY

In this chapter we have introduced the idea of *models* for signals and systems. Signals can be of many different forms including *continuous-time, discrete-time,* and *quantized*. The concept of *periodic* versus *aperiodic* signals was introduced; the values of a periodic signal repeat each fundamental period, and multiples thereof, for *all* time. A rotating phasor signal is a useful model in that it permits the introduction of the concept of the *amplitude* and *phase spectra* of a signal. This *frequency-domain* viewpoint of a signal is one that we adopt for a large portion of this book. Another class of useful signal models is the *singularity functions*, which include the *unit impulse, unit step,* and *unit ramp*. Many simple signals can be built up of ramps and steps. An important property of the unit impulse is the *sifting property*. Two broad classes of signals are *energy* and *power signals*, which include the vast majority of, but not all, signals. For example, a unit impulse is neither an energy nor a power signal.

Several properties of systems were defined. These included *lumped* versus *distributed* or *nonlumped, order* of a lumped system, *fixed* versus *time-varying, causal* versus *noncausal,* and *linear* versus *nonlinear*. The student should write definitions for each of these terms in his own words and be prepared to illustrate them by example.

FURTHER READING

R. J. SCHWARZ and B. FRIEDLAND, *Linear Systems*. New York: McGraw-Hill, 1965. Many of the concepts presented in this chapter are treated at a more advanced level in this reference.

M. E. VAN VALKENBURG, *Network Analysis*, 3rd ed. Englewood Cliffs, N.J.: Prentice-Hall, 1974. Many elementary circuits texts treat signal and system modeling, of which this and the following are representative.

W. H. HAYT and J. E. KEMMERLY, *Engineering Circuit Analysis*. New York: McGraw-Hill, 1978.

R. E. ZIEMER and W. H. TRANTER, *Principles of Communications: Systems, Modulation, and Noise*. Boston: Houghton Mifflin, 1976. The spectrum ideas introduced in this chapter are treated in more detail in this reference.

R. A. GABEL and R. A. ROBERTS, *Signals and Linear Systems,* 2nd ed. New York: Wiley, 1980. Another recent, well-written text which treats continuous- and discrete-time signal and system theory.

C. D. MCGILLEM and G. R. COOPER, *Continuous and Discrete Signal and System Analysis*. New York: Holt, Rinehart and Winston, 1974. A book which in many ways parallels the treatment of this book; in particular, both continuous-time and discrete-time systems are treated.

The last two references include continuous-time and discrete-time systems together, whereas we have chosen to treat continuous-time systems first (Chapters 1 through 6) followed by discrete-time systems (Chapters 7 through 9).

PROBLEMS

SECTION 1-2

1-1. Sketch the liquid level in the tank as a function of time for Example 1-1. Dimension your graph completely.

1-2. Suppose that the pump in Example 1-1 is capable of pumping liquid out of the tank at a flow rate of 15 liters/s. Sketch the liquid level versus time, assuming that the flow rate into the tank is still 5 liters/s.

1-3. In the satellite communications example of Section 1-1, suppose that a unit-amplitude square pulse signal of duration 1 s is transmitted and that $\alpha\beta = 0.1$. (a) Sketch $s(t) = x(t) + \alpha\beta \, x(t - 2\tau)$, where $2\tau = 0.26$ s. (b) If the transmitted signal is 0.1 s in duration, sketch $s(t)$.

SECTION 1-3

1-4. Show by substitution into (1-2) that a unit-amplitude pulse signal of width τ can be represented as $\Pi(t/\tau)$.

1-5. Sketch the following signals:
(a) $\Pi(0.1t)$
(b) $\Pi(10t)$
(c) $\Pi(t - 1/2)$
(d) $\Pi[(t - 2)/5]$
(e) $\Pi[(t - 1)/2] + \Pi(t - 1)$

1-6. What are the fundamental periods of the signals given below? (Assume units of seconds.)
(a) $4 \sin 100\pi t$
(b) $2 \cos 120\pi t$
(c) $\sin 100\pi t + 2 \sin 200\pi t$
(d) $3 \sin 100\pi t + \cos 120\pi t$

1-7. Which of the following signals are periodic? Give their fundamental periods.
  (a) sin 150πt
  (b) 2 cos (120πt + π/3)
  (c) 6 cos 377t
  (d) 2 cos (120πt + π/3) + 6 cos 377t

1-8. Write the signals given in Problem 1-6 as the real parts of rotating phasors.

1-9. Write the signals given in Problem 1-6 as one-half the sum of a rotating phasor and its complex conjugate.

1-10. Sketch the single-sided amplitude and phase spectra of the signals given in Problem 1-6.

1-11. Sketch the double-sided amplitude and phase spectra of the signals given in Problem 1-6.

1-12. Write the signals given in Problem 1-7 as the real parts of rotating phasors.

1-13. Write the signals given in Problem 1-7 as one-half the sum of a rotating phasor and its complex conjugate.

1-14. Sketch the single-sided amplitude and phase spectra of the signals given in Problem 1-7.

1-15. Sketch the double-sided amplitude and phase spectra of the signals given in Problem 1-7.

1-16. Sketch the following signals. Dimension the ordinate and abscissa carefully.
  (a) $x_1(t) = u(t) + r(t) - 2r(t - 1) + r(t - 2) - u(t - 2)$
  (b) $x_2(t) = u(t) u(1 - t)$
  (c) $x_3(t) = u(t) r(2 - t)$
  (d) $x_4(t) = 4u(t) u(2 - t) \delta(t - 1)$
  (e) $x_5(t) = 4u(t) u(2 - t) \delta(t - 3)$

1-17. Write the signals shown in terms of singularity functions.

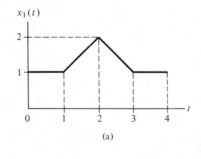

(a)

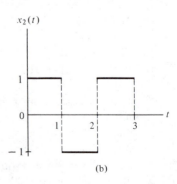

(b)

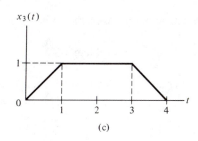

(c)

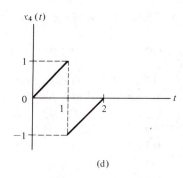

(d)

1-18. Evaluate the following integrals.

(a) $\displaystyle\int_5^{10} \cos 2\pi t \, \delta(t - 2) \, dt$

(b) $\displaystyle\int_0^5 \cos 2\pi t \, \delta(t - 2) \, dt$

(c) $\displaystyle\int_0^5 \cos 2\pi t \, \delta(t - 0.5) \, dt$

(d) $\displaystyle\int_{-\infty}^{\infty} (t - 2)^2 \delta(t - 2) \, dt$

(e) $\displaystyle\int_{-\infty}^{\infty} t^2 \delta(t - 2) \, dt$

1-19. Evaluate the following integrals.

(a) $\displaystyle\int_{-\infty}^{\infty} e^{-2t} \ddot{\delta}(t - 1) \, dt$

(b) $\displaystyle\int_0^{10} \cos 2\pi t \, \ddot{\delta}(t - 5) \, dt$

(c) $\displaystyle\int_{-\infty}^{\infty} (e^{-5t} + \sin 10\pi t) \, \dot{\delta}(t) \, dt$

1-20. Find the unspecified constants, denoted as $C_1, C_2, \ldots$, in the following expressions.

(a) $10\delta(t) + C_1\dot{\delta}(t) + (2 + C_2) \ddot{\delta}(t) = (3 + C_3)\delta(t) + 5\dot{\delta}(t) + 6\ddot{\delta}(t)$

(b) $(3 + C_1)\delta^{(4)}(t) + C_2\ddot{\delta}(t) + C_3\dot{\delta}(t) = C_4\delta^{(3)}(t) + C_5\delta(t)$

1-21. Sketch the following signals and calculate their energies.

(a) $e^{-10t}u(t)$

(b) $u(t) - u(t - 15)$

(c) $\cos 10\pi t \, u(t)u(2 - t)$

(d) $r(t) - 2r(t - 1) + r(t - 2)$

1-22. Obtain the energies of the signals in Problem 1-17.

1-23. Which of the signals given in Problem 1-16 are energy signals? Justify your answers.

1-24. Obtain the average powers of the signals given in Problem 1-6.

1-25. Obtain the average powers of the signals given in Problem 1-7.

1-26. Which of the following signals are power signals and which are energy signals? Which are neither? Justify your answers.
(a) $u(t) + 5u(t - 1) - 2u(t - 2)$
(b) $u(t) + 5u(t - 1) - 6u(t - 2)$
(c) $e^{-5t} u(t)$
(d) $(e^{-5t} + 1)u(t)$
(e) $(1 - e^{-5t})u(t)$
(f) $r(t)$
(g) $r(t) - r(t - 1)$
(h) $r(t) - r(t - 1) - r(t - 2) + r(t - 3)$

1-27. Given the following signals:
(1) $\cos 5\pi t + \sin 6\pi t$
(2) $\sin 2t + \cos \pi t$
(3) $e^{-10t}u(t)$
(4) $e^{2t}u(t)$
(a) Which are periodic? Give their periods.
(b) Which are power signals? Compute their average powers.
(c) Which are energy signals? Compute their energies.

1-28. (a) Sketch the following signals:
(1) $x_1(t) = r(t + 2) - 2r(t) + r(t - 2)$
(2) $x_2(t) = u(t)u(10 - t)$
(3) $x_3(t) = 2u(t) + \delta(t - 2)$
(4) $x_4(t) = 2u(t)\delta(t - 2)$
(b) For the signal shown, write an equation in terms of singularity functions.

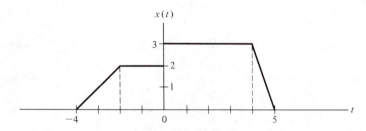

1-29. For the signal shown, write an equation in terms of singularity functions.

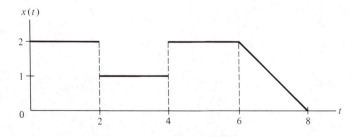

1-30.   Evaluate the following integrals.

(a) $\int_{-\infty}^{\infty} t^2 \delta(t - 10) \, dt$

(b) $\int_{-\infty}^{10} (5 + \cos t) \delta(t - 20) \, dt$

(c) $\int_{-\infty}^{\infty} e^{-5t} \dot{\delta}(t - 5) \, dt$

1-31.   Write the following signals in terms of singularity functions.

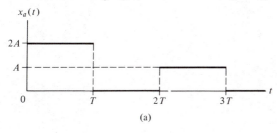

(a)

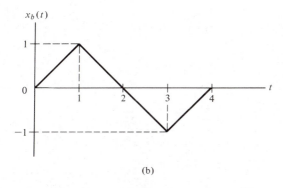

(b)

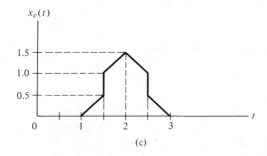

(c)

1-32.   Prove Equation (1-29) by starting with (1-28)

SECTION 1-4

1-33.   A system is defined by the set of algebraic equations

$$y_1(t) = 2x_1(t) - x_2(t)$$

$$y_2(t) = 5x_2(t) + 3x_3(t)$$

Represent this system in terms of a matrix equation of the form

$$\mathbf{y}(t) = \mathbf{H} \cdot \mathbf{x}(t)$$

That is, give $\mathbf{y}(t)$, $\mathbf{H}$, and $\mathbf{x}(t)$ explicitly.

1-34. What is the order of each system defined by the following equations?

(a) $2\dfrac{dy(t)}{dt} + 3y(t) = \dfrac{d^2x(t)}{dt^2} + x(t)$

(b) $3y(t) + \displaystyle\int_{-\infty}^{t} y(\lambda)\, d\lambda = x(t)$

(c) $4y(t) + 10 = \dfrac{dx(t)}{dt} + 5x(t)$

(d) $\dfrac{dy(t)}{dt} + t^2 y(t) = \displaystyle\int_{-\infty}^{t} x(\lambda)\, d\lambda$

(e) $\dfrac{d^2y(t)}{dt^2} + y(t)\dfrac{dy(t)}{dt} + y(t) = 5x(t)$

1-35. Which of the systems defined by the equations of Problem 1-34 are fixed? Justify your answers.

1-36. Which of the systems defined by the equations of Problem 1-34 are nonlinear? Justify your answers.

1-37. A system is defined by the input–output relationship

$$y(t) = x(t + 10)$$

Is this system causal or noncausal? Justify your answer by choosing a specific pair of inputs that will or will not satisfy (1-36).

1-38. A system is defined by the input–output relationship

$$y(t) = 10x(t) + 5$$

Is this system linear? Prove your answer.

1-39. A system is defined by the input–output relationship

$$y(t) = x(t^2)$$

Is this system:
(a) Linear?
(b) Causal?
(c) Fixed?
Prove your answers.

1-40. Is the system with input and output as shown causal? Why?

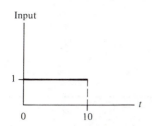
Input

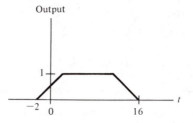
Output

1-41. A system has input and output as shown. Can you say anything about
(a) Linearity?
(b) Causality?
(c) Fixed versus time-varying?
(d) Zero memory?
For which of the above is insufficient information given?

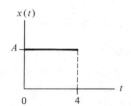

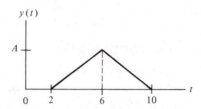

1-42. (a) Write a differential equation relating $y(t)$ to $x(t)$ for the circuit shown.
(b) Show that this system is linear.
(c) Show that this system is fixed.
(d) Convert the differential equation found in part (a) to an integral equation of the form (1-33). Assume that the input $x(t)$ is applied at $t = 0$, and that the current through the inductor is initially zero.

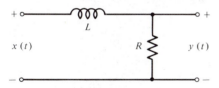

REVIEW QUESTION

1-43. State whether the following signals are periodic or aperiodic, energy, power, or neither energy nor power. Fill in the table with your answers.
(a) $u(t)$
(b) $u(t) - u(t - 10)$
(c) $\cos 20\pi t + \cos 40\pi t$
(d) $\cos 20\pi t + \cos 120t$

(e) $\exp(-10t)$
(f) $r(t) - r(t - 1)$
(g) $r(t) - r(t - 1) - r(t - 2) + r(t - 3)$
(h) $\cos 20\pi t + \exp(-10t)$

If true, check the appropriate box.

Property	Signal							
	a	b	c	d	e	f	g	h
Periodic								
Aperiodic								
Energy								
Power								
Neither energy nor power								

REVIEW QUESTION

1-44. State whether the following systems are linear or nonlinear, causal or noncausal, fixed or time-varying, dynamic or instantaneous. As usual, $x(t)$ denotes the input and $y(t)$ the output. Fill in the table with your answers.

(a) $\dfrac{dy(t)}{dt} + 10y(t) = x(t)$

(b) $\dfrac{dy(t)}{dt} + 10y(t) + 5 = x(t)$

(c) $\dfrac{dy(t)}{dt} + t^2y(t) = x(t)$

(d) $\dfrac{dy(t)}{dt} + y^2(t) = x(t)$

(e) $\dfrac{dy(t)}{dt} + y(t) = x(t + 10)$

(f) $y(t) = 10x^2(t) + x(t)$

(g) $y(t) = x(t + 10) + x^2(t)$

If true, check the appropriate box.

Property	Signal						
	a	b	c	d	e	f	g
Linear							
Causal							
Fixed							
Dynamic							

REVIEW QUESTION

1-45. Classify the following input–output relations for systems as to linearity, order, causality, and time invariance.

(a) $\dfrac{dy}{dt} + \dfrac{1}{RC} y(t) = x(t); \quad RC = \text{constant}$

(b) $\dfrac{dy}{dt} + t^2 y(t) = x(t)$

(c) $\dfrac{d^2y}{dt^2} + y(t) \dfrac{dy}{dt} + y(t) = x(t)$

(d) $y(t) = \displaystyle\int_{-\infty}^{t} x(t - \lambda) \dfrac{e^{-\lambda/RC}}{RC} \, d\lambda$

(e) $y(t) = x^2(t) + 1$

CHAPTER 2

System Analysis in the Time Domain

2-1

INTRODUCTION

The systems analysis problem is *given a system and an input; what is the output?* Basically, the three approaches that can be taken in finding the response of a fixed, linear system to a given input are: (1) Obtain a solution to the modeling equations through standard methods of solving differential equations; (2) carry out a solution in the time domain using the superposition integral; (3) use frequency-domain analysis by means of the Fourier or Laplace transforms. The first method should be familiar to the student from circuits courses. The objective of this chapter is to examine the second method. The third method is dealt with in later chapters.

The reader may well ask why the concentration on linear systems? Fortunately, many physical systems are adequately modeled as linear systems since the analysis of nonlinear systems is difficult. Indeed, analysis techniques for nonlinear systems is an active research area.

In the development of time-domain analysis techniques, an integral known as the *convolution integral* arises. If $x(t)$ and $h(t)$ are two signals, the convolution of $x(t)$ with $h(t)$ is a new signal $y(t)$ given by the operation

$$y(t) = \int_{-\infty}^{\infty} x(\lambda)h(t - \lambda) \, d\lambda, \qquad -\infty < t < \infty \qquad (2\text{-}1)$$

Later, in Section 2.3, we show that if $x(t)$ is the system input and $h(t)$ is a system-characterizing function called the impulse response, the operation expressed by (2-1) gives the system output. This was alluded to in Section 1.3.

Alternatively, $y(t)$ can be found by convolving $h(t)$ with $x(t)$, which is expressed as

$$y(t) = \int_{-\infty}^{\infty} h(\eta)x(t - \eta)\, d\eta, \qquad -\infty < t < \infty \qquad (2\text{-}2)$$

That (2-1) and (2-2) are equivalent can readily be proven by making the change of variables $\eta = t - \lambda$ in (2-1).

Equations (2-1) and (2-2) have applications other than time-domain analysis of systems. Since they are so widely useful, we begin this chapter by examining them more closely through examples.

2-2

EXAMPLES ILLUSTRATING EVALUATION OF THE CONVOLUTION INTEGRAL

The convolution of two signals, $x(t)$ and $h(t)$, is a new function of time, $y(t)$, which is given by (2-1) or (2-2). A useful symbolic notation often employed to denote (2-1) and (2-2), respectively, is

$$y(t) = x(t) * h(t) \qquad (2\text{-}3)$$

$$y(t) = h(t) * x(t) \qquad (2\text{-}4)$$

The integrand of (2-1) is found by three operations: (1) reversal in time, or folding, to obtain $h(-\lambda)$; (2) shifting to obtain $h(t - \lambda)$; and (3) multiplication of $x(\lambda)$ and $h(t - \lambda)$ to obtain the integrand.

Two examples will be given to illustrate these three operations and the subsequent evaluation of the integral to form $y(t)$.

EXAMPLE 2-1
Consider the convolution of the two rectangular pulse signals

$$x(t) = 2\Pi\left(\frac{t - 5}{2}\right)$$

and

$$h(t) = \Pi\left(\frac{t - 2}{4}\right)$$

These signals are sketched in Figure 2-1a and b.

We base our evaluation of the convolution of $x(t)$ and $h(t)$ on (2-2). Thus $x(t)$ is reversed and the variable changed to η, which results in $x(-\eta)$, also shown in Figure 2-1. Finally, $x(-\eta)$ is shifted so that what was the origin now appears at $\eta = t$. It is sometimes helpful to think of $x(t - \eta)$ as $x[-(\eta - t)]$; that is, t in $x(t)$ is replaced by $\eta - t$, with η thought of as the independent

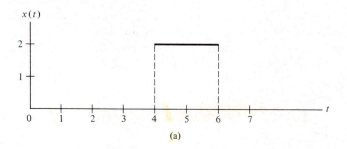

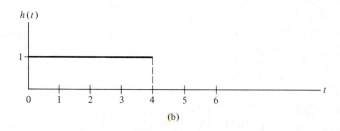

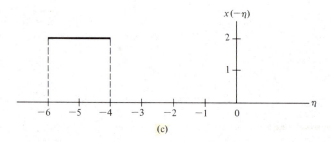

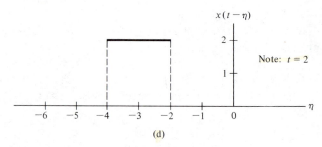

FIGURE 2-1. Signals to be convolved in Example 2.1.

variable, whereupon the function $x(\eta - t)$ is reversed or folded (its mirror image about the ordinate is taken). The result of these operations is $x(t - \eta)$, which is shown in Figure 2-1d for $t = 2$.

The product of $x(t - \eta)$ and $h(\eta)$ is formed, point by point, and this product integrated to form $y(t)$. Clearly, the product will be zero if $t < 4$ in this example. If $t \geq 4$, the rectangles forming $h(\eta)$ and $x(t - \eta)$ overlap until $t > 10$, whereupon the product is again zero.

After a little thought, we see that three different cases of overlap result. These are illustrated in Figure 2-2. In each case, the area of integration is shaded; because the height of $h(t)$ is unity, the area is simply the width of the overlap times the height of $x(t - \eta)$. Thus the area for Case I is $2(t - 4)$; that is, it is linearly increasing with t. For Case II, it is $2[(t - 4) - (t - 6)] = 4$. Finally, for Case III, it is $2[4 - (t - 6)] = 2(10 - t)$; that is, the area linearly decreases with t. The result for $y(t)$ is sketched in Figure 2-2d. ∎

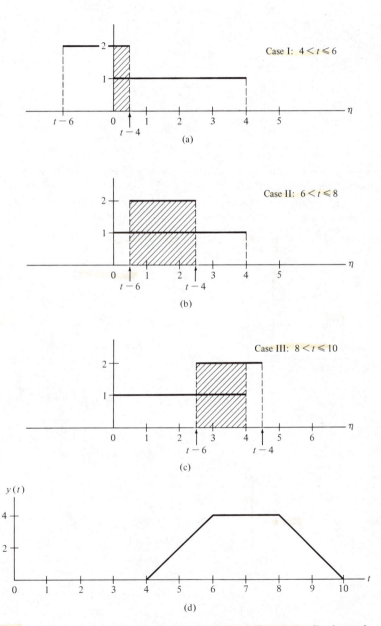

FIGURE 2-2. Steps in the convolution of Example 2.2 and the final result.

EXAMPLE 2-2

As a second example of the convolution operation, we consider the signals shown in Figure 2-3a. We use the convolution integral (2-1) in obtaining the result. This example differs from Example 2-1 in that part of the area under the product $x(\lambda)h(t - \lambda)$ is negative. Our first step is to express $h(t - \lambda)$ mathematically. Since

$$h(t) = \begin{cases} \dfrac{t}{2}, & 0 \le t \le 2 \\ 0, & \text{otherwise} \end{cases}$$

we obtain

$$h(t - \lambda) = \begin{cases} \dfrac{t - \lambda}{2}, & 0 \le t - \lambda \le 2 \quad \text{or} \quad t - 2 \le \lambda \le t \\ 0, & \text{otherwise} \end{cases}$$

The signals $x(\lambda)$ and $h(t - \lambda)$ are sketched in Figure 2-3b. Some thought on the part of the student should result in the four cases of nonzero overlap illustrated in Figure 2-3c through f. The resulting convolutions for these cases are expressed mathematically by the following equations:

$$y(t) = \begin{cases} 0, & t < -1 \quad \text{or} \quad t > 3 \\[2mm] \displaystyle\int_{-1}^{t} (t - \lambda)\, d\lambda, & -1 \le t < 0 \\[2mm] \displaystyle\int_{-1}^{0} (t - \lambda)\, d\lambda - \int_{0}^{t} (t - \lambda)\, d\lambda, & 0 \le t < 1 \\[2mm] \displaystyle\int_{t-2}^{0} (t - \lambda)\, d\lambda - \int_{0}^{1} (t - \lambda)\, d\lambda, & 1 \le t < 2 \\[2mm] \displaystyle -\int_{t-2}^{1} (t - \lambda)\, d\lambda, & 2 \le t < 3 \end{cases}$$

Integration of these expressions results in the following for $y(t)$:

$$y(t) = \begin{cases} 0, & t < -1 \quad \text{or} \quad t > 3 \\[1mm] \tfrac{1}{2}(t + 1)^2, & -1 \le t < 0 \\[1mm] \tfrac{1}{2}(-t^2 + 2t + 1), & 0 \le t < 1 \\[1mm] \tfrac{1}{2}(-t^2 - 2t + 1) + 2, & 1 \le t < 2 \\[1mm] \tfrac{1}{2}(t^2 - 2t + 1) - 2, & 2 \le t < 3 \end{cases}$$

The resulting signal, which is the convolution of $x(t)$ with $h(t)$, is shown in Figure 2-3g. ∎

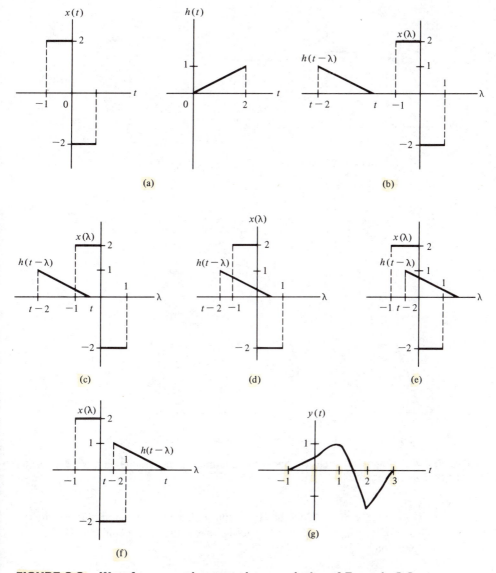

FIGURE 2-3. Waveforms pertinent to the convolution of Example 2.2.

SUPERPOSITION INTEGRAL FOR FIXED, LINEAR SYSTEMS

In this section we show that (2-1) and (2-2) can be used to obtain the output of a fixed, linear system in response to an input, $x(t)$, which we assume for the time being to be a continuous function of time in the time interval $T_1 \leq t \leq T_2$ and zero elsewhere. Figure 2-4a illustrates an arbitrary input, $x(t)$, in the time

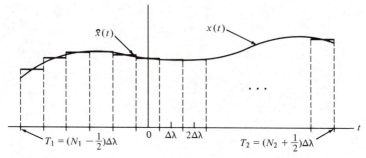

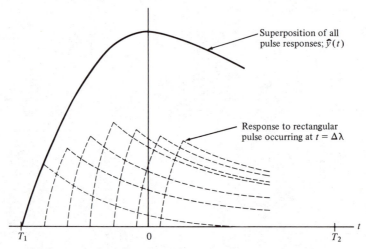

(a) Input to a fixed, linear system and a stairstep approximation

(b) Output of a fixed, linear system due to the stairstep approximate input.

FIGURE 2-4.   Input and output signals for a fixed, linear system used to derive the superposition integral.

interval $T_1 \leq t \leq T_2$ which has been approximated by a stairstep function $\hat{x}(t)$. That is, $\hat{x}(t)$ can be expressed in terms of the unit pulse function $\Pi(t)$ as the summation

$$\hat{x}(t) = \sum_{n=N_1}^{N_2} x(n\,\Delta\lambda)\Pi\left(\frac{t - n\,\Delta\lambda}{\Delta\lambda}\right)$$

$$= \sum_{n=N_1}^{N_2} x(n\,\Delta\lambda)\,\frac{1}{\Delta\lambda}\,\Pi\left(\frac{t - n\,\Delta\lambda}{\Delta\lambda}\right)\Delta\lambda, \qquad T_1 \leq t \leq T_2$$

(2-5)

where $T_1 = (N_1 - \frac{1}{2})\,\Delta\lambda$ and $T_2 = (N_2 + \frac{1}{2})\,\Delta\lambda$. The reason for multiplying and dividing by $\Delta\lambda$ to obtain the second equation will be apparent shortly. Clearly, the approximation of $\hat{x}(t)$ to $x(t)$ improves as $\Delta\lambda$ gets smaller. In fact, in the limit as $\Delta\lambda \to 0$, the second equation of (2-5) becomes the sifting integral for the unit impulse function since

$$\lim_{\Delta\lambda\to 0} \frac{\Pi\left[(t - n\,\Delta\lambda)/\Delta\lambda\right]}{\Delta\lambda} = \delta\,(t - \lambda)$$

where $n\,\Delta\lambda$ has been replaced by the continuous variable λ.

Now consider the response of a fixed linear system to the summation of pulses (2-5). Let the response of the system to $\Pi(t/\Delta\lambda)/\Delta\lambda$ be $\hat{h}(t)$. That is, in terms of the operator notation introduced in Chapter 1,

$$\hat{h}(t) = \mathcal{H}\left[\frac{\Pi(t/\Delta\lambda)}{\Delta\lambda}\right] \tag{2-6}$$

where $\mathcal{H}[\cdot]$ signifies the operation that produces the system output in response to a particular input. We want the output of a fixed, linear system in response to the input (2-5). In terms of the operator $\mathcal{H}[\cdot]$, it is

$$\hat{y}(t) = \mathcal{H}[\hat{x}(t)] = \mathcal{H}\left[\sum_{n=N_1}^{N_2} x(n\,\Delta\lambda)\,\frac{1}{\Delta\lambda}\,\Pi\left(\frac{t - n\Delta\lambda}{\Delta\lambda}\right)\Delta\lambda\right] \tag{2-7}$$

Because $\mathcal{H}[\cdot]$ represents a linear system, superposition holds and (2-7) can be written as

$$\hat{y}(t) = \sum_{n=N_1}^{N_2} x(n\,\Delta\lambda)\,\mathcal{H}\left[\frac{1}{\Delta\lambda}\,\Pi\left(\frac{t - n\,\Delta\lambda}{\Delta\lambda}\right)\right]\Delta\lambda \tag{2-8}$$

where we assume that all initial conditions are zero. More will be said about nonzero initial conditions shortly. Although the superposition property is stated by (1-38) for inputs of the form $x(t) = \alpha_1 x_1(t) + \alpha_2 x_2(t)$, we may identify α_1 as $x(N_1\,\Delta\lambda)\,\Delta\lambda$, $x_1(t)$ as $\Pi[(t - N_1\,\Delta\lambda)/\Delta\lambda]/\Delta\lambda$, and $x_2(t)$ as $\sum_{n=N_1+1}^{N_2} x(n\,\Delta\lambda)$ $\cdot\,\Delta\lambda\,\Pi[(t - n\,\Delta\lambda)/\Delta\lambda]\,\Delta\lambda$ with $\alpha_2 = 1$. The superposition property is applied $N_2 - N_1 - 1$ times in this manner to arrive at (2-8).

Because the system is assumed to be fixed, it follows that

$$\mathcal{H}\left[\frac{1}{\Delta\lambda}\,\Pi\left(\frac{t - n\Delta\lambda}{\Delta\lambda}\right)\right] = \hat{h}(t - n\,\Delta\lambda) \tag{2-9}$$

where $\hat{h}(t)$ was defined as the response of the system to $\Pi(t/\Delta\lambda)/\Delta\lambda$ in (2-6). That is, the system's response to a pulse centered at $t = n\Delta\lambda$ is simply $\hat{h}(t)$ shifted by $t = n\Delta\lambda$. Using (2-9) in (2-8) we can express the system output, $\hat{y}(t)$, in response to the stairstep approximation to $x(t)$, $\hat{x}(t)$, as

$$\hat{y}(t) = \sum_{n=N_1}^{N_2} x(n\,\Delta\lambda)\hat{h}(t - n\,\Delta\lambda)\,\Delta\lambda \tag{2-10}$$

which shows that $\hat{y}(t)$ is the *superposition* of $N = N_2 - N_1$, elementary responses, $\hat{h}\,(t - n\,\Delta\lambda)$, of the system to $\Pi[(t - n\,\Delta\lambda)/\Delta\lambda]/\Delta\lambda$, each of which is weighted by the value of the input signal at the time $t = n\,\Delta\lambda$. A typical superposition of elementary responses is shown in Figure 2-4b corresponding to the stairstep input of Figure 2-4a. Note that $\hat{h}(t)$ is a characteristic of the system in the sense that it tells us how the system responds to the test pulse $\Pi(t/\Delta\lambda)/\Delta\lambda$ of width $\Delta\lambda$ and height $1/\Delta\lambda$. To remove the dependence of $\hat{h}(t)$ on $\Delta\lambda$, we *define* the *impulse response* of the system as

$$h(t) = \lim_{\Delta\lambda \to 0} \hat{h}(t) = \lim_{\Delta\lambda \to 0} \mathcal{H}\left[\frac{\Pi(t/\Delta\lambda)}{\Delta\lambda}\right]$$

where we recall that zero initial conditions were assumed. By interchanging the order of the limit and the operator $\mathcal{H}[\cdot]$, which is a linear operator by assumption, and recalling that $\lim_{\Delta\lambda \to 0} \Pi(t/\Delta\lambda)/\Delta\lambda$ has the properties of a unit impulse function, it is seen that *the impulse response of a system is its response to a unit impulse applied at time $t = 0$ with all initial conditions of the system zero.* That is,

$$h(t) \triangleq \mathcal{H}[\delta(t)] \qquad \text{(zero initial conditions)} \qquad (2\text{-}11)$$

Returning to (2-10) and taking the limit as $\Delta\lambda \to 0$ and $n\Delta\lambda \to \lambda$, a continuous variable, we recognize (2-10) as an approximation to the integral

$$y(t) = \int_{T_1}^{T_2} x(\lambda)h(t - \lambda) \, d\lambda, \qquad T_1 \le t \le T_2 \qquad (2\text{-}12)$$

Assuming that the input may have been present since the infinite past and may last indefinitely into the future, we have, in the limit as $T_1 \to -\infty$ and $T_2 \to \infty$,

$$y(t) = \int_{-\infty}^{\infty} x(\lambda)h(t - \lambda) \, d\lambda \qquad (2\text{-}13)$$

for the response of a system characterized by an impulse response $h(t)$ and having an input $x(t)$.

With the lower limit of $t = -\infty$ on (2-13), we can include the response due to initial conditions by identifying it as the response due to $x(t)$ from $t = -\infty$ to an appropriately chosen starting instant, t_0, usually chosen as $t_0 = 0$. In Chapter 4 we consider another way to handle initial conditions. Making the substitution $\sigma = t - \lambda$, we obtain the equivalent result

$$y(t) = \int_{-\infty}^{\infty} x(t - \sigma)h(\sigma) \, d\sigma \qquad (2\text{-}14)$$

Because these equations were obtained by superposition of a number of elementary responses due to each individual impulse, they are referred to as *superposition integrals*. A simplification results if the system under consideration is causal. In Chapter 1 we defined such a system as being one that did not anticipate its input. For a causal system, $h(t - \tau) = 0$ for $t < \tau$ and the upper limit of (2-13) can be set equal to t. Furthermore, if $x(t) = 0$ for $t < 0$, the lower limit becomes zero, and the resulting integral is given by

$$y(t) = \int_0^t x(\lambda)h(t - \lambda)\, d\lambda, \qquad t \geq 0 \qquad (2\text{-}15)$$

(causal system with input zero for $t < 0$ and zero initial conditions).

EXAMPLE 2-3

Consider the system with impulse response

$$h(t) = \frac{1}{RC}\, e^{-t/RC} u(t) \qquad (2\text{-}16)$$

Let the input be a unit step:

$$x(t) = u(t)$$

Applying (2-15), the output is given by

$$a(t) = \int_0^t \frac{1}{RC}\, e^{-(t-\lambda)/RC} d\lambda, \qquad t \geq 0$$

$$= \frac{1}{RC}\, e^{-t/RC} \int_0^t e^{\lambda/RC} d\lambda$$

$$= e^{-t/RC} e^{\lambda/RC} \Big|_{\lambda=0}^{\lambda=t}$$

$$= 1 - e^{-t/RC}, \qquad t \geq 0$$

Since $a(t) = 0$ for $t < 0$, this output, referred to as the *step response* of the system, can be written compactly as

$$a(t) = \left[1 - \exp\left(\frac{-t}{RC}\right) \right] u(t) \qquad (2\text{-}17)$$

Note that $a(t) = \int_{-\infty}^t h(\lambda)\, d\lambda.$ ∎

2-4

IMPULSE RESPONSE OF A FIXED, LINEAR SYSTEM

We have shown in Section 2.3 that the response of a fixed, linear system with impulse response $h(t)$ to an input $x(t)$ is the convolution of $h(t)$ and $x(t)$. In this section we therefore consider the impulse response of a linear system. Although the impulse response of a time-varying system can be defined also, we limit our consideration to *fixed* linear systems.

We have defined the *impulse response,* denoted $h(t)$, of a fixed, linear system, assumed initially unexcited, to be *the response of the system to a unit impulse applied at time $t = 0$.* Since the system is assumed to be fixed, the response to an impulse applied at some time other than zero, say $t = \tau$, is simply $h(t - \tau)$. We now illustrate by example two techniques for obtaining the impulse response of a system.

EXAMPLE 2-4

Find the impulse response of a system modeled by the differential equation

$$\tau_0 \frac{dy(t)}{dt} + y(t) = x(t), \qquad -\infty < t < \infty \qquad (2\text{-}18)$$

where $x(t)$ is the input and $y(t)$ the output.

Solution: Setting $x(t) = \delta(t)$ results in the response $y(t) = h(t)$. For $t > 0$, $\delta(t) = 0$, so that the governing differential equation for the impulse response is

$$\tau_0 \frac{dh(t)}{dt} + h(t) = 0, \qquad t > 0$$

Assuming a solution of the form

$$h(t) = Ae^{pt}$$

and substituting into the equation for $h(t)$, we obtain

$$(\tau_0 p + 1)Ae^{pt} = 0$$

which is satisfied if

$$p = \frac{-1}{\tau_0}$$

Thus

$$h(t) = Ae^{-t/\tau_0}, \qquad t > 0 \qquad (2\text{-}19)$$

To fix A, we require an initial condition for $h(t)$. The system is unexcited for $t < 0$. Therefore, from the definition of the impulse response,

$$h(t) = 0, \qquad t < 0,$$

From (2-18) it follows that the impulse response for $t \geq 0$ obeys the differential equation

$$\tau_0 \frac{dh(t)}{dt} + h(t) = \delta(t), \qquad t \geq 0 \qquad (2\text{-}20)$$

For the left-hand side to be identically equal to the right-hand side, one of the terms on the left-hand side must contain an impulse at $t = 0$. It cannot be $h(t)$, for then $\tau_0 [dh(t)/dt]$ would contain a doublet,[†] and there is no doublet on the right-hand side. Thus $h(t)$ is discontinuous at $t = 0$, but no impulse is present in $h(t)$. Therefore, its integral through $t = 0$ must be zero,

$$\int_{t=0^-}^{0^+} h(t)\, dt = 0$$

[†]A doublet was defined as $d\delta(t)/dt$ in Chapter 1.

and the integral of (2-20) from $t = 0^-$ to $t = 0^+$ gives

$$\int_{t=0^-}^{0^+} \tau_0 \frac{dh(t)}{dt}\, dt + \int_{t=0^-}^{0^+} h(t)\, dt = \int_{t=0^-}^{0^+} \delta(t)\, dt$$

$$\tau_0[h(0^+) - h(0^-)] + \qquad 0 \qquad = \qquad 1$$

Since $h(0^-) = 0$, it follows that

$$h(0^+) = \frac{1}{\tau_0}$$

Setting this result equal to (2-19) with $t = 0^+$ shows that

$$A = \frac{1}{\tau_0}$$

Thus

$$h(t) = \begin{cases} 0, & t < 0 \\ \dfrac{1}{\tau_0}\, e^{-t/\tau_0}, & t \geq 0 \end{cases} \qquad (2\text{-}21)$$

is the impulse response of the system. ∎

EXAMPLE 2-5

Considering next the RC circuit shown in Figure 2-5, we see that the governing differential equation is (2-18) with $\tau_0 = RC$. We will find the impulse response by considering the physical properties of the resistor and capacitor.

At $t = 0^-$, the capacitor is uncharged and acts as a short circuit (infinite charge sink) to any current flowing through R. With $x(t) = \delta(t)$, the current through R at $t = 0$ is given by

$$i(t) = \frac{\delta(t)}{R}, \qquad 0^- \leq t \leq 0^+$$

The charge stored in C due to this initial current flow is

$$q(0^+) = \int_{0^-}^{0^+} \frac{\delta(t)}{R}\, dt = \frac{1}{R}$$

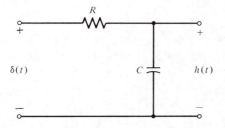

FIGURE 2-5. *RC circuit to illustrate the calculation of impulse response.*

Thus the capacitor voltage is $v_c(0^+) = 1/RC$. For $t > 0$, $x(t) = 0$; that is, the input is a short circuit and C discharges. The voltage across C exponentially decays according to

$$v_c(t) = Ae^{-t/RC}, \qquad t > 0$$

Setting $v_c(0^+) = A = 1/RC$ and noting that $h(t) = v_c(t)$, we obtain the same result for $h(t)$ as in Example 2.4 with $\tau_0 = RC$. ∎

2-5

SUPERPOSITION INTEGRALS IN TERMS OF STEP RESPONSE

We saw in Example 2-3 that the step response of the RC circuit shown in Figure 2-5 is simply the integral of the impulse response of the circuit. We now ask: Can the response of a system to an arbitrary input be expressed in terms of its step response? The answer is yes.

To find the appropriate relationship, consider the superposition integral in terms of impulse response. Repeating (2-14) here for convenience, we have

$$y(t) = \int_{-\infty}^{\infty} x(t - \lambda)h(\lambda)\, d\lambda$$

We employ the formula for integration by parts:

$$\int_a^b u\, dv = uv\,\Big|_a^b - \int_a^b v\, du \qquad (2\text{-}22)$$

with $u = x(t - \lambda)$ and $dv = h(\lambda)\, d\lambda$. Thus

$$v(\lambda) = \int_{-\infty}^{\lambda} h(\zeta)\, d\zeta \triangleq a(\lambda)$$

is simply the step response, and

$$du(\lambda) = -\dot{x}(t - \lambda)\, d\lambda$$

where the overdot denotes differentiation with respect to the argument. Substituting these expressions into (2-22), we obtain

$$y(t) = a(\lambda)x(t - \lambda)\,\Big|_{\lambda = -\infty}^{\infty} + \int_{-\infty}^{\infty} \dot{x}(t - \lambda)a(\lambda)\, d\lambda \qquad (2\text{-}23)$$

The system is initially unexcited, so that $a(-\infty) = 0$ and $x(t - \lambda)|_{\lambda = \infty} = 0$. Thus the first term is zero when the limits are substituted, and the final result is

$$y(t) = \int_{-\infty}^{\infty} \dot{x}(t - \lambda)a(\lambda)\, d\lambda \qquad (2\text{-}24)$$

The change of variables $\eta = t - \lambda$ results in the alternative form

$$y(t) = \int_{-\infty}^{\infty} \dot{x}(\eta)a(t - \eta)\, d\eta \qquad (2\text{-}25)$$

Thus in terms of the step response $a(t)$, the response of a system to an input $x(t)$ is the *convolution of the derivative of the input with the step response*. Equations (2-24) and (2-25) are known as *DuHamel's integrals*. Note the similarity to (2-13) and (2-14).

EXAMPLE 2-6

Consider a system with a ramp input for which

$$x(t) = tu(t)$$

$$\dot{x}(t) = u(t)$$

Applying (2-25), we obtain the response to a ramp as

$$y_R(t) = \int_{-\infty}^{\infty} u(t - \lambda)a(\lambda)\, d\lambda$$

$$= \int_{-\infty}^{t} a(\lambda)\, d\lambda \qquad (2\text{-}26)$$

Thus the response of a system to a unit ramp, which is the integral of the unit step, is the integral of the step response. ∎

Generalizing, we conclude that *for a fixed, linear system, any linear operation on the input produces the same linear operation on the output.*

EXAMPLE 2-7

Find the response of the *RC* circuit of Figure 2-5 to the triangular signal

$$x_\Delta(t) = r(t) - 2r(t - 1) + r(t - 2)$$

Solution: We first substitute (2-17) into (2-26) to obtain the ramp response. Thus

$$y_R(t) = \int_{-\infty}^{t} \left[1 - \exp\left(\frac{-\lambda}{RC} \right) \right] u(\lambda)\, d\lambda$$

$$= r(t) + RC\left[\exp\left(\frac{-\lambda}{RC} \right) \right]_0^t u(t)$$

$$= r(t) - RC\left[1 - \exp\left(\frac{-t}{RC} \right) \right] u(t)$$

By superposition, the response to the triangle $x_\Delta(t)$ is $y_\Delta(t) = y_R(t) - 2y_R(t - 1) + y_R(t - 2)$. This input and output are compared in Figure 2-6. Note that for $RC \ll 1$, the output closely approximates the input, whereas if $RC = 1$, the output does not resemble the input. ∎

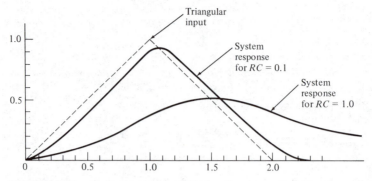

FIGURE 2-6. Response of an *RC* circuit to a triangular input signal.

EXAMPLE 2-8

We again consider the *RC* circuit of Figure 2-5 and compute its impulse response by first finding its step response and then differentiating it. The differential equation relating input and output is

$$RC \frac{dy(t)}{dt} + y(t) = x(t)$$

which is (2-18) with $\tau_0 = RC$. With a unit step input, this equation can be written as

$$RC \frac{da(t)}{dt} + a(t) = 1, \qquad t \geq 0$$

where $a(t)$ is the step response. We have already considered the solution of this differential equation in Example 1-9 for an arbitrary input. As mentioned in that example, the solution consists of the sum of a homogeneous solution, which is

$$a_h(t) = A \exp\left(\frac{-t}{RC}\right)$$

plus a particular solution for the forcing function $u(t)$. This particular solution is simply $a_p(t) = 1$, which can be seen by direct substitution into the nonhomogeneous differential equation for $a(t)$. Thus

$$a(t) = a_h(t) + a_p(t) = A \exp\left(\frac{-t}{RC}\right) + 1, \qquad t \geq 0$$

The initial charge on the capacitor is zero for $t < 0$. Since the forcing function does not contain impulses (it is a step and therefore cannot change the capacitor charge instantaneously), it follows that the charge on the capacitor at $t = 0^+$ is zero. Therefore, $a(0^+) = 0$ or

$$a(0^+) = A \exp\left(\frac{-t}{RC}\right)_{t=0} + 1 = A + 1 = 0$$

and

$$a(t) = 1 - \exp\left(\frac{-t}{RC}\right), \qquad t \geq 0$$

Using the fact that $a(t) = 0$ for $t < 0$, we may write this as

$$a(t) = \left[1 - \exp\left(\frac{-t}{RC}\right)\right]u(t)$$

Since $\delta(t) = du(t)/dt$ and using the principle stated after Example 2.6 that for a fixed, linear system, any linear operation on the input produces the same linear operation on the output, we can obtain the impulse response by differentiating $a(t)$ with respect to t. The result is

$$h(t) = \frac{da(t)}{dt} = \left[1 - \exp\left(\frac{-t}{RC}\right)\right]\delta(t) + \frac{1}{RC}\exp\left(\frac{-t}{RC}\right)u(t)$$

$$= \frac{1}{RC}\exp\left(\frac{-t}{RC}\right)u(t)$$

which follows by using the chain rule for differentiation and noting that $[1 - \exp(-t/RC)] = 0$ for $t = 0$. ∎

EXAMPLE 2-9

Our final example involves finding the response of the circuit of Figure 2-5 to the rectangular pulse input

$$\delta_\epsilon(t) = \frac{1}{2\epsilon}[u(t + \epsilon) - u(t - \epsilon)]$$

and taking the limit of the output as $\epsilon \to 0$. As illustrated by Figure 1-12, $\delta_\epsilon(t)$ is a suitable approximation for $\delta(t)$ as $\epsilon \to 0$. From Example 2.8 and superposition, it follows that the response to $\delta_\epsilon(t)$ is

$$y_\epsilon(t) = \frac{1}{2\epsilon}\{[1 - e^{-(t+\epsilon)/RC}]u(t + \epsilon)$$

$$- [1 - e^{-(t-\epsilon)/RC}]u(t - \epsilon)\}$$

which is zero for $t < -\epsilon$. If $t > \epsilon$, the response is

$$y_\epsilon(t) = \frac{1}{2\epsilon}[e^{-(t-\epsilon)/RC} - e^{-(t+\epsilon)/RC}], \qquad t > \epsilon$$

Factoring out the term $\exp(-t/RC)$, we may write $y_\epsilon(t)$ as

$$y_\epsilon(t) = \left[\frac{\exp(\epsilon/RC) - \exp(-\epsilon/RC)}{2\epsilon}\right]\exp\left(\frac{-t}{RC}\right), \qquad t > \epsilon$$

In the limit as $\epsilon \to 0$, the expression in brackets assumes the form $0/0$. We may use l'Hospital's rule to find its limit as $\epsilon \to 0$ or expand the bracketed term to first-order terms in ϵ. Using the Taylor series $\exp(u) = 1 + u/1! + u^2/2! +$

$u^3/3! + \cdots$, we find that

$$y_\epsilon(t) = \left(\frac{1}{RC} + \frac{\epsilon^2}{6RC} + \cdots \right) \exp \left(\frac{-t}{RC} \right), \qquad t > \epsilon$$

Taking the limit as $\epsilon \to 0$, we obtain $\lim_{\epsilon \to 0} y_\epsilon(t) = (1/RC) \exp(-t/RC)$, $t > 0$, which is the same result as obtained previously for the impulse response. ■

2-6

STABILITY OF LINEAR SYSTEMS

One of the considerations in any system design is the question of stability. Although there are various definitions of stability in common usage, the one we shall use for the present is referred to as *bounded-input, bounded-output* (BIBO) *stability*. By definition, *a system is BIBO stable if and only if every bounded input results in a bounded output*. Clearly, stability is a desirable property of most systems.

For a fixed, linear system we may obtain a condition on the impulse response which guarantees BIBO stability. To derive this condition, consider (2-13), which relates the input and output of a fixed, linear system:

$$y(t) = \int_{-\infty}^{\infty} x(\lambda) h(t - \lambda) \, d\lambda$$

It follows that

$$|y(t)| = \left| \int_{-\infty}^{\infty} x(\lambda) h(t - \lambda) \, d\lambda \right| \le \int_{\infty}^{\infty} |x(\lambda)| \, |h(t - \lambda)| \, d\lambda \qquad (2\text{-}27)$$

If the input is bounded, then

$$|x(\lambda)| \le M < \infty$$

where M is a finite constant. Replacing $|x(\lambda)|$ by M in (2-27), we have the inequality

$$|y(t)| \le M \int_{-\infty}^{\infty} |h(t - \lambda)| \, d\lambda$$

or

$$|y(t)| \le M \int_{-\infty}^{\infty} |h(\eta)| \, d\eta$$

which follows by the change of variables $\eta = t - \lambda$. Thus the output is bounded provided that

$$\int_{-\infty}^{\infty} |h(t)| \, dt < \infty \qquad (2\text{-}28)$$

That is, (2-28) is a sufficient condition for stability.

To show that it is also a necessary condition, consider the input

$$x(\lambda) = \begin{cases} +1 & \text{if } h(t - \lambda) > 0 \\ 0 & \text{if } h(t - \lambda) = 0 \\ -1 & \text{if } h(t - \lambda) < 0 \end{cases}$$

For this input it follows that

$$|y(t)| = \left| \int_{-\infty}^{\infty} |h(t - \lambda)| \, d\lambda \right| = \int_{-\infty}^{\infty} |h(\eta)| \, d\eta$$

for any fixed value of t since the integrand is always nonnegative. Thus the output will be unbounded if (2-28) is not satisfied. That is, (2-28) is also a necessary condition for BIBO stability.

EXAMPLE 2-10

The system of Example 2-5 is BIBO stable since

$$\int_{-\infty}^{\infty} \frac{1}{RC} e^{-t/RC} u(t) \, dt = \int_{0}^{\infty} \exp(-v) \, dv$$

$$= -\exp(-v) \Big|_{0}^{\infty}$$

$$= 1 < \infty$$

Thus the system is BIBO stable ∎

We will return to the idea of stability of a system in Chapter 5, where additional methods for determining the stability of systems describable by constant coefficient, linear differential equations, are discussed.

*2-7

SYSTEMATIC PROCEDURES FOR WRITING GOVERNING EQUATIONS FOR LUMPED SYSTEMS

In the analysis of simple systems composed of two or three components, it is usually possible to write down the governing equations by inspection. In more complicated cases, however, we require systematic procedures for obtaining the system equations. For lumped systems, this procedure consists of the following basic steps:

1. Identify the important system elements or components and determine their individual *describing equations*; examples are Ohm's law, $v = iR$, for a resistor, or Coulomb's law, $v = q/C$, for a capacitor.
2. Write down the *combining equations,* which relate the variables of the individual elements to each other; examples are Kirchhoff's voltage and current laws, and application of d'Alembert's principle.

*Optional section.

3. Eliminate all variables that are not of interest by means of the combining equations. The variables of interest will include, but may not be limited to, the input and output variables. Quite often, other auxiliary variables will be involved.
4. Examine the model closely to determine if it is an adequate representation of the physical system. If stray capacitance, lead inductance, or nonlinear effects are of importance, these may be added to the model. Quite often, the first solution for the desired quantities will be carried out with the simplest model that is deemed adequate and refinements will be made in subsequent solutions.
5. Solve the governing equations of the system using a systematic method of solution.

We now discuss further the procedures involved in each of these steps. In order to be able to solve the resulting system equations analytically, it is almost a necessity (except in a few specific cases) that the describing equations be linear. Furthermore, many physical components are well approximated by linear relationships. Therefore, we limit our attention to system components that are described by linear relationships. The resulting system equations then satisfy the superposition principle.

Lumped, Linear Components and Their Describing Equations

The student is already familiar with the describing equations for lumped, linear circuit elements. These are the basic relations or laws relating the terminal currents and voltages for resistance, inductance, and capacitance. Each of these equations constitutes a model of the physical device by mathematically relating two variables that can be measured at the device terminals; in particular, the voltage *across* the device and the curent *through* it. To generalize such describing equations to other physical components, such as mechanical or hydraulic components, it is useful to think of the terminal variables as either *across variables* or *through variables*. An *across* variable is one that is measured by placing an appropriate meter *across* the terminals of the device; a *through* variable is measured by placing an appropriate meter in series with the device so that the measured quantity is transmitted *through* the meter. Thus voltage is an across variable and current is a through variable. For a mechanical device such as a spring, the appropriate variables are force and displacement (or, possibly, the velocity) of one end with respect to the other. To measure the force transmitted by the spring to another object, we would need to place a force meter in series with the spring so that force is a *through* variable. Measurement of the displacement of one end relative to the other, on the other hand, would require that a ruler be placed across the spring. Thus displacement (and relative velocity, since it is the derivative of the displacement) is an *across* variable.

In writing down the describing equations for a device, some convention regarding sign must be adopted. Each device is represented by a schematic diagram symbol with reference directions shown for the through and across variables. Consider, as an example, a resistance of value R, the schematic representation of which, together with reference directions for through (current) and across (voltage) variables, is shown in Figure 2-7. The describing equation is Ohm's law, given by

$$v = iR \qquad (2-29)$$

The convention we will adopt regarding signs for the through and across variables is the following. To measure the through variable (current), we must insert an instantaneous reading meter in series with the device. If the actual through variable flow is in the direction of the reference arrow on the schematic representation of the device, the meter will read a positive value for the through variable. For the across variable (voltage), the convention is that the meter is connected so that the product of instantaneous across- and through-variable meter readings is *positive* when power is *delivered* to the device.

The describing equations for several types of components are summarized in Table 2-1. The terminology and conventions regarding signs is that described above. Symbols that are used in schematic diagrams showing interconnection of various devices are also shown.

In addition to the passive elements defined in Table 2-1, ideal sources are also useful models. These include both independent and dependent sources. For electrical systems, the two types of independent sources are constant-voltage and constant-current sources; for translational mechanical systems, the two types of independent sources are constant-velocity and constant-force sources, with analogous sources defined for rotational mechanical systems. By this terminology, of course, we do not mean that a source delivers a time-independent quantity, be it electrical or mechanical, but rather that the delivered quantity does not depend on the load placed on the source. For example, an ideal velocity source given by

$$\dot{\Delta}_s(t) = A \cos \omega_0 t$$

maintains the velocity $A \cos \omega_0 t$ regardless of the forces applied.

A controlled source produces a terminal quantity whose value is proportional to some other system variable but is independent of the load placed on the source. For example an ideal angular velocity source that is controlled by the current $i_1(t)$ can be represented by the equation

$$\dot{\theta}_s(t) = KF[i_1(t)]$$

independent of the load placed on the source, where $F(\cdot)$ is a specified function and K is a constant of proportionality.

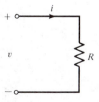

FIGURE 2-7. Schematic representation of a resistor, together with reference directions for the associated through and across variables.

TABLE 2-1 Electrical, Translational Mechanical, and Rotational Mechanical Elements

Component; Symbol (Units)	Schematic Symbol	Across Variable; Units	Through Variable; Units	Describing Equations
Resistance, R (ohms)		Voltage (volts)	Current (amperes)	$v = iR$ $i = \dfrac{1}{R} v$
Inductance, L (henries)		Voltage (volts)	Current (amperes)	$v = L \dfrac{di}{dt}$ $i = \dfrac{1}{L} \displaystyle\int_{-\infty}^{t} v(\lambda)\, d\lambda$
Capacitance, C (farads)		Voltage (volts)	Current (amperes)	$v = \dfrac{1}{C} \displaystyle\int_{-\infty}^{t} i(\lambda)\, d\lambda$ $i = C \dfrac{dv}{dt}$
Translational damper, B (kg/s)		Velocity (m/s)	Force (newtons)	$\dot{\Delta} = \dfrac{1}{B} f$ $f = B \dot{\Delta}$
Translational spring, K (kg/s^2)		Velocity (m/s)	Force (newtons)	$\dot{\Delta} = \dfrac{1}{K} \dfrac{df}{dt}$ $f = K\Delta$ $= K \displaystyle\int_{-\infty}^{t} \dot{\Delta}(\lambda)\, d\lambda$
Mass, M (kg)		Velocity (m/s)	Force (newtons)	$\dot{\Delta} = \dfrac{1}{M} \displaystyle\int_{-\infty}^{t} f(\lambda)\, d\lambda$ $f = M \dfrac{d\dot{\Delta}}{dt}$
Rotational damper, B (kg-m^2/s)		Angular velocity (rad/s)	Torque (N−m)	$\dot{\theta} = \dfrac{1}{B} T$ $T = B \dot{\theta}$
Rotational spring, K (N − m)		Angular velocity (rad/s)	Torque (N−m)	$\dot{\theta} = \dfrac{1}{K} \dfrac{dT}{dt}$ $T = K \displaystyle\int_{-\infty}^{t} \dot{\theta}(\lambda)\, d\lambda$
Rotational mass, J (kg − m^2) (moment of inertia)		Angular velocity (rad/s)	Torque (N−m)	$\dot{\theta} = \dfrac{1}{J} \displaystyle\int_{-\infty}^{t} T(\lambda)\, d\lambda$ $T = J \dfrac{d\dot{\theta}}{dt}$

EXAMPLE 2-11

To illustrate some of the ideas introduced here and demonstrate, at least in a particular case, that the analogies given in Table 2-1 are valid, consider the simple mechanical system shown in Figure 2-8. The mass has force F_s acting on it and is assumed to move on rollers without friction. Its displacement from equilibrium is Δ as shown. From d'Alembert's principle we have

$$M \frac{d\dot{\Delta}}{dt} + B\dot{\Delta} + K\Delta = F_s$$

where B is the coefficient of viscous friction describing the damper as shown in Table 2-1. We wish to show that by suitable association of elements and variables, this equation may be represented by an electrical network. First we rewrite the equation as

$$M \frac{d\dot{\Delta}}{dt} + B\dot{\Delta} + K \int \dot{\Delta} \, dt = F_s \qquad (2\text{-}30)$$

We have chosen to write the equation in terms of velocity $\dot{\Delta}$ since, as we shall see, it is analogous to voltage. Now consider the parallel RLC network shown in Figure 2-9. From KCL and the element relations we obtain the differential equation for the voltage v as

$$C \frac{dv}{dt} + \frac{v}{R} + \frac{1}{L} \int v \, dt = I_s \qquad (2\text{-}31)$$

Examination of (2-30) and (2-31) shows that the equations are exactly the same if we make the following associations:

$$M \longleftrightarrow C$$

$$B \longleftrightarrow \frac{1}{R}$$

$$K \longleftrightarrow \frac{1}{L}$$

$$\dot{\Delta} \longleftrightarrow v$$

$$F_s \longleftrightarrow I_s$$

■

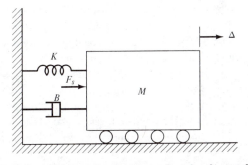

FIGURE 2-8. Schematic diagram of a simple mechanical system.

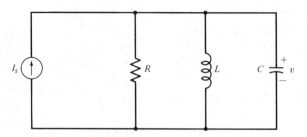

FIGURE 2-9.   Schematic diagram of a parallel *RLC* network.

Since the mechanical and electrical systems are described by the same differential equation, we say that the electrical network and the mechanical system are *analogous* to each other; in particular:

1. Mass M is analogous to capacitance C.
2. Coefficient of friction B is analogous to conductance $G = B = 1/R$.
3. Spring constant K is analogous to the inverse of inductance $1/L = K$.
4. Across-variable velocity is analogous to across-variable voltage.
5. Through-variable force is analogous to through-variable current.

Analogies also exist for other types of systems, as suggested by Table 2-1. Clearly, such analogies can aid electrical engineers in understanding and analyzing other kinds of systems. The power of the analogous-network concept lies in the fact that the analysis of electrical networks is particularly well established. Therefore, complicated networks can be handled with relative ease. If the concept of analogous networks can be extended to more complex systems, we will indeed have a powerful tool for systems analysis. To do this we need systematic procedures for obtaining the analogous network or, perhaps more conveniently, the system graph. These topics are examined in more detail in the following section.

System Graphs

The concept of a system graph, although not absolutely essential for the writing of equilibrium equations for many lumped systems, nevertheless provides a systematic procedure that one may resort to in cases where these equations are not readily written down by inspection. The graph of a system of two-terminal components is obtained by simply representing each component by a directed line segment. The direction, indicated by an arrow, shows the positive reference sense for the across and through variables of the component. Each line segment is terminated by two nodes. An example is given in Figure 2-10, which shows the schematic representation of an electrical system and its graph. The system graph of an electrical network bears a resemblance to the shape of the schematic diagram of the original circuit, a characteristic that is not true of mechanical or other systems.

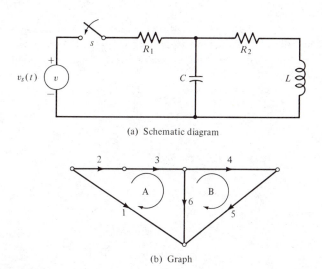

(a) Schematic diagram

(b) Graph

FIGURE 2-10. Example that illustrates the concept of a system graph of an electrical circuit.

The describing equations† for the components of the system of Figure 2-10 are:

$$v_1(t) = v_s(t)$$

$$i_2(t) = 0, \quad \text{switch open}$$

$$v_2(t) = 0, \quad \text{switch closed}$$

$$v_3(t) = R_1 i_3(t)$$

$$v_4(t) = R_2 i_4(t)$$

$$v_5(t) = L \frac{di_5}{dt}$$

and

$$v_6(t) = \frac{1}{C} \int_{-\infty}^{t} i_6(\lambda) \, d\lambda$$

Note that the component describing equations may be linear or nonlinear; there is no restriction from the standpoint of constructing the system graph.

As a second example, consider the translational mechanical system shown schematically in Figure 2-11, which represents the suspension system on an automobile. The mass M_1 represents the mass of the automobile, which is coupled to the axle and wheels by a spring and damper system. The smaller mass, M_2, of the axle and wheels is coupled to the ground by the tires, which are represented by a second spring and damper system. The automobile frame is

†Subscripts refer to the numbering of the graph branches.

driven by a force generator $f_s(t)$. The system graph is shown in Figure 2-11b, where it is noted that the component graph for the masses has nodes located by the mass and "earth." The terminal equations for the system components are, from Table 2-1, given by

$$f_1(t) = f_s(t)$$

$$f_2(t) = B_1 \dot{\Delta}_2(t)$$

$$f_3(t) = K_1 \Delta_3(t)$$

$$f_4(t) = M_1 \frac{d\dot{\Delta}_4(t)}{dt}$$

$$f_5(t) = B_2 \dot{\Delta}_5(t)$$

$$f_6(t) = K_2 \Delta_6(t)$$

and

$$f_7(t) = M_2 \frac{d\dot{\Delta}_7(t)}{dt}$$

Node and Circuit Postulates

We next need a procedure for combining component equations to obtain the system equations. A basis for this is provided by graph theory, a few basic definitions for which are given below.

1. Each line element of a system graph is called a *branch*. An indicated direction on a branch makes it an *oriented branch*.
2. A *node* is the end point of a branch.
3. An *oriented graph* is a set of oriented branches, no two of which have a point in common which is not a node.
4. A *subgraph* is any subset of the branches of a graph.
5. A *circuit* is a subgraph such that two and only two branches are incident on each node of the subgraph.
6. The *complement* $\overline{B}$ of a subgraph G is the subgraph remaining in G when the branches of $\overline{B}$ are removed.
7. A connected subgraph of a graph is called a *separate part* if it contains no nodes in common with its complement.

With these definitions stated, we can now give two fundamental properties that relate across and through variables of the elements of a system graph.

1. *The Node Postulate.* Let the system graph of a physical system be composed of n oriented branches. Let $u_j(t)$ be the through variable of the jth element. Then at the kth node of the graph

$$\sum_{j=1}^{n} a_j u_j(t) = 0 \qquad (2\text{-}32)$$

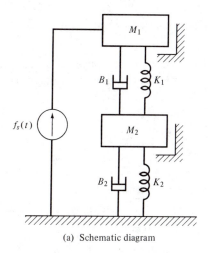

(a) Schematic diagram

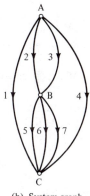

(b) System graph

FIGURE 2-11. Schematic representation and system graph for an automobile suspension system.

where

$$
a_j = \begin{cases}
0 & \text{if the } j\text{th branch is not incident at the } k\text{th node} \\
1 & \text{if the } j\text{th branch is oriented away away from the } k\text{th node} \\
-1 & \text{if the } j\text{th branch is oriented toward the } k\text{th node}
\end{cases}
$$

We recognize Postulate 1 simply as corresponding to a general statement of Kirchhoff's current law (KCL) for electrical systems. The physical principle on which it is based is the conservation of charge. For translational mechanical systems, Postulate 1 states that the force components along a coordinate direction of each system element connected at a common point must sum algebraically to zero. For example, in Figure 2-11, the damper B_1, spring K_1, mass M_1, and force generator $f_s(t)$ all must exert a net force of zero at their common juncture. We recognize this as d'Alembert's principle.

To state the second fundamental property of a system graph, we need to define the concept of *circuit orientation,* which is assigned by an oriented arrow paralleling the circuit. Examples of oriented arrows are shown in Figure 2-10b labeled A and B.

2. *The Circuit Postulate.* Let the linear graph of a physical system contain n oriented branches and let $v_j(t)$ represent the across variable of the jth branch; then for the kth circuit

$$\sum_{j=1}^{n} b_j v_j(t) = 0 \qquad (2\text{-}33)$$

where

$$b_j = \begin{cases} 0 & \text{if the } j\text{th branch is not included in the } k\text{th circuit} \\ 1 & \text{if the orientation of the } j\text{th branch is the same as the orientation chosen for the } k\text{th circuit} \\ -1 & \text{if the orientation of the } j\text{th branch is opposite to that of the } k\text{th circuit} \end{cases}$$

Postulate 2 corresponds to Kirchhoff's voltage law (KVL) for electrical systems. The physical principle on which it is based is conservation of energy. For translational mechanical systems, Postulate 2 states that the oriented sum of each component of the vectorial velocity around a circuit of the system graph must sum to zero.

As an example of writing node equations (Postulate 1), we consider the mechanical system and system graph of Figure 2-11. Considering node A, and letting the corresponding through variables be $f_1(t)$, $f_2(t)$, . . ., we have

$$f_1(t) + f_2(t) + f_3(t) + f_4(t) = 0 \qquad \text{(node } A\text{)}$$

For node B, we have

$$f_5(t) + f_6(t) + f_7(t) - f_2(t) - f_3(t) = 0 \qquad \text{(node } B\text{)}$$

and for node C, we have

$$-f_1(t) - f_5(t) - f_6(t) - f_7(t) - f_4(t) = 0 \qquad \text{(node } C\text{)}$$

We note that the three equations are not independent; for example, adding the third to the second gives the negative of the first. Therefore, we choose the first two equations. We may eliminate the through variables in each equation in favor of the across variables by using the describing equations. Note that by doing this we do not decrease the number of unknowns to be equal to the number of equations. However, suppose that we define the velocities of nodes A, B, and C relative to a common reference. Since we have eliminated node C as a Node Postulate equation, we choose its velocity as our reference and let it be $v_c(t) = 0$. Then

$$\dot{\Delta}_2(t) = v_A(t) - v_B(t) = \dot{\Delta}_3(t)$$

$$\dot{\Delta}_4(t) = v_A(t)$$

$$\dot{\Delta}_5(t) = v_B(t) = \dot{\Delta}_6(t) = \dot{\Delta}_7(t)$$

In terms of $v_A(t)$ and $v_B(t)$, the terminal equations become

$$f_1(t) = -f_s(t) \quad \text{(unknown)}$$

$$f_2(t) = B_1(v_A - v_B)$$

$$f_3(t) = K_1 \int_{-\infty}^{t} (v_A - v_B) \, d\lambda$$

$$f_4(t) = M_1 \frac{dv_A}{dt}$$

$$f_5(t) = B_2 v_B$$

$$f_6(t) = K_2 \int_{-\infty}^{t} v_B \, d\lambda$$

$$f_7(t) = M_2 \frac{dv_B}{dt}$$

Substituting these into the node A and node B equations, we obtain

$$-f_s + B_1(v_A - v_B) + K_1 \int_{-\infty}^{t} (v_A - v_B) \, d\lambda + M_1 \frac{dv_A}{dt} = 0$$

$$B_2 v_B + K_2 \int_{-\infty}^{t} v_B \, d\lambda + M_2 \frac{dv_B}{dt}$$
$$- B_1(v_A - v_B) - K_1 \int_{-\infty}^{t} (v_A - v_B) \, d\lambda = 0$$

Rearranging and differentiating, we have

$$M_1 \frac{d^2 v_A}{dt^2} + B_1 \frac{dv_A}{dt} + K_1 v_A \qquad (2\text{-}34a)$$

$$- B_1 \frac{dv_B}{dt} - K_1 v_B = \frac{df_s}{dt}$$

$$-B_1 \frac{dv_A}{dt} - K_1 v_A + M_2 \frac{d^2 v_B}{dt^2} \qquad (2\text{-}34b)$$

$$+ (B_1 + B_2) \frac{dv_B}{dt} + (K_1 + K_2) v_B = 0$$

If we wish, we may eliminate $v_A(t)$ or $v_B(t)$ to obtain a differential equation involving one unknown variable alone. The elimination is not trivial, however. A much easier method for elimination and solution is discussed in Chapter 5, which deals with applications of the Laplace transform.

As an example of writing circuit equations (Postulate 2), we consider the circuit and system graph of Figure 2-10. The circuit equation for A is

$$-v_1 + v_2 + v_3 + v_6 = 0$$

and for B the circuit equation is

$$v_4 + v_5 - v_6 = 0$$

where $v_1, v_2, \ldots$ are the across variables for branches $1, 2, \ldots$. We may eliminate the branch across variables in terms of through variables to obtain

$$-v_s + 0 + R_1 i_3 + \frac{1}{C} \int_{-\infty}^{t} i_6 \, d\lambda = 0$$

$$R_2 i_4 + L \frac{di_5}{dt} - \frac{1}{C} \int_{-\infty}^{t} i_6 \, d\lambda = 0$$

where it is assumed that the switch is closed. We note that

$$i_2 = i_A = i_3, \qquad i_4 = i_B = i_5$$

and

$$i_6 = i_A - i_B$$

When written in terms of the circuit currents, the Circuit Postulate equations are

$$R_1 \frac{di_A}{dt} + \frac{1}{C} i_A - \frac{1}{C} i_B = \frac{dv_s}{dt} \qquad (2\text{-}35a)$$

$$-\frac{1}{C} i_A + L \frac{d^2 i_B}{dt^2} + R_2 \frac{di_B}{dt} + \frac{1}{C} i_B = 0 \qquad (2\text{-}35b)$$

where each equation has been differentiated to eliminate the integrals. As with the mechanical system example, either i_A or i_B can be eliminated to obtain an equation involving one unknown variable. The order of the resulting equation would be second. The elimination of variables is left to the problems.

Topology of System Graphs

In order to select an independent set of node or circuit equations for a system, we introduce some basic concepts of graph topology. A subgraph has been defined as any subset of the branches of a graph. A subgraph that is of basic importance in determining a set of independent node or circuit equations for a system is a *tree*. A tree of a connected graph with N_v nodes has the following properties:

1. All nodes of the graph are included in the tree, and no nodes are left in isolated positions.
2. A tree contains $N_v - 1$ branches.
3. A tree includes no closed paths.

Those branches removed from a graph in forming a tree are called *chords* or *links*.

There are a number of different possible trees for a given graph. Several examples of trees for the graph of Figure 2-10b are shown in Figure 2-12.

The number of branches in a tree is obviously related to the number of nodes in the graph (and the tree), N_v, by

$$\text{number of branches in a tree} = N_v - 1 \qquad (2\text{-}36)$$

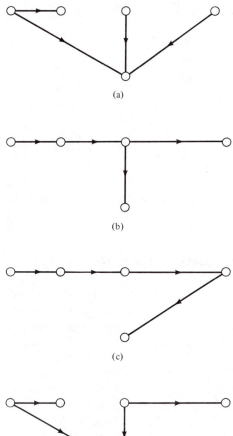

(a)

(b)

(c)

(d)

FIGURE 2-12.   Trees corresponding to the graph of Figure 2-10b.

The number of chords, N_c, in a graph is the difference between the total number, N_b, of branches in the graph and the number of branches in a tree. Thus, from (2-36), we obtain

$$N_c = N_b - (N_v - 1)$$

$$= N_b - N_v + 1$$

(2-37)

For a graph with P separate parts,

$$N_c = N_b - N_v + P$$

(2-38)

EXAMPLE 2-12

The number of branches in the graph of Figure 2-10b is

$$N_b = 6$$

and the number of nodes is

$$N_v = 5$$

The number of chords is

$$N_c = 6 - 5 + 1 = 2$$

The number of branches in the tree is

$$N_{tb} = 6 - 2 = 4 = N_v - 1$$

which checks with (2-36). ■

We now use these ideas from the topology of graphs to establish the number of independent equations that can be written for a system.

Choosing Independent Equations for a System

We assume that each element of a schematic diagram of a system is represented by one branch in the system graph. Actually, there is no need to represent the switch of Figure 2-10a by a branch if $t > 0$ (switch closed). Also, we assume that across-variable sources (voltage sources in the case of electrical systems) appear in series with other elements and that through-variable sources (current sources for electrical systems) appear in parallel with other elements.

The procedure we may follow is to use the Node Postulate (1) to write equations for $N_v - 1$ nodes after one has been chosen as a reference. We can then write, in addition, $N_b - (N_v - 1)$ equations using Circuit Postulate (2), giving a total of N_b equations. Since the Node Postulate equations involve through variables and the Circuit Postulate equations involve across variables, we will have $2N_b$ variables and only N_b equations. However, the through and across variables for each system element are related by a describing equation and we may reduce the number of variables from $2N_b$ to N_b. Hence, if the N_b equations obtained by applying Postulates 1 and 2 are independent, we may, in principle, solve for the N_b unknown through or across variables, depending on which ones remain after the describing equations have been applied.

It is perhaps a good thing at this point to review what is meant by independence of a set of equations. The equations

$$x + y = 5$$
$$2x - 2y = 10$$

are independent, whereas the pair of equations

$$x + y = 5$$
$$2x + 2y = 10$$

are not. After some reflection, the reason for the latter set not being independent is clear. The second equation can be obtained from the first by multiplying through by two. Geometrically, both sets of equations represent lines in the $x - y$

plane, the latter pair representing a pair of colinear lines, whereas the former set of equations represent a pair of lines that are perpendicular. Thus, for the first set, we may obtain a solution for x and y that corresponds to the crossing point; a unique solution is impossible to obtain for the second set.

We now turn to the problem of choosing a set of independent variables in terms of which to express the system equations. To keep the solution of these equations as simple as possible, we want the minimum number of independent variables that can be chosen. For simplicity, discussion will center on the choice of voltage and current variables for electrical systems although it will apply equally well to the choice of independent through and across variables for mechanical systems.

To proceed, consider Figure 2-13, which is the tree shown in Figure 2-12d for the network of Figure 2-10. Assume that node e is chosen as the reference, or datum, node. The voltages of nodes a, b, c, and d measured with respect to node e are defined as v_a, v_b, v_c, and v_d, respectively. Also shown in Figure 2-13 are the branch voltages (across variables), with polarities indicated by plus and minus signs to emphasize the meaning of the reference arrows shown on the tree branches. Now each node voltage can be expressed in terms of tree branch voltages. For example, v_b is given by

$$v_b = -v_1 + v_2$$

and v_c by

$$v_c = -v_3 + v_4$$

Conversely, if we know the node voltages, we may obtain the tree branch voltages and, as a matter of fact, any branch voltage. Hence a KCL equation (Postulate 1) written for each of $N_v - 1$ nodes in terms of node voltages by using the describing equation of each element will provide a set of $N_v - 1$ independent equations in terms of the $N_v - 1$ node voltages relative to the reference node. Note that $N_v - 1 < N_b$. Other choices are also available for determining the branch voltages, but these will not be discussed here.

We now investigate the possibility for the choice of current variables fewer in number than the branch currents. From Figure 2-13 we note that insertion of the link labeled 5 will allow the link current i_5 to flow. The loop current i_A is thereby established. Similarly, insertion of the link labeled 6 will allow link current i_6 to flow and the loop current i_B is thereby established. A little thought

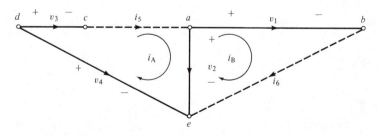

FIGURE 2-13. Tree showing appropriate voltage and current variables to establish independent through and across variables.

results in the conclusion that any of the branch currents can be expressed in terms of the loop currents i_A and i_B. For example,

$$i_2 = i_A - i_B$$

Thus the link currents and the loop currents defined thereby constitute an independent set of variables. We may write

$$N_c = N_b - N_v + 1$$

KVL (Postulate 2) equations in terms of these N_c loop currents by employing the describing equations to eliminate the across variables for each element in favor of their through variables (branch currents in this discussion) and, in turn, express these branch currents in terms of loop currents defined by the link currents. Thus we will have $N_c = N_b - N_v + 1$ independent equations involving N_c loop current variables.

Usually, it will not be necessary to construct a tree to obtain an independent set of KCL or KVL equations. For example, we note that the circuit (KVL) equations written in terms of loop currents defined by the windows or meshes of a planar network will always give an independent set of equations. The node equations (KCL) written in terms of the $N_v - 1$ node voltages with one node used as a reference also result in a set of independent equations. Whether one chooses loop current equations or node voltage equations usually depends on whether $N_c = N_b - N_v + 1$ or $N_v - 1$ is smaller.

We close this chapter with two examples that illustrate the use of the concepts introduced in this section. The first example involves a circuit. The second is an electromechanical system that involves controlled sources.

EXAMPLE 2-13

Consider the cubic array of 1-Ω resistors shown in Figure 2-14a and its associated graph shown in Figure 2-14b. Figure 2-14c shows one possible tree. The links are the branches numbered 2, 8, 9, 11, 12, and 14. Insertion of these links establishes loop currents i_A, i_B, . . ., i_F, respectively.

Suppose that we desire v_o with $v_I = 2$ V. To minimize the chance for error, we need a systematic procedure to solve for V_o. In order to conveniently relate branch currents and loop currents, we construct Table 2-2, which is called a *tie-set schedule*. The method of construction is the following. If a loop current established by the insertion of a link includes the ith branch of the graph, a $+1$

TABLE 2-2. Tie-Set Schedule for Circuit of Figure 2-14

Loops	Branch Number													
	1	2	3	4	5	6	7	8	9	10	11	12	13	14
A	−1	1	1	0	0	0	0	0	0	0	0	0	0	0
B	0	0	−1	0	−1	0	1	1	0	0	0	0	0	0
C	0	0	0	−1	0	0	1	0	1	−1	0	0	0	0
D	0	0	−1	1	0	−1	0	0	0	0	1	0	0	0
E	0	0	0	0	1	−1	0	0	0	0	0	1	−1	0
F	0	0	−1	1	0	−1	0	0	0	1	0	0	−1	1

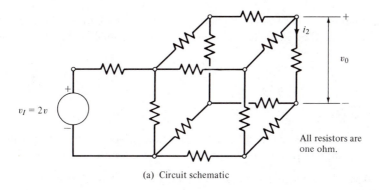

(a) Circuit schematic

$v_I = 2v$

All resistors are one ohm.

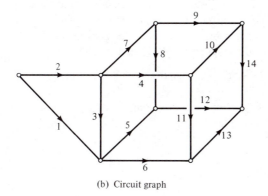

(b) Circuit graph

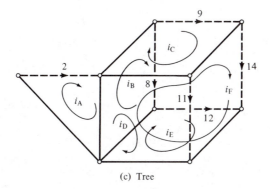

(c) Tree

FIGURE 2-14. Circuit showing the selection of loop currents using a tree.

is entered in column i of the row corresponding to that loop current if branch and loop currents are in the same direction; a -1 is entered if they are in opposite directions; a 0 is entered if the loop does not include the ith branch. The rows of the tie-set schedule give us the KVL equations around each loop in terms of branch voltages. For example, row A of Table 2-2 tells us that

$$-v_1 + v_2 + v_3 = 0$$

The columns of the tie-set schedule yield the branch currents in terms of the loop currents. For example, column 1 of Table 2-2 tells us that

$$i_1 = -i_A$$

while column 5 gives us

$$i_5 = -i_B + i_E$$

The procedure in obtaining the KVL equations around each circuit defined by the link currents is then the following:

1. Write down the voltage-law equations for each loop in terms of the branch voltages.
2. Substitute for each branch voltage the describing equation for each branch with the branch current as the independent variable.
3. Substitute for the branch currents the expressions involving loop currents obtained from the columns of the tie-set schedule.

The result is $N_b - N_v + 1$ equations in terms of the $N_b - N_v + 1$ loop current variables.

In the present example, all components are 1-Ω resistors, so that these steps are particularly simple. Thus since $v_1 = 2$ V, the row A equation becomes

$$v_2 + v_3 = 2$$

or, in terms of branch currents,

$$i_2 + i_3 = 2$$

But columns 2 and 3 give us

$$i_2 = i_A \qquad \text{and} \qquad i_3 = i_A - i_B - i_D - i_F$$

Thus

$$i_A + (i_A - i_B - i_D - i_F) = 2$$

or

$$2i_A - i_B - i_D - i_F = 2$$

The student should obtain the remaining five equations.   ∎

EXAMPLE 2-14

As a final example, consider the electromechanical system shown in Figure 2-15, which represents a dc motor with separately excited field damped by a viscous friction load. The applied voltage to the motor armature is assumed to be a unit step. An expression for the angular velocity of the motor as a function of time is desired. Except for the fact that household motors are not normally dc machines, this example could represent the pumping cycle for the example on the liquid-level controller of Chapter 1.

The torque produced by the armature is proportional to the armature current.

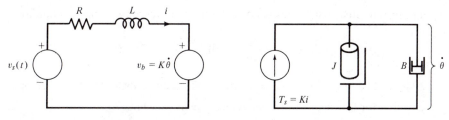

FIGURE 2-15. Schematic representation for a separately field-excited dc motor with viscous friction loading.

This is represented by the controlled source defined by the equation

$$T_s = Ki$$

where K is a constant of proportionality. The armature, when rotating in the separately produced magnetic field, delivers a back-electromotive force to the armature circuit which is proportional to the angular velocity of the armature. This is represented by the controlled source

$$v_b = K\dot{\theta}$$

where it can be shown that the constant K is the same as for the torque equation. The schematic diagram is so simple that we do not have to construct a system graph. On the electrical side KVL gives

$$v_s = Ri + L\frac{di}{dt} + K\dot{\theta}, \qquad -\infty < t < \infty$$

while the mechanical side yields

$$T_s = Ki = J\frac{d\dot{\theta}}{dt} + B\dot{\theta}, \qquad -\infty < t < \infty$$

where the time dependence is not shown explicitly in order to simplify the notation. We wish to eliminate i from the second equation in order to solve for $\dot{\theta}$ as a function of time. We may accomplish this by solving the second equation for i, differentiating and substituting into the first equation for i and di/dt. The result is

$$v_s = \frac{LJ}{K}\frac{d^2\dot{\theta}}{dt^2} + \left(\frac{RJ + LB}{K}\right)\frac{d\dot{\theta}}{dt} + \left(\frac{RB}{K} + K\right)\dot{\theta}, \qquad -\infty < t < \infty$$

Let $LJ/K = 1$, $(RJ + LB)/K = 2$, and $RB/K + K = 5$. Also assume a unit-step input voltage. The equation for $\dot{\theta}$ then becomes

$$\frac{d^2\dot{\theta}}{dt^2} + 2\frac{d\dot{\theta}}{dt} + 5\dot{\theta} = u(t)$$

The solution to this equation is composed of two parts, the transient (natural) and forced (steady-state) responses. The transient response is the solution to

$$\frac{d^2\dot{\theta}_{tr}}{dt^2} + 2\frac{d\dot{\theta}_{tr}}{dt} + 5\dot{\theta}_{tr} = 0, \qquad t > 0$$

Assume that $\dot{\theta}_{tr}(t)$ is of the form Ae^{pt}. Substituting this into the differential equation for $\dot{\theta}_{tr}$ yields the *characteristic equation*, which is

$$p^2 + 2p + 5 = 0$$

The solutions for p are complex conjugates:

$$p_1 = -1 + j2$$

and

$$p_2 = -1 - j2$$

Thus the transient solution is of the form

$$\dot{\theta}_{tr}(t) = A_1 e^{p_1 t} + A_2 e^{p_2 t}$$

$$= A_1 e^{(-1+j2)t} + A_2 e^{(-1-j2)t}$$

$$= e^{-t}(A_3 \cos 2t + A_4 \sin 2t)$$

where A_3 and A_4 (or A_1 and A_2) are constants to be fixed by the initial conditions.

The forced response, for a constant input, is itself a constant, for substitution of $\dot{\theta}_s = A_5$ into the differential equation results in

$$\dot{\theta}_s = \tfrac{1}{5}$$

Thus the total solution, transient plus steady state, is

$$\dot{\theta}(t) = [\tfrac{1}{5} + e^{-t}(A_3 \cos 2t + A_4 \sin 2t)]u(t)$$

To determine A_3 and A_4, we need two initial conditions. We assume the initial conditions

$$\dot{\theta}(0) = \frac{d\dot{\theta}}{dt}\bigg|_{t=0} = 0$$

in order to get two independent equations for A_3 and A_4. Letting $t = 0$ in the solution yields

$$\tfrac{1}{5} + A_3 = 0 \qquad \text{or} \qquad A_3 = -\tfrac{1}{5}$$

Differentiating, letting $t = 0$, and setting the result equal to zero gives

$$-A_3 + 2A_4 = 0 \qquad \text{or} \qquad A_4 = \frac{A_3}{2} = -\tfrac{1}{10}$$

Thus the angular velocity as a function of time is

$$\dot{\theta}(t) = \tfrac{1}{5}[1 - e^{-t}(\cos 2t + \tfrac{1}{2}\sin 2t)]u(t) \qquad \text{rad/s}$$

Note that as $t \to \infty$, $\dot{\theta} \to \tfrac{1}{5}$ rad/s, the steady-state value.

For a step input of magnitude greater than 1 V, $\dot{\theta}(t)$ would simply be multiplied by a corresponding factor since the system is linear. ∎

SUMMARY

In this chapter the problem of determining the response of a fixed, linear system to an arbitrary input has been considered. The basic tool for accomplishing this is the superposition integral, which provides the output of a system as the convolution of the input signal with the impulse response of the system. The impulse response is defined as the system response to a unit impulse input applied at time zero. The convolution operation consists of time-reversing one signal to be convolved, shifting or delaying it, multiplication point by point with the second signal, and integrating the product. Successive time shifts of the time reversal of the first function, multiplication, and integration then provides the convolution of the two signals for all time. Although superposition in terms of the impulse response to obtain a system output is usually employed, superposition of unit step responses weighted by the *derivatives* of the input signal can also be employed. An important characteristic of a fixed, linear system is the following. Suppose that a given input produces a given output. Now modify the original input through a linear operation, such as integration or differentiation. The new system output is related to the previous system output by the same linear operation performed on the original input to produce the new input. As an example, the ramp response of a fixed, linear system is the integral of its step response since the integral, with respect to time, of a unit step is a unit ramp.

A bounded-input, bounded-output (BIBO) stable linear system is defined as one that produces a bounded output in response to any bounded input. A necessary and sufficient condition that a system be BIBO stable is that the integral over all time of the absolute value of its impulse response be finite.

This chapter was concluded with a discussion of systematic techniques for obtaining governing equations for lumped systems. Although the graph theory concepts introduced work for virtually any fixed, linear system, electrical or mechanical, they are usually not necessary for the simple systems involved in the larger share of problems encountered. However, if a relatively complex system is to be analyzed, the graph theory approach is almost a necessity to ensure that an independent set of equations is obtained.

FURTHER READING

In addition to the references listed in Chapter 1, the following book provides a unified and complete approach to the analysis of linear systems:

W. A. BLACKWELL, *Mathematical Modeling of Physical Networks,* New York: Macmillan, 1968.

PROBLEMS

SECTION 2-2

2-1. Find and sketch the signal $y(t)$, which is the convolution of the following pairs of signals.
(a) $x(t) = 2 \exp(-10t)u(t)$ and $h(t) = \Pi(t/2)$
(b) $x(t) = \Pi[(t - 1)/2]$ and $h(t) = u(t - 10)$
(c) $x(t) = 2 \exp(-10t)u(t)$ and $h(t) = u(t - 2)$

2-2. Show that (2-1) and (2-2) are equivalent.

2-3. If $x_1(t)$, $x_2(t)$, and $h(t)$ are arbitrary signals, and α is a constant, show the following:
(a) $h(t) * [x_1(t) + x_2(t)] = h(t) * x_1(t) + h(t) * x_2(t)$
(b) $h(t) * [x_1(t) * x_2(t)] = [h(t) * x_1(t)] * x_2(t)$
(c) $h(t) * [\alpha x_1(t)] = \alpha h(t) * x_1(t)$

SECTIONS 2-3 AND 2-4

2-4. Assuming the superposition property for the sum of two signals, extend this expression to the sum of N components as suggested by the discussion accompanying Equation (2-8).

2-5. Obtain the impulse response of the system shown by:
(a) Using the method of Example 2.4
(b) Using the method of Example 2.5

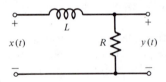

2-6. Obtain the impulse response of the system shown.

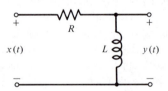

2-7. Obtain the impulse response of the system shown.

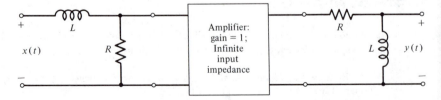

2-8. Show that the impulse response of the system of Problem 2-7 is the convolution of the impulse responses of the systems of Problems 2-5 and 2-6.

2-9. Obtain the impulse response of the system shown.

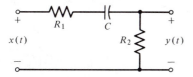

SECTION 2-5

2-10. Obtain the output of the system of Problem 2-5 in response to the input

$$x(t) = \Pi(t - 0.5)$$

Sketch for $R/L = 0.1$ and 1.0. Use Equation (2-13).

2-11. Obtain the output of the system of Problem 2-6 in response to the input

$$x(t) = \Pi(t - 0.5)$$

Sketch for $R/L = 0.1$ and 1.0.

2-12. Obtain the step responses of the systems of:
(a) Problem 2-5
(b) Problem 2-6
(c) Problem 2-7

2-13. Using Equation (2-25), obtain the output of the system of Problem 2-5 in response to the input

$$x(t) = \Pi(t - 0.5)$$

Sketch for $R/L = 0.1$ and 1.0.

2-14. Given that the RL filter shown has impulse response

$$h(t) = \delta(t) - \frac{R}{L} e^{-(R/L)t} u(t)$$

(a) Find the step response.
(b) Find the ramp response.

2-15. Obtain and sketch the response of the filter of Problem 2-14 to the input shown for:
(a) $R/L = 0.1$
(b) $R/L = 1.0$.
(*Hint*: Use the answers obtained for Problem 2-14 together with superposition.)

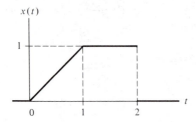

$x(t)$

1

0 1 2 t

2-16. Using the superposition property find the response of the *RC* circuit of Example 2-5 to the input

$$x(t) = r(t) - r(t - 1) - u(t - 2)$$

Sketch the input and the output for $RC = 0.1$.

SECTION 2-6

2-17. Show that the system of Problem 2-5 is BIBO stable.

2-18. Show that the system of Problem 2-6 is BIBO stable.

2-19. Show that the system of Problem 2-7 is BIBO stable.

SECTION 2-7

2-20. Assume an applied sinusoidal force of the form $f(t) = A \cos \omega_0 t$ for the translational mechanical elements of Table 2-1. The average power delivered to the device is

$$P_{av} = \frac{1}{T_0} \int_0^{T_0} f(t) \frac{d\Delta(t)}{dt} dt \qquad \text{where } T = \frac{2\pi}{\omega_0}.$$

(a) Show that the average power delivered to the translational damper is $A^2/2B$ watts.

(b) Show that the average power delivered to the spring and mass elements is zero.

2-21. Assume an applied sinusoidal torque of the form $T(t) = A \cos \omega_0 t$ for the rotational mechanical elements of Table 2-1. The average power delivered to the device is

$$P_{av} = \frac{1}{T_0} \int_0^{T_0} T(t) \frac{d\theta(t)}{dt} dt \qquad \text{where } \omega_0 = \frac{2\pi}{T_0}$$

(a) Show that the average power delivered to the rotational damper is $A^2/2B$ watts.

(b) Show that P_{av} for the rotational spring and mass is zero.

2-22. Eliminate $i_B(t)$ between (2-35a) and 2-35b) so that one equation in $i_A(t)$ is obtained.

2-23. For each of the system schematics shown, draw the system graph, choose a tree, and write down the Node and Circuit Postulate equations, (2-32) and (2-33).

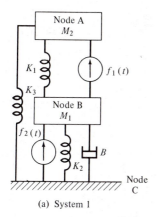

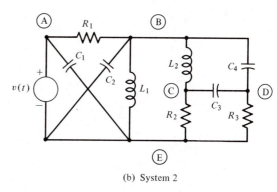

(a) System 1

(b) System 2

2-24. For System 1 of Problem 2-23, let C be the reference node with velocity zero. Write terminal equations for all elements in terms of V_A and V_B, the velocities of nodes A and B relative to node C. Express the Node Postulate equations in terms of V_A and V_B. How many independent equations are there?

2-25. For System 2 of Problem 2-23, write terminal equations for all elements in terms of appropriately chosen loop currents. Substitute into the Circuit Postulate equations obtained in Problem 2-23. How many independent Circuit Postulate equations are there.

2-26. **(a)** Sketch trees for each of the system graphs shown.

(b) Assuming that electrical networks are represented, compare the number of independent Postulate 1 (KCL) equations that can be written in terms of node voltages with the number of independent Postulate 2 (KVL) equations that can be written in terms of loop currents.

(c) Assume that link 1 for both parts (a) and (b) corresponds to a voltage source $v_s(t)$, and that all other links are 1-Ω resistors. Write down the KVL equations for each network in terms of loop currents determined by your choice of trees.

(d) Assume that link 1 for both parts (a) and (b) represents a current source $i_s(t)$, and that all other links are 1-Ω resistors. Write down the KCL equations in terms of node voltages with node C chosen as reference.

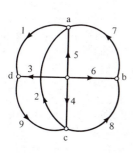

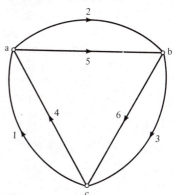

CHAPTER 3

Fourier Series and the Fourier Transform

INTRODUCTION

After considering some simple signal models in Chapter 1, we turned to systems analysis in the time domain in Chapter 2. The principal tool developed there was the superposition integral which resulted from resolving the input signal into a continuum of unit impulses and superimposing the elemental system outputs due to each impulsive input—thus the name "superposition integral." The application of the superposition integral involves first finding the impulse response of the system and then carrying out the actual integration—tasks that often are not simple.

In this chapter we consider procedures for resolving certain classes of signals into superpositions of sines and cosines, or equivalently, complex exponential signals of the form exp $(j\omega t)$. For periodic power signals this resolution results in the Fourier series coefficients of the signal, and the resulting representation of such signals is known as the Fourier series of the signal. For energy signals of finite or infinite extent, the Fourier integral provides the desired resolution with the signal representation for such signals given by the inverse Fourier transform integral. Signals of the *exponential class,* of which energy and power signals are subclasses, may be resolved with complex exponential signals of the form exp $[(\sigma + j\omega)t]$ by means of the Laplace transform integral. We deal with the Laplace transform in Chapter 4.

Advantages of Fourier series and Fourier transform representations for signals are twofold. First, in the analysis and design of systems it is often useful to characterize signals in terms of *frequency-domain* parameters such as bandwidth or spectral content. Second, the superposition property of linear systems, and the fact that the steady-state response of a fixed, linear system to a sinusoid of a given frequency is itself a sinusoid of the same frequency, provides a means for solving for the response of such systems. Indeed, the Fourier or Laplace transform, as appropriate, allows one to convert the constant-coefficient linear differential equation representation for a lumped system to an algebraic expression which considerably simplifies the solution for the system output. We consider this method of solution later in this chapter and in Chapters 4 and 5.

We begin this chapter with some simple examples in order to convince ourselves that a sine–cosine series is a useful representation for a periodic signal.

TRIGONOMETRIC SERIES†

Recalling Example 1-2, parts (d) and (e), we note that the sum of two sinusoids is periodic provided that their frequencies are commensurable. Stated another way, the sum of two sinusoids is periodic provided that their frequencies are integer multiples of a fundamental frequency.

To examine the fruitfulness of representing periodic signals by sums of sinusoids whose frequencies are harmonics, or integer multiples, of a fundamental frequency, we consider two such series and plot their partial sums.

EXAMPLE 3-1

As a first example, consider the trigonometric series, assumed to hold for all t, given by

$$x(t) = \sin \omega_0 t + \tfrac{1}{3} \sin 3\omega_0 t + \tfrac{1}{5} \sin 5\omega_0 t + \cdots \tag{3-1}$$

where $2\pi/\omega_0$ is the period. Its partial sums are

$$s_1 = \sin \omega_0 t$$

$$s_2 = \sin \omega_0 t + \tfrac{1}{3} \sin 3\omega_0 t$$

and so on. Several of them are plotted in Figure 3-1 with $\omega_0 t$ as the independent variable. Stretching our imagination, we see that a square wave appears to be taking shape as more terms are included in the partial sum. As each harmonic is added, the ripple in the flat-top approximation of the square wave becomes smaller and possesses a frequency equal to that of the highest harmonic included in the series.

Somewhat disconcerting, however, is the appearance of "ears" (or overshoot) in the plot of the partial sums where the square wave is discontinuous. This phenomenon, referred to as the Gibbs phenomenon, is examined in more detail in Section 3-19.

†An alternative approach to Fourier series is provided in Section 3-22 in terms of generalized vector-space concepts.

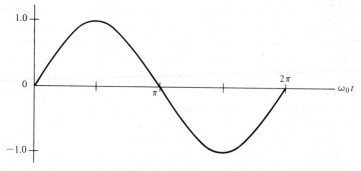

(a) The first partial sum is a sine wave

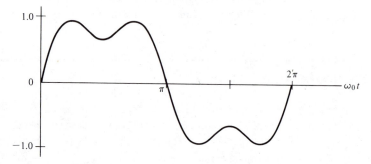

(b) The second partial sum begins to approximate a square wave

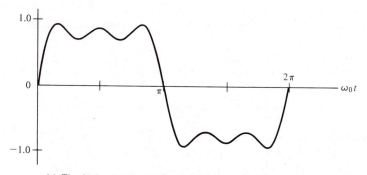

(c) The third partial sum provides a better approximation to a square wave

FIGURE 3-1. The series of Example 3-1 shows how a trigonometric series approximates a square wave (only one period shown).

The convergence of the series at any particular point may be examined by substituting the appropriate value of $\omega_0 t$. For example, setting $\omega_0 t = \pi/2$ in (3-1), we obtain the alternating series

$$1 - \frac{1}{3} + \frac{1}{5} - \frac{1}{7} + \cdots = \frac{\pi}{4} \tag{3-2}$$

which may be checked by consulting a table of sums of series.† We can nor-

†See, for example, *Mathematical Tables from Handbook of Chemistry and Physics* (Boca Raton, Fla.: CRC Press, 1962), p. 325.

malize the series (3-1) to have a value of unity at $\omega_0 t = \pi/2$ by multiplying it by $4/\pi$. Later we will show that a unity-amplitude square wave can be represented by the series

$$x_{sq}(t) = \frac{4}{\pi} (\sin \omega_0 t + \tfrac{1}{3} \sin 3\omega_0 t + \tfrac{1}{5} \sin 5\omega_0 t + \cdots),$$
$$(3-3)$$
$$-\infty < t < \infty$$
■

EXAMPLE 3-2

As a second example, consider the partial sums of the trigonometric series

$$y(t) = \sin \omega_0 t - \tfrac{1}{2} \sin 2\omega_0 t + \tfrac{1}{3} \sin 3\omega_0 t \cdots, \qquad -\infty < t < \infty \quad (3-4)$$

Plots showing its partial sums are given in Figure 3-2. Again, with a little imagination, we can see the beginnings of a sawtooth waveform. Each term added decreases the ripple in magnitude. The frequency of the ripple is equal to that of the highest-frequency term included in the series. The overshoot phenomenon, present at the discontinuities of the square wave considered previously, is also present with the sawtooth waveform. ■

Having convinced ourselves of the possibility of building up arbitrary periodic waveforms from sums of harmonically related sinusoidal terms, we turn to the crux of the matter: *Given a specific periodic waveform, how do we find its trigonometric-series representation?* The resulting series, which is unique for each periodic signal, is called the *trigonometric Fourier series* of the signal.

3-3

OBTAINING TRIGONOMETRIC FOURIER SERIES REPRESENTATIONS FOR PERIODIC SIGNALS

We begin by writing down the general form that the trigonometric Fourier series representation of a periodic signal may have:

$$x(t) = a_0 + a_1 \cos \omega_0 t + a_2 \cos 2\omega_0 t + \cdots$$
$$+ b_1 \sin \omega_0 t + b_2 \sin 2\omega_0 t + \cdots, \qquad -\infty < t < \infty \quad (3-5)$$

This may be written compactly in terms of summations as

$$x(t) = a_0 + \sum_{n=1}^{\infty} a_n \cos n\omega_0 t + \sum_{n=1}^{\infty} b_n \sin n\omega_0 t \qquad (3-6)$$

where $-\infty < t < \infty$.† Since the right-hand side is a sum of harmonically related sinusoids, which themselves are periodic functions, the left-hand side is periodic. The problem that faces us is to find $a_0, a_1, a_2, \ldots, b_1, b_2, \ldots$ for a given $x(t)$.

†Since most signals in this chapter are defined over the entire t-axis, we dispense with giving the range of definition unless it is other than $-\infty < t < \infty$.

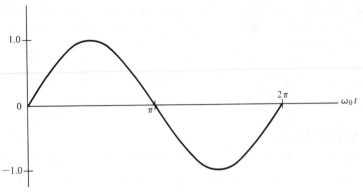

(a) The first partial sum is a sine wave

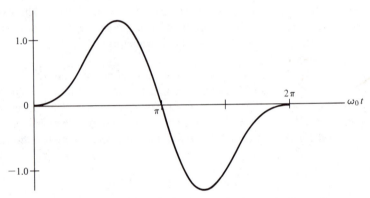

(b) The second partial sum begins to approximate a saw tooth waveform

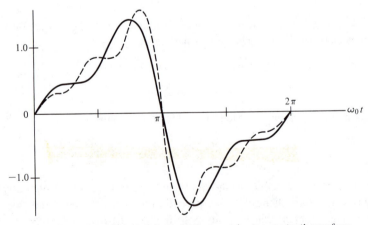

(c) The third and fifth partial sums better approximate a saw tooth waveform

FIGURE 3-2.   The trigonometric series of Example 3-2 illustrates approximation of a sawtooth waveform (only one period shown).

To do this, we first note some rather interesting properties of integrals involving products of sines and cosines. Clearly, the integral of $\sin n\omega_0 t$ or $\cos n\omega_0 t$ over an integer number of periods is zero, for as much area appears above the t-axis as below it in one period. However, what about products involving $\sin m\omega_0 t$ and $\cos m\omega_0 t$? Three possible cases occur. The resulting integrals are

$$I_1 = \int_0^{T_0} \sin m\omega_0 t \sin n\omega_0 t \, dt = \begin{cases} 0, & m \neq n \\ T_0/2, & m = n \neq 0 \end{cases} \qquad (3\text{-}7)$$

$$I_2 = \int_0^{T_0} \cos m\omega_0 t \cos n\omega_0 t \, dt = \begin{cases} 0, & m \neq n \\ T_0/2, & m = n \neq 0 \end{cases} \qquad (3\text{-}8)$$

and

$$I_3 = \int_0^{T_0} \sin m\omega_0 t \cos n\omega_0 t \, dt = 0, \qquad \text{all } m, n \qquad (3\text{-}9)$$

where $T_0 = 2\pi/\omega_0$ is a period of the fundamental and n and m are integers.[†] We note that with $n = 0$ in (3-8) and (3-9), we have the integral of $\cos m\omega_0 t$ and $\sin m\omega_0 t$, respectively. We also note that the product of two sinusoids with harmonically related frequencies is periodic, so that the integrals could be taken over *any* period.

With the integrals given by (3-7) through (3-9), we may now obtain relationships for finding the a_n's and b_n's for an arbitrary (periodic) $x(t)$. We begin with a_0. Integrating the series (3-5) term by term over one period of $x(t)$, we obtain

[†]To show (3-7) through (3-9), we use the trigonometric identities

$$\sin u \sin v = \tfrac{1}{2} \cos (u - v) - \tfrac{1}{2} \cos (u + v)$$
$$\cos u \cos v = \tfrac{1}{2} \cos (u - v) + \tfrac{1}{2} \cos (u + v)$$

and

$$\sin u \cos v = \tfrac{1}{2} \sin (u - v) + \tfrac{1}{2} \sin (u + v)$$

Using the first trigonometric identity in (3-7), we obtain

$$I_1 = \tfrac{1}{2} \int_0^{T_0} \cos (m - n)\omega_0 t \, dt - \tfrac{1}{2} \int_0^{T_0} \cos (m + n)\omega_0 t \, dt$$

Both integrals may be easily evaluated. It is simpler, however, to note that both involve the integration of a cosine over an integer number of periods if $m \neq n$; therefore, the net area between the integrand and the t-axis is zero. If $m = n$, this observation still holds for the second integral, but now the first integrand is $\cos 0 = 1$. Hence the value of the first integral is

$$I_1 = \tfrac{1}{2} T_0 \qquad (m = n)$$

which proves (3-7).

Application of the second identity to (3-8) and use of similar reasoning to that just used to prove (3-7) results in a proof of (3-8).

When the last trigonometric identity is applied to (3-9), the integral becomes

$$I_3 = \tfrac{1}{2} \int_0^{T_0} \sin (m - n)\omega_0 t \, dt + \tfrac{1}{2} \int_0^{T_0} \sin (m + n)\omega_0 t \, dt$$

which is clearly zero for all integer values of m and n, including $m = n$.

$$\int_{T_0} x(t)\, dt = a_0 \int_{T_0} dt$$

$$+ a_1 \int_{T_0} \cos \omega_0 t\, dt + a_2 \int_{T_0} \cos 2\omega_0 t\, dt + \cdots \quad (3\text{-}10)$$

$$+ b_1 \int_{T_0} \sin \omega_0 t\, dt + b_2 \int_{T_0} \sin 2\omega_0 t\, dt + \cdots$$

where $\int_{T_0} (\cdot)\, dt$ denotes integration over any period. All terms except the first involve the integration of a sine or cosine over an integral number of periods and are therefore zero. This first term yields $a_0 T_0$. Thus the coefficient a_0 is given by

$$a_0 = \frac{1}{T_0} \int_{T_0} x(t)\, dt \quad (3\text{-}11)$$

which is the average value of the waveform.

By applying (3-8) and (3-9) to (3-5), we see how the same procedure may be used to find a_1, a_2, or any specific a_n. Using (3-6), however, we may derive a general expression valid for any of the a_n's. The derivation proceeds as follows for $m \neq 0$.

Multiplying both sides of (3-6) by $\cos m\omega_0 t$ (the use of m allows us to choose any a_n coefficient), and integrating over a period of $x(t)$, we obtain

$$\int_{T_0} x(t) \cos m\omega_0 t\, dt = a_0 \int_{T_0} \cos m\omega_0 t\, dt$$

$$+ \int_{T_0} \left(\sum_{n=1}^{\infty} a_n \cos n\omega_0 t \right) \cos m\omega_0 t\, dt$$

$$+ \int_{T_0} \left(\sum_{n=1}^{\infty} b_n \sin n\omega_0 t \right) \cos m\omega_0 t\, dt \quad (3\text{-}12)$$

The first term integrates to zero. We multiply each term of the series in parentheses by $\cos m\omega_0 t$ and integrate term by term to obtain

$$\int_{T_0} x(t) \cos m\omega_0 t\, dt = \sum_{n=1}^{\infty} a_n \int_{T_0} \cos n\omega_0 t \cos m\omega_0 t\, dt$$

$$+ \sum_{n=1}^{\infty} b_n \int_{T_0} \sin n\omega_0 t \cos m\omega_0 t\, dt \quad (3\text{-}13)$$

Applying (3-9), we see that each term of the second series on the right-hand side of (3-13) is zero. Applying (3-8), we see that all terms of the first series on the right-hand side of (3-13) are also zero except for the term for which $n = m$. For $n = m$, the integral gives $T_0/2$, which yields

$$\int_{T_0} x(t) \cos m\omega_0 t\, dt = a_m \left(\frac{T_0}{2} \right) \quad (3\text{-}14)$$

Multiplying both sides of (3-14) by $2/T_0$ gives

$$a_m = \frac{2}{T_0} \int_{T_0} x(t) \cos m\omega_0 t \, dt, \, m \neq 0 \tag{3-15}$$

In a similar manner, by employing (3-6), (3-7), and (3-9), we may show that

$$b_m = \frac{2}{T_0} \int_{T_0} x(t) \sin m\omega_0 t \, dt \tag{3-16}$$

Without concerning ourselves about the validity of the steps in obtaining (3-11), (3-15), and (3-16), we *define* a Fourier series to be any trigonometric series of the form (3-4), where the coefficients are found according to the formulas just derived. To find the coefficients, we require only that the integrals involved exist.†

In going from (3-12) to (3-13) it was necessary to interchange the order of summation and integration. We did not worry about the validity of this step. Indeed, the whole development so far has been based on the assumption that (3-6) is truly an equality; that is, an arbitrary periodic function can be represented in terms of a sum of sines and cosines. There is no reason to believe that this is always the case. However, we noted in the previous paragraph that the coefficients a_n and b_n will exist if $x(t)$ is integrable. We can therefore find the coefficients for an appropriate $x(t)$ and then examine the convergence properties of the series formed with these coefficients. It is too much to expect that the Fourier series will converge to a periodic, but otherwise arbitrary, $x(t)$ for every t. For example, in Example 3-1 it is apparent that each partial sum is zero at $\omega_0 t = n\pi$, where n is an integer. If $x(t)$ has been defined to be $\pi/4$ at these points (or any other desired value) the series cannot represent $x(t)$ at $\omega_0 t = n\pi$. As another example, consider two periodic signals $x(t)$ and $y(t)$ which are identical except at a single point, where they differ by a finite amount. This finite difference will not affect the value of the integrals for the coefficients given by (3-15) and (3-16). Therefore, both $x(t)$ and $y(t)$ have identical Fourier series and we conclude that the Fourier series cannot represent both functions at the point in question. In general, functions that are equal ''almost everywhere''—that is, everywhere except at a number of isolated points—will have identical Fourier series representations. Fourier series, therefore, do not converge in a *pointwise* sense.

The question remains as to what conditions are required to ensure that a Fourier series *will* converge to $x(t)$. Indeed, very little is required to guarantee convergence. It turns out that if $x(t)$ has period equal to T_0, and has continuous first and second derivatives for all t, *except possibly at a finite number of points where it may have finite jump discontinuities,* the Fourier series for $x(t)$ con-

---

†We give only an informal discussion of convergence here. Conditions concerning the possibility of expanding a function in a Fourier series are given in a theorem proved by Dirichlet in 1829. These sufficient conditions are that $x(t)$ be defined and bounded on the range $(t_0, t_0 + T_0)$ and have only a finite number of maxima and minima and a finite number of discontinuities on this range. These conditions, being *sufficient conditions,* are more restrictive than necessary. Fejér, in 1904, published two theorems giving conditions for the convergence of Fourier series which are less restrictive than the Dirichlet conditions.

verges uniformly to $x(t)$ for every t except at the points of discontinuity.† Even at points of discontinuity, the behavior of a Fourier series is very reasonable. The limit of the Fourier series sum at a point of discontinuity is simply the average of the left- and right-hand limits. This is illustrated in Figure 3-3.

Although we have discussed only periodic functions so far, it should be noted that the basic coefficient formulas *use* only the values of $x(t)$ in a T_0-second interval. Thus, if $x(t)$ is given only in some interval of length T_0, the corresponding Fourier series can be formed and converges uniformly to $x(t)$ within this interval at all points of continuity. Outside this interval, the Fourier series converges to a signal which is the *periodic extension of $x(t)$*. This is illustrated in Figure 3-4.

We close this section with an example that involves computation of the Fourier series coefficients for a square wave.

EXAMPLE 3-3

Consider the square wave defined by

$$
x(t) = \begin{cases} A, & 0 < t < \dfrac{T_0}{2} \\[2mm] -A, & \dfrac{T_0}{2} < t < T_0 \end{cases}
$$

and periodically extended outside this interval. The average value is zero, so $a_0 = 0$. From (3-15), the a_m's are

$$
a_m = \frac{2}{T_0} \int_0^{T_0/2} A \cos m\omega_0 t \, dt + \frac{2}{T_0} \int_{T_0/2}^{T_0} (-A) \cos m\omega_0 t \, dt
$$

$$
= \frac{2A}{T_0} \left[\frac{\sin m\omega_0 t}{m\omega_0} \bigg|_0^{T_0/2} - \frac{\sin m\omega_0 t}{m\omega_0} \bigg|_{T_0/2}^{T_0} \right]
$$

Recalling that $\omega_0 = 2\pi/T_0$, we see that

$$
\sin \frac{m\omega_0 T_0}{2} = \sin m\pi = 0
$$

---

†By uniform convergence of an infinite series whose terms are functions of a variable, it is meant that the numerical value of the remainder after the first n terms is as small as desired *throughout the given interval* for n greater than a sufficiently large chosen number.

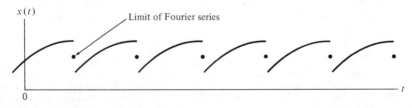

$x(t)$

Limit of Fourier series

0

t

FIGURE 3-3. The Fourier series of a waveform converges to the mean of the left- and right-hand limits at a point of discontinuity.

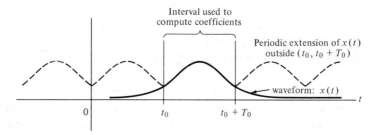

Interval used to
compute coefficients

FIGURE 3-4. Expansion of a nonperiodic signal in terms of a Fourier series.

and

$$\sin m\omega_0 T_0 = \sin 2m\pi = 0$$

Thus all the a_m coefficients are zero. The b_m coefficients result by applying (3-16). This gives

$$b_m = \frac{2}{T_0} \int_0^{T_0/2} A \sin m\omega_0 t \, dt + \frac{2}{T_0} \int_{T_0/2}^{T_0} (-A) \sin m\omega_0 t \, dt$$

$$= \frac{2A}{T_0} \left[-\frac{\cos m\omega_0 t}{m\omega_0} \Big|_0^{T_0/2} + \frac{\cos m\omega_0 t}{m\omega_0} \Big|_{T_0/2}^{T_0} \right]$$

Again, recalling that $\omega_0 = 2\pi/T_0$, we find upon substitution of the limits that

$$b_m = \frac{2A}{m\pi} (1 - \cos m\pi)$$

which results because $\cos 2m\pi = 1$ for all m. Finally, noting that $\cos m\pi = -1$ for m odd and $\cos m\pi = 1$ for m even, we obtain

$$b_m = \begin{cases} \dfrac{4A}{m\pi}, & m \text{ odd} \\ 0, & m \text{ even} \end{cases}$$

The Fourier series of a square wave of amplitude A with odd symmetry is therefore

$$x(t) = \frac{4A}{\pi} (\sin \omega_0 t + \tfrac{1}{3} \sin 3\omega_0 t + \tfrac{1}{5} \sin 5\omega_0 t + \cdots) \qquad (3\text{-}17)$$

We note that only odd harmonic terms are present. This is a consequence of $x(t)$ having a special type of symmetry, referred to as half-wave odd symmetry. More will be said about special symmetry cases later. If $A = 1$, (3-17) becomes the series first shown in (3-3) and verifies the previous assertion that (3-3) represents a unity amplitude square wave. ∎

ANOTHER FORM OF THE TRIGONOMETRIC FOURIER SERIES

Except for special cases involving odd or even symmetry,† as in Example 3-3, both sine and cosine terms will be present in the Fourier series defined by (3-6). The series (3-6) may be expressed in terms of cosines alone, however, if a phase angle is introduced in the argument of the cosine. This is accomplished by noting that

$$a_n \cos n\omega_0 t + b_n \sin n\omega_0 t = A_n \cos (n\omega_0 t + \theta_n) \qquad (3\text{-}18)$$

with

$$A_n = \sqrt{a_n^2 + b_n^2} \qquad (3\text{-}19)$$

and

$$\theta_n = -\tan^{-1} \frac{b_n}{a_n} \qquad (3\text{-}20)$$

That (3-18) is an identity can be proved by expanding $A_n \cos (n\omega_0 t + \theta_n)$ as

$$A_n \cos \theta_n \cos \omega_0 t - A_n \sin \theta_n \sin n\omega_0 t$$

and matching $A_n \cos \theta_n$ with a_n and $-A_n \sin \theta_n$ with b_n in (3-18).

When (3-18) is used in (3-4), we obtain an alternative form for the Fourier series given by

$$x(t) = A_0 + \sum_{n=1}^{\infty} A_n \cos (n\omega_0 t + \theta_n) \qquad (3\text{-}21)$$

where $A_0 = a_0$ has been used to make the notation consistent.

This form of the Fourier series is convenient for providing a *spectral representation* of a signal $x(t)$, similar to the spectral plots for the sinusoidal signal $A \cos (\omega_0 t + \theta)$ shown in Figure 1-8. To obtain the *single-sided amplitude spectral representation or amplitude spectrum* of a periodic signal $x(t)$, we simply plot $A_0, A_1, A_2, \ldots$ at the frequency locations $0, f_0, 2f_0, \ldots$, respectively. To obtain the *single-sided* phase spectrum, we plot $\theta_1, \theta_2, \theta_3, \ldots$ at the frequency locations $f_0, 2f_0, \ldots$, respectively [$\theta_0 = 0$ because A_0, the average value of $x(t)$, is a constant with no phase associated with it].‡

Double-sided spectra may be obtained by representing $\cos (n\omega_0 t + \theta_n)$ in (3-21) in exponential form. We will not consider them until the exponential Fourier series is introduced, however.

We now consider an example to illustrate the calculation of Fourier series for the purpose of making spectral plots later.

---

†A signal, $x(t)$, has *even symmetry* if $x(t) = x(-t)$, whereas it has *odd symmetry* if $x(t) = -x(-t)$.

‡It is assumed that $A_0 \geq 0$. If not, let $x'(t) = -x(t)$ and obtain its Fourier series.

EXAMPLE 3-4

Find the Fourier series for the half-rectified sine wave defined by

$$x(t) = \begin{cases} A \sin \omega_0 t, & 0 \le t \le \dfrac{T_0}{2} \\ 0, & \dfrac{T_0}{2} \le t \le T_0 \end{cases}$$

with $x(t) = x(t + T_0)$. Express in the form (3-21).

Solution: From (3-15) and (3-16),

$$a_m = \frac{2}{T_0} \int_0^{T_0/2} A \sin \omega_0 t \cos m\omega_0 t \, dt$$

and

$$b_m = \frac{2}{T_0} \int_0^{T_0/2} A \sin \omega_0 t \sin m\omega_0 t \, dt$$

Using trigonometric identities, these expressions become

$$a_m = \frac{A}{T_0} \left[\int_0^{T_0/2} \sin (1 - m)\omega_0 t \, dt + \int_0^{T_0/2} \sin (1 + m)\omega_0 t \, dt \right] \quad (3\text{-}22)$$

and

$$b_m = \frac{A}{T_0} \left[\int_0^{T_0/2} \cos (1 - m)\omega_0 t \, dt - \int_0^{T_0/2} \cos (1 + m)\omega_0 t \, dt \right] \quad (3\text{-}23)$$

Integrating the expression for b_m, we find that

$$b_m = \frac{A}{T_0} \left[\frac{\sin (1 - m)\omega_0 t}{(1 - m)\omega_0} \Big|_0^{T_0/2} - \frac{\sin (1 + m)\omega_0 t}{(1 + m)\omega_0} \Big|_0^{T_0/2} \right], \quad m \ne 1$$

$$= \frac{A}{T_0} \left[\frac{\sin (1 - m) \pi}{(1 - m)\omega_0} - \frac{\sin (1 + m)\pi}{(1 + m)\pi} \right], \quad m \ne 1$$

where $\omega_0 = 2\pi/T_0$ has been used. Noting that the sine of an integer multiple of π is zero, we obtain

$$b_m = 0, \quad m \ne 1$$

The case $m = 1$ must be treated separately because of the factor $(1 - m)$ in the denominator after integration. Returning to (3-23) and letting $m = 1$, we obtain

$$b_1 = \frac{A}{T_0} \left(\int_0^{T_0/2} dt - \int_0^{T_0/2} \cos 2\omega_0 t \, dt \right)$$

$$= \frac{A}{2}$$

Considering next the a_m's, we integrate (3-22) to obtain

$$a_m = \frac{A}{2\pi} \left\{ \frac{1}{1-m} [1 - \cos(1-m)\pi] + \frac{1}{1+m} [1 - \cos(1+m)\pi] \right\}$$

where $\omega_0 = 2\pi/T_0$ has been used as well as $\cos 0 = 1$. Now

$$\cos(1+m)\pi = \cos(1-m)\pi = \begin{cases} -1, & m = 2, 4, 6, \ldots \\ 1, & m = 3, 5, 7, \ldots \end{cases}$$

Substituting the values of $\cos(1 \pm m)\pi$, we may reduce the expression for a_m to

$$a_m = \begin{cases} \dfrac{A}{2\pi} \left(\dfrac{2}{1-m} + \dfrac{2}{1+m} \right) = \dfrac{2A}{\pi(1-m^2)}, & m = 2, 4, 6, \ldots \\ 0, & m = 3, 5, 7, \ldots \end{cases}$$

As before for the b_m's, the case $m = 1$ must be considered separately because of the $(1 - m)$ in the denominator after integration. Returning to (3-22) with $m = 1$, we find a_1 to be

$$a_1 = \frac{A}{T_0} \int_0^{T_0/2} \sin 2\omega_0 t \, dt = 0$$

It remains to calculate a_0, the dc value. By definition

$$a_0 = \frac{1}{T_0} \int_0^{T_0} x(t) \, dt = \frac{A}{T_0} \int_0^{T_0/2} \sin \omega_0 t \, dt$$

$$= - \frac{A \cos \omega_0 t}{\omega_0 T_0} \Big|_0^{T_0/2}$$

$$= \frac{A}{\pi}$$

Summarizing, we have found that

$$a_m = \begin{cases} \dfrac{A}{\pi}, & m = 0 \\ \dfrac{2A}{\pi(1-m^2)}, & m \text{ even} \\ 0, & m \text{ odd} \end{cases} \tag{3-24}$$

$$b_m = \begin{cases} \dfrac{A}{2}, & m = 1 \\ 0, & m > 1 \end{cases} \tag{3-25}$$

Employing (3-19) and (3-20), we obtain

$$A_m = \begin{cases} \dfrac{A}{\pi}, & m = 0 \\[2mm] \dfrac{A}{2}, & m = 1 \\[2mm] \dfrac{2A}{\pi\,|1 - m^2|}, & m \text{ even} \\[2mm] 0, & m > 1 \text{ and odd} \end{cases} \qquad (3\text{-}26)$$

and

$$\theta_m = \begin{cases} 0, & m = 0 \\[2mm] -\dfrac{\pi}{2}, & m = 1 \\[2mm] \pm\pi, & m, \text{ even} \quad (\text{because } a_m \text{ is negative}) \\[2mm] 0, & m > 1 \text{ and odd } (\text{because } A_m = 0) \end{cases} \qquad (3\text{-}27)$$

The sign on θ_m for m even is arbitrary. ∎

We postpone making spectral plots for the half-rectified sine wave until after we discuss the exponential Fourier series.

3-5

THE EXPONENTIAL FOURIER SERIES

As mentioned in Section 3-4, there is yet a third form of the Fourier series which involves complex exponential functions. A way of obtaining it would be to substitute the complex exponential form of $\cos(n\omega_0 t + \theta_n)$ in (3-21):

$$\cos(n\omega_0 t + \theta_n) = \tfrac{1}{2} e^{j(n\omega_0 t + \theta_n)} + \tfrac{1}{2} e^{-j(n\omega_0 t + \theta_n)} \qquad (3\text{-}28)$$

However, as we saw in Example 3.4, the calculation of the Fourier coefficients, A_n, and phases, θ_n, is often inconvenient if one is forced to employ (3-19) and (3-20). Thus we seek an alternative procedure in deriving the series.

Upon substituting (3-28) into (3-21), we see that terms involving both $\exp(j n\omega_0 t)$ and $\exp(-j n\omega_0 t)$ will be present in the series. This is in keeping with our observation in Chapter 1 that the sum of two complex-conjugate phasors is required to produce a real, sinusoidal signal. Thus we assume that a periodic signal $x(t)$ can be represented by a complex exponential series of the form

$$x(t) = \cdots X_{-2}e^{-j2\omega_0 t} + X_{-1}e^{-j\omega_0 t} + X_0$$
$$+ X_1 e^{j\omega_0 t} + X_2 e^{j2\omega_0 t} + \cdots \qquad (3\text{-}29)$$

$$= \sum_{n=-\infty}^{\infty} X_n e^{jn\omega_0 t}$$

where the X_n's are, in general, complex constants. In order to find an expression for computing the X_n's, we multiply both sides of (3-29) by $e^{-jm\omega_0 t}$ and integrate over any period of $x(t)$. This results in

$$\int_{T_0} x(t)e^{-jm\omega_0 t}\, dt = \int_{T_0} \left(\sum_{n=-\infty}^{\infty} X_n e^{jn\omega_0 t} \right) e^{-jm\omega_0 t}\, dt \tag{3-30}$$

We next multiply each term in the sum by $e^{-jm\omega_0 t}\, dt$ and integrate the result term by term without worrying about the validity of these steps at this point.†
This results in

$$\int_{T_0} x(t)e^{-jm\omega_0 t}\, dt = \sum_{n=-\infty}^{\infty} X_n \int_{T_0} e^{j(n-m)\omega_0 t}\, dt \tag{3-31}$$

If $m \neq n$, the integral

$$\int_{T_0} e^{j(n-m)\omega_0 t}\, dt$$

is zero because

$$e^{j(n-m)\omega_0 t} = \cos(n-m)\omega_0 t + j\sin(n-m)\omega_0 t$$

is a periodic function with period T_0 and is symmetric about the t-axis. For $n = m$, $e^{j(n-m)\omega_0 t} = 1$, and the integral evaluates to T_0. Thus every term in the sum on the right-hand side of (3-31) is zero except the one for $n = m$; (3-31) therefore reduces to

$$\int_{T_0} x(t)e^{-jm\omega_0 t}\, dt = T_0 X_m$$

or, solving for X_m, we obtain

$$X_m = \frac{1}{T_0} \int_{T_0} x(t)e^{-jm\omega_0 t}\, dt \tag{3-32}$$

As with the trigonometric Fourier series, we *define* an exponential Fourier series to be a series of the form (3-29) with the coefficients calculated according to (3-32). The computation of the coefficients is usually considerably simpler than the computation of the trigonometric Fourier series coefficients. In addition, the exponential Fourier series provides us with a direct means of plotting the two-sided amplitude and phase spectra of a signal.

We note that the coefficients of the trigonometric Fourier series (3-21) and the complex coefficients (3-32) are related by

$$A_n = 2|X_n| \tag{3-33a}$$

and

$$\theta_n = \underline{/X_n} = \tan^{-1}\frac{\text{Im } X_n}{\text{Re } X_n} \tag{3-33b}$$

†A discussion of convergence for this form of the series would be identical to the one given for the trigonometric form.

The proof is left to the problems. Thus the single-sided amplitude spectrum of a signal is twice its double-sided amplitude spectrum for positive frequencies. The phase spectra are identical for positive frequencies.

This section will be concluded with an example to illustrate the calculation of complex Fourier series coefficients.

EXAMPLE 3-5

Consider the periodic sequence of asymmetrical pulses (referred to as a pulse train)

$$x(t) = \sum_{m=-\infty}^{\infty} A\Pi \left(\frac{t - t_0 - mT_0}{\tau} \right), \qquad \tau < T_0$$

whose period is T_0. The exponential Fourier series coefficients are calculated by evaluating the integral

$$X_n = \frac{1}{T_0} \int_{t_0-\tau/2}^{t_0+\tau/2} A e^{-jn\omega_0 t} \, dt \tag{3-34}$$

where $\omega_0 = 2\pi/T_0$. The evaluation is straightforward. Integration gives

$$X_n = \frac{-A}{jn\omega_0 T_0} e^{-jn\omega_0 t} \Big|_{t_0-\tau/2}^{t_0+\tau/2}$$

$$= \frac{2A}{n\omega_0 T_0} e^{-jn\omega_0 t_0} \left(\frac{e^{jn\omega_0\tau/2} - e^{-jn\omega_0\tau/2}}{2j} \right), \qquad n \neq 0$$

$$= \frac{2A}{n\omega_0 T_0} e^{-jn\omega_0 t_0} \sin \frac{n\omega_0\tau}{2}, \qquad n \neq 0 \tag{3-35a}$$

Letting $n = 0$ in (3-34), we find X_0 to be

$$X_0 = \frac{A\tau}{T_0} \tag{3-35b}$$

Substituting $\omega_0 = 2\pi f_0$ into (3-35a), where f_0 is the frequency in hertz, we obtain the form

$$X_n = \frac{A\tau}{T_0} e^{-j2\pi n f_0 t_0} \frac{\sin \pi n f_0 \tau}{\pi n f_0 \tau} \tag{3-35c}$$

where the numerator and denominator have been multiplied by τ. ∎

It is convenient to define.

$$\operatorname{sinc} z = \frac{\sin \pi z}{\pi z} \tag{3-36}$$

which is referred to as the *sinc function*. Values for sinc z and $\operatorname{sinc}^2 z$ are tabulated in Appendix D. Examining (3-36), we see that the sinc function is

TABLE 3-1. Coefficients for the Complex Exponential Fourier Series of Several Signals

Signal	Waveform	Coefficients
1. Half-rectified sine wave	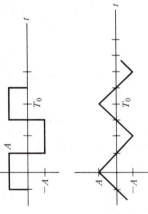	$X_n = \begin{cases} \dfrac{A}{\pi(1 - n^2)}, & n = 0, \pm 2, \pm 4, \ldots \\ 0, & n \text{ odd and } \neq 1 \\ -\dfrac{1}{4} jnA, & n = \pm 1 \end{cases}$
2. Full-rectified sine wave		$X_n = \begin{cases} \dfrac{2A}{\pi(1 - n^2)}, & n \text{ even} \\ 0, & n \text{ odd} \end{cases}$
3. Pulse-train signal		$X_n = \dfrac{A\tau}{T_0} \operatorname{sinc} n f_0 \tau \, e^{-j2\pi n f_0 t_0}, \quad f_0 = T_0^{-1}$
4. Square wave		$X_n = \begin{cases} \dfrac{2A}{|n|\pi}, & n = \pm 1, \pm 5, \ldots \\ \dfrac{-2A}{|n|\pi}, & n = \pm 3, \pm 7, \ldots \\ 0, & n \text{ even} \end{cases}$
5. Triangular wave		$X_n = \begin{cases} \dfrac{4}{\pi^2 n^2}, & n \text{ odd} \\ 0, & n \text{ even} \end{cases}$

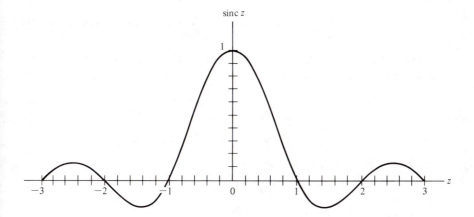

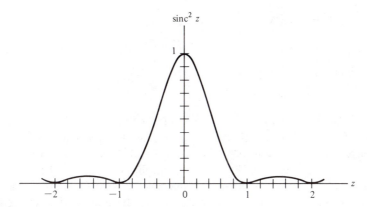

FIGURE 3-5. Sinc z and sinc² z.

zero whenever

$$\sin \pi z = 0$$

This happens for $z = \pm 1, \pm 2, \pm 3, \ldots$. Furthermore, the sinc function is even (substitute $-z$ for z and obtain the same function) and has a maximum of unity at $z = 0$. The latter may be shown by using l'Hospital's rule. Also, sinc z is oscillatory with the amplitude of the oscillations decreasing as $|z| \to \infty$. Thus a sketch of sinc z appears as shown in Figure 3-5.

It is straightforward to express (3-35c) in terms of the sinc function. Table 3-1 summarizes the Fourier coefficients for the pulse train as well as those for the half-rectified sine wave and several other waveforms of interest.

3-6

SYMMETRY PROPERTIES OF THE FOURIER COEFFICIENTS

The expression for the coefficients of the complex exponential Fourier series (3-32), through use of Euler's theorem, can be written as

$$X_m = \frac{1}{T_0} \int_{T_0} x(t) \cos m\omega_0 t \, dt - \frac{j}{T_0} \int_{T_0} x(t) \sin m\omega_0 t \, dt \qquad (3\text{-}37)$$

If $x(t)$ is real, the first term is the real part of X_m and the second term is the imaginary part of X_m. Comparing this with (3-15) and (3-16), we see that

$$X_m = \begin{cases} \frac{1}{2}(a_m - jb_m), & m > 0 \\ \frac{1}{2}(a_{-m} + jb_{-m}), & m < 0 \end{cases} \qquad (3\text{-}38)$$

Solving (3-38) for a_m and b_m, we obtain

$$a_m = 2 \operatorname{Re} X_m \quad \text{and} \quad b_m = -2 \operatorname{Im} X_m, \qquad m > 0 \qquad (3\text{-}39)$$

If we replace m by $-m$ in (3-37) or (3-38), it is clear that

$$X_m = X^*_{-m} \qquad (3\text{-}40)$$

for $x(t)$ real. Writing X_m in polar form as

$$X_m = |X_m| e^{j\theta_m} \qquad (3\text{-}41)$$

we conclude from (3-40) that

$$|X_m| = |X_{-m}| \quad \text{and} \quad \theta_m = -\theta_{-m} \qquad (3\text{-}42)$$

That is, for real signals, the magnitude of the Fourier coefficients is an even function of the index, m, and the argument is an odd function of m.

Considering the Fourier coefficients for a real, even signal, that is $x(t) = x(-t)$, we see from (3-37) that the imaginary part will be zero since $x(t) \sin m\omega_0 t$ is an odd function that integrates to zero over an interval symmetrically placed about $t = 0$. That is, the complex exponential Fourier series coefficients X_m of a real, even signal are real (see Example 3-5 with $t_0 = 0$). Furthermore, since the dependence on m is through the function $\cos m\omega_0 t$, they are also even functions of m. From (3-37) and (3-39) it follows that the b_m's are zero for the trigonometric series of real, even signals.

Similar reasoning shows that the X_m's are imaginary and odd functions of m if $x(t)$ is odd, that is, if $x(t) = -x(-t)$. Also $a_m = 0$, all m, if $x(t)$ is odd, so that the trigonometric Fourier series of an odd function consists only of sine terms (see Example 3-3). The proofs of these statements are left to the problems.

Another type of symmetry is *half-wave* (odd) symmetry defined as

$$x\left(t \pm \frac{T_0}{2}\right) = -x(t) \qquad (3\text{-}43)$$

where T_0 is a period of $x(t)$.† For signals with half-wave symmetry, it turns out that

$$X_n = 0, \qquad n = 0, \pm 2, \pm 4, \ldots$$

†To illustrate half-wave symmetry, consider waveform 4 in Table 3-1. Pick an arbitrary point on the t-axis. The value of the signal at this time is equal in magnitude, but opposite in sign, to the value of the signal one-half period earlier or later. The same argument holds for waveform 5 in Table 3-1.

The proof is left to the problems. Thus the Fourier series of a half-wave (odd) symmetrical signal contains only odd harmonics (see Example 3-3 and Table 3-1, waveforms 4 and 5). We note the half-wave even symmetry, defined by

$$x\left(\frac{t \pm T_0}{2}\right) = x(t) \tag{3-44}$$

simply means that the fundamental period of the waveform is $T_0/2$ rather than T_0.

Table 3-2 summarizes the properties of the three types of Fourier series that we have considered.

3-7

PARSEVAL'S THEOREM

In Chapter 1 the average normalized power of a periodic waveform $x(t)$ was written as

$$P_{av} = \frac{1}{T_0} \int_{T_0} |x(t)|^2 \, dt = \frac{1}{T_0} \int x(t)x^*(t) \, dt \tag{3-45}$$

We can express P_{av} in terms of the Fourier coefficients of $x(t)$ by replacing $x^*(t)$ in (3-45) with its Fourier series representation (3-29). The substitution yields

$$P_{av} = \frac{1}{T_0} \int_{T_0} x(t) \left(\sum_{n=-\infty}^{\infty} X_n^* e^{-jn\omega_0 t}\right) dt$$

$$= \sum_{n=-\infty}^{\infty} X_n^* \left[\frac{1}{T_0} \int_{T_0} x(t)e^{-jn\omega_0 t} \, dt\right]$$

where the order of integration and summation have been interchanged. We recognize the term in brackets as X_n [see (3-32)]. Thus we may write

$$P_{av} = \frac{1}{T_0} \int_{T_0} |x(t)|^2 \, dt = \sum_{n=-\infty}^{\infty} |X_n|^2 \tag{3-46a}$$

$$= X_0^2 + 2 \sum_{n=1}^{\infty} |X_n|^2 \tag{3-46b}$$

where (3-46b) results by applying (3-40). In words, (3-46a) simply states that the average power of a periodic signal $x(t)$ is the sum of the powers in the phasor components of its Fourier series (recall Example 1.8, in which the power of a phasor was computed). Equivalently, (3-46b) states that the average power in a periodic signal is the sum of the power in its dc component plus the powers in its harmonic components (recall that the average power of a sinusoid is one-half the square of its amplitude).

TABLE 3-2. Summary of Fourier Series Properties[a]

Series	Coefficients[b]	Symmetry Properties
1. Trigonometric sine–cosine $$x(t) = a_0 + \sum_{n=1}^{\infty} (a_n \cos n\omega_0 t + b_n \sin n\omega_0 t)$$	$$a_0 = \frac{1}{T_0} \int_{T_0} x(t)\, dt$$ $$a_n = \frac{2}{T_0} \int_{T_0} x(t) \cos n\omega_0 t\, dt$$ $$b_n = \frac{2}{T_0} \int_{T_0} x(t) \sin n\omega_0 t\, dt$$	$a_0 = $ Average value of $x(t)$ $a_n = 0$ for $x(t)$ odd, $b_n = 0$ for $x(t)$ even $a_n, b_n = 0$, n even, for $x(t)$ odd, half-wave symmetrical
2. Trigonometric cosine $$x(t) = A_0 + \sum_{n=1}^{\infty} A_n \cos (n\omega_0 t + \theta_n)$$	$$A_0 = a_0; \quad A_n = \sqrt{a_n^2 + b_n^2}$$ $$\theta_n = -\tan^{-1} \frac{b_n}{a_n}$$	$A_0 = $ Average value of $x(t)$ $A_n = 0$, n even, for $x(t)$ odd, half-wave symmetrical
3. Complex exponential $$x(t) = \sum_{n=-\infty}^{\infty} X_n e^{jn\omega_0 t}$$	$$X_n = \frac{1}{T_0} \int_{T_0} x(t) e^{-jn\omega_0 t}\, dt$$ $$X_n = \begin{cases} \frac{1}{2}(a_n - jb_n), & n > 0 \\ \frac{1}{2}(a_{-n} + jb_{-n}), & n < 0 \end{cases}$$ $$X_n = X_{-n}^* \quad \text{for } x(t) \text{ real}$$	$X_0 = $ Average value of $x(t)$ X_n real for $x(t)$ even X_n imaginary for $x(t)$ odd $X_n = 0$, n even, for $x(t)$ odd half-wave symmetrical

[a] $x(t)$ even means that $x(t) = x(-t)$; $x(t)$ odd means that $x(t) = -x(-t)$; $x(t)$ odd half-wave symetrical means that $x(t) = -x(t \pm T_0/2)$.

[b] $\int_{T_0} (\cdot)\, dt$ means integration over any period T_0 of $x(t)$.

EXAMPLE 3-6

The average power of the sine wave $x(t) = 4 \sin 50\pi t \; dt$ is

$$P_{av} = 25 \int_0^{0.04} 16 \sin^2 50\pi t \; dt = \frac{16}{2} = 8 \text{ W}$$

computed from (3-45). Its exponential Fourier series coefficients are $X_{-1} = -X_1 = 2j$ and its power, when computed from (3-46a), is $P_{av} = 4 + 4 = 8$ W. ∎

EXAMPLE 3-7

Consider the use of a square wave to test the fidelity of an amplifier. Suppose that the amplifier ideally passes all Fourier components of the input with frequencies less than 51 kHz. By ideal, it is meant that the individual Fourier components of the input with frequencies less than 51 kHz suffer zero phase shift in passing through the amplifier and all are multiplied by the same factor, called the *voltage gain*, G_v, while all Fourier components above 51 kHz do not appear at the output.

Suppose that $G_v = 10$ and the input, $x(t)$, is a 10-kHz square wave with amplitudes ± 1 V. (a) What is the average power P_x of the input signal? (b) What is the average power P_y of the output signal, $y(t)$? (c) What would the average power of the output signal be if the amplifier passed all frequency components at its input?

Solution: (a) The average power of the input is found by using (1-29) for the average power of a periodic signal. Letting T_0 be the period of the square wave and A its amplitude, we obtain

$$P_x = \frac{1}{T_0} \int_{t_0}^{t_0+T_0} x^2(t) \; dt$$

$$= \frac{1}{T_0} \int_{t_0}^{t_0+T_0} A^2 \; dt$$

The immediately preceding equation follows by virtue of the square-wave amplitude being either A or $-A$. Thus the squared amplitude of $x(t)$ is A^2 and the average power beomes

$$P_x = A^2 \qquad \text{(square wave of amplitudes } \pm A\text{)}$$

$$= 1 \text{ W}$$

Parseval's theorem could be used to obtain P_x by summing the powers in the spectral components provided that one has the patience to sum an infinite series!

(b) The exponential Fourier series of an even-symmetry square wave, $x(t)$, with amplitude $A = 1$ V is

$$x(t) = \frac{2}{\pi} (\cdots - \tfrac{1}{7} e^{-j7\omega_0 t} + \tfrac{1}{5} e^{-j5\omega_0 t} - \tfrac{1}{3} e^{-j3\omega_0 t} + e^{-j\omega_0 t}$$

$$+ e^{j\omega_0 t} - \tfrac{1}{3} e^{j3\omega_0 t} + \tfrac{1}{5} e^{j5\omega_0 t} - \tfrac{1}{7} e^{j7\omega_0 t} + \cdots)$$

where $\omega_0 = 2\pi/T_0 = 2\pi f_0$ with $f_0 = 10$ kHz. The amplifier multiplies all spectral components with frequencies less than 51 kHz by the factor $G_v = 10$ and blocks all other spectral components. Thus the Fourier series of the output of the amplifier is

$$y(t) = \frac{20}{\pi} \quad (\tfrac{1}{5} e^{-j5\omega_0 t} - \tfrac{1}{3} e^{-j3\omega_0 t} + e^{-j\omega_0 t}$$

$$+ e^{j\omega_0 t} - \tfrac{1}{3} e^{j3\omega_0 t} + \tfrac{1}{5} e^{j5\omega_0 t})$$

$$\equiv \sum_{n=-\infty}^{\infty} Y_n e^{jn\omega_0 t}$$

The amplitudes of the Fourier components of the output are therefore

$$Y_n = 0, \quad |n| > 5$$

$$Y_n = 0, \quad n \text{ even}$$

$$Y_{-1} = Y_1 = \frac{20}{\pi}$$

$$Y_{-3} = Y_3 = -\frac{20}{3\pi}$$

$$Y_{-5} = Y_5 = \frac{20}{5\pi}$$

Using Parseval's theorem, the output power is

$$P_y = Y_0^2 + 2 \sum_{n=1}^{\infty} |Y_n|^2 = 2 \left[\left(\frac{20}{\pi}\right)^2 + \left(\frac{20}{3\pi}\right)^2 + \left(\frac{20}{5\pi}\right)^2 \right]$$

$$= \frac{8 \times 10^2}{\pi^2} (1 + \tfrac{1}{9} + \tfrac{1}{25})$$

$$= 93.31 \text{ W}$$

(c) If the amplifier passes $x(t)$ perfectly, then

$$\hat{y}(t) = G_v x(t)$$

and

$$P_{\hat{y}} = G_v^2 P_x$$

$$= 10^2 \text{ W}$$

Note that $P_y/P_{\hat{y}} = 93.31\%$. This result tells us that only the first through fifth harmonics of the square wave need to be included to obtain 93.3% of the average signal power possible at the amplifier output. The student should compute the percent of average signal power possible at the amplifier output if its cutoff frequency is such that the seventh, ninth, eleventh, and so on, harmonics are passed by the amplifier but all others rejected. (*Answers:* 94.96%; 95.96%; 96.63%.) ■

LINE SPECTRA

The complex exponential Fourier series (3-29) of a signal consists of a summation of rotating phasors. In Chapter 1 we showed how sums of rotating phasors can be characterized in the frequency domain by two plots, one showing their amplitudes as a function of frequency and one showing their phases. This observation allows a periodic signal to be characterized graphically in the frequency domain by making two plots. The first, showing amplitudes of the separate phasor components versus frequency, is known as the *amplitude spectrum* of the signal. The second, showing the relative phases of each component versus frequency, is called the *phase spectrum* of the signal. Since these plots are obtained from the complex exponential Fourier series, spectral components, or *lines* as they are called, are present at both positive and negative frequencies. Such spectra are referred to as *two-sided*. From (3-42) it follows that, for a real signal, the amplitude spectrum is even and the phase spectrum is odd, which simply is a result of the necessity to add complex-conjugate phasors to get a real sinusoidal signal. Figure 3-6 shows the two-sided spectra for a half-rectified sine wave. The complex exponential Fourier coefficients are found from (3-26) and (3-27) by applying (3-33). In order that the phase be odd, X_n is represented as

$$X_n = - \left| \frac{A}{\pi(1 - n^2)} \right| = \frac{A}{\pi(n^2 - 1)} e^{-j\pi}$$

for $n = 2, 4, \ldots$, and as

$$X_n = - \left| \frac{A}{\pi(1 - n^2)} \right| = \frac{A}{\pi(n^2 - 1)} e^{j\pi}$$

for $n = -2, -4, \ldots$. For $n = \pm 1$, the amplitude spectrum is $A/4$ with a phase shift of $-\pi/2$ for $n = 1$ and a phase shift of $\pi/2$ for $n = -1$, which again assures that the phase is odd.

The *single-sided line spectra* are obtained by plotting the amplitudes and phase angles of the terms in the trigonometric Fourier series (3-21) versus nf_0 for n positive since the series (3-21) has nonnegative frequency terms only. From (3-33) it is readily apparent that the single-sided phase spectrum of a periodic signal is identical to its double-sided phase spectrum for $nf_0 \geq 0$, and zero for $nf_0 < 0$. The single-sided amplitude spectrum is obtained from the double-sided amplitude spectrum by doubling the amplitude of all lines for $nf_0 > 0$. The line at $nf_0 = 0$ stays the same. The student should plot the single-sided spectra for the half-rectified sine wave.

As a second example, consider the periodic pulse train shown in Table 3-1 as waveform 3. We consider the special case of $t_0 = \tau/2$. From (3-35), the Fourier coefficients are

$$X_n = \frac{A\tau}{T_0} \operatorname{sinc} nf_0\tau \, e^{-j\pi nf_0\tau}$$

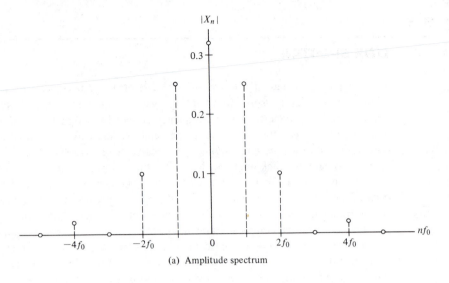

(a) Amplitude spectrum

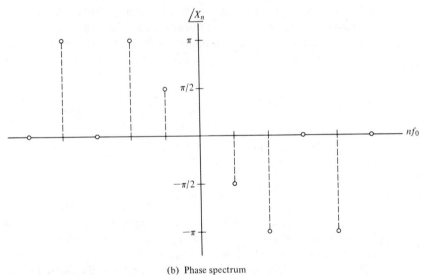

(b) Phase spectrum

FIGURE 3-6. Two-sided line spectra for a half-rectified sine wave.

The Fourier coefficients can be put in the form $|X_n| \exp(j\underline{/X_n})$, where

$$|X_n| = \frac{A\tau}{T_0} |\text{sinc } nf_0\tau| \qquad (3\text{-}47a)$$

and

$$\underline{/X_n} = \begin{cases} -\pi nf_0\tau & \text{if sinc } nf_0\tau > 0 \\ -\pi nf_0\tau + \pi & \text{if } nf_0 > 0 \text{ and sinc } nf_0\tau < 0 \quad (3\text{-}47b) \\ -\pi nf_0\tau - \pi & \text{if } nf_0 < 0 \text{ and sinc } nf_0\tau < 0 \end{cases}$$

The $\pm\pi$ on the right-hand side of (3-47b) accounts for $|\text{sinc } nf_0\tau| = -\text{sinc } nf_0\tau$ whenever sinc $nf_0 < 0$. Since the phase spectrum must have odd symmetry if $x(t)$ is real, π is subtracted if $nf_0 < 0$ and added if $nf_0 > 0$. The two-sided amplitude and phase spectra are shown in Figure 3-7 for several choices of τ and T_0. Note that whenever a phase line is 2π or greater in magnitude, 2π is subtracted. Two important observations can be made from Figure 3-7. First, comparing Figure 3-7a and b we note that the width of the envelope of the amplitude spectrum increases as the pulse width decreases. That is, *the pulse width of the signal and its corresponding spectral width are inversely proportional.* Second, comparing Figure 3-7a and c, we note that the separation between lines in the spectrum is $1/T_0$, and therefore *the density of the spectral lines with frequency increases as the period of $x(t)$ increases.*

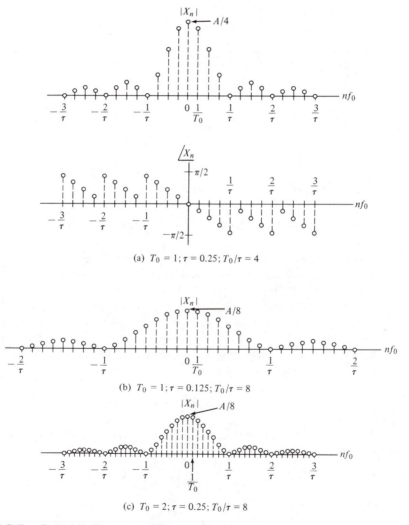

(a) $T_0 = 1; \tau = 0.25; T_0/\tau = 4$

(b) $T_0 = 1; \tau = 0.125; T_0/\tau = 8$

(c) $T_0 = 2; \tau = 0.25; T_0/\tau = 8$

FIGURE 3-7. Spectra for a periodic pulse-train signal.

TRANSFER FUNCTION OF A FIXED, LINEAR SYSTEM; DISTORTIONLESS SYSTEMS

If the input to a fixed, linear system is a sinusoid of frequency ω rad/s, the steady-state output is also a sinusoid of the same frequency, but with amplitude multiplied by a factor $A(\omega)$ and phase shifted by $\theta(\omega)$ radians. The complex function of frequency

$$\tilde{H}(\omega) = A(\omega) \exp [j\theta(\omega)] \qquad (3\text{-}48)$$

is called the *transfer function* or *frequency-response function* of the system; $A(\omega)$ and $\theta(\omega)$ are referred to as the *amplitude-response* and *phase-response functions* of the system, respectively.

Applying the superposition principle, we conclude that the steady-state output of a system resulting from a periodic input with Fourier series

$$x(t) = \sum_{n=0}^{\infty} (a_n \cos n\omega_0 t + b_n \sin n\omega_0 t) \qquad (3\text{-}49)$$

is

$$y(t) = \sum_{n=0}^{\infty} A(n\omega_0) \{a_n \cos [n\omega_0 t + \theta(n\omega_0)] $$

$$+ b_n \sin [n\omega_0 t + \theta(n\omega_0)]\} \qquad (3\text{-}50)$$

When written in terms of a complex exponential Fourier series, $y(t)$ becomes

$$y(t) = \sum_{n=-\infty}^{\infty} A(n\omega_0) X_n \exp \{j[n\omega_0 t + \theta(n\omega_0)]\} \qquad (3\text{-}51)$$

The frequency-response function, (3-48), completely characterizes the response of a fixed, linear system to a sinusoid or, equivalently, a complex exponential $e^{j\omega t}$. When the input $e^{j\omega t}$ is applied to the system, its steady-state response will be $\tilde{H}(\omega)e^{j\omega t}$. That is, the steady-state output of a fixed, linear system is of the same form as its input when its input is $e^{j\omega t}$. In mathematical terms, $e^{j\omega t}$ is said to be an *eigenfunction* of the system. That this is the case can be shown by considering the superposition integral (2-2) with the input $x(t) = e^{j\omega t}$. Then the output is

$$y(t) = \int_{-\infty}^{\infty} e^{j\omega(t-\lambda)}h(\lambda) \, d\lambda$$

where $h(\lambda)$ is the impulse response. Since $e^{j\omega t}$ can be factored out of the integral, the output can be written as

$$y(t) = e^{j\omega t} \int_{-\infty}^{\infty} h(\lambda)e^{-j\omega\lambda} \, d\lambda \qquad (3\text{-}52)$$

We identify the complex function $\int_{-\infty}^{\infty} h(\lambda)e^{-j\omega\lambda} \, d\lambda$ as the frequency response

$\tilde{H}(\omega)$. Later we shall see that $\tilde{H}(\omega)$ corresponds to the *Fourier transform* of the impulse response $h(t)$.

If the system is to have no effect on the input except, possibly, for an amplitude change and delay which are the same for all frequency components, the system is called *distortionless*. For a distortionless system with periodic input of the form (3-49), the output is of the form

$$y(t) = K \sum_{n=0}^{\infty} [a_n \cos n\omega_0(t - \tau_0) + b_n \sin n\omega_0(t - \tau_0)]$$

$$= Kx(t - \tau_0)$$

Writing this in the form of (3-50), we see that

$$A(n\omega_0) = K \qquad \text{and} \qquad \theta(n\omega_0) = -\tau n\omega_0$$

or

$$A(\omega) = K \qquad \text{and} \qquad \theta(\omega) = -\tau_0\omega \tag{3-53}$$

respectively. That is, the amplitude response of a distortionless system is the same, or *constant,* for all frequency components of the input and the phase response is a *linear* function of frequency. In other words, the frequency response function of a distortionless system is of the form

$$\tilde{H}_d(\omega) = K \exp(-j\tau_0\omega) \tag{3-54}$$

In general, three major types of distortion can be introduced by a system. First, if a system is linear but its amplitude response is not constant with frequency, the system introduces *amplitude distortion*. Second, if a system is linear but its phase shift is not a linear function of frequency, it introduces *phase,* or *delay, distortion*. Third, if the system is not linear, *nonlinear* distortion results. These three types of distortion may occur in combination with each other.

EXAMPLE 3-8

To illustrate the ideas of amplitude and phase distortion, consider a system with amplitude response and phase response as shown in Figure 3-8, and the following three inputs:

1. $x_1(t) = 2 \cos 10\pi t + \sin 12\pi t$
2. $x_2(t) = 2 \cos 10\pi t + \sin 26\pi t$
3. $x_3(t) = 2 \cos 26\pi t + \sin 34\pi t$

Although this system has idealized frequency response characteristics, it can be used to illustrate various combinations of amplitude and phase distortion. Using (3-49), we find the outputs for the given inputs to be:

1. $y_1(t) = 4 \cos(10\pi t - \pi/6) + 2 \sin(12\pi t - \pi/5)$
   $= 4 \cos 10\pi(t - 1/60) + 2 \sin 12\pi(t - 1/60)$
2. $y_2(t) = 4 \cos(10\pi t - \pi/6) + \sin(26\pi t - 13\pi/30)$
   $= 4 \cos 10\pi(t - 1/60) + \sin 26\pi(t - 1/60)$
3. $y_3(t) = 2 \cos(26\pi t - 13\pi/30) + \sin(34\pi t - \pi/2)$
   $= 2 \cos 26\pi(t - 1/60) + \sin 34\pi(t - 1/68)$

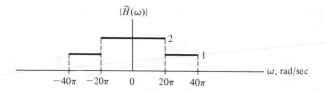

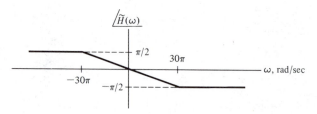

FIGURE 3-8 Amplitude and phase response of the filter for Example 3-8.

Comparison with (3-53) shows that only the input $x_1(t)$ is passed without distortion by the system. The system imposes amplitude distortion on $x_2(t)$ and phase (delay) distortion on $x_3(t)$. ∎

EXERCISE

A second-order low-pass Butterworth filter is one having frequency-response function

$$H(\omega) = \frac{\omega_c^2}{\omega_c^2 - \omega^2 + j\sqrt{2}\,\omega_c\omega}$$

where ω_c is a filter parameter.

(a) Show that its amplitude- and phase-response functions are

$$A(\omega) = \left[1 + \left(\frac{\omega}{\omega_c}\right)^4\right]^{-1/2}$$

and

$$\theta(\omega) = \begin{cases} -\tan^{-1}\dfrac{\sqrt{2}(\omega/\omega_c)}{1 - \omega^2/\omega_c^2}, & \omega < \omega_c \\[2ex] -\pi + \tan^{-1}\dfrac{\sqrt{2}(\omega/\omega_c)}{\omega^2/\omega_c^2 - 1}, & \omega > \omega_c \end{cases}$$

respectively, where the value of the arctangent function is taken as its principal value.

(b) If $\omega_c = 300$ rad/s above and $\omega_0 = 100$ rad/s in (3-17), obtain the attenuations and phase shifts introduced by the filter in the first three terms of the Fourier series of a square-wave input to the filter. (*Answer:* attenuations: 0.994, 0.707, 0.339; phase shifts: $-27.9°$, $-90°$, $-127°$.)

(c) Write out the first three nonzero terms of the Fourier series of the output and compare with the input by plotting.

(*Answer:* $y(t) = 4/\pi [0.994 \sin(200\pi t - 27.9°) + 0.707 \sin(600\pi t - 90°) + 0.339 \sin (1000\pi t - 127°)]$.)

FREQUENCY GROUPS

As a means toward the goal of obtaining a frequency-domain representation for signals that are not necessarily periodic and yet are defined over $-\infty < t < \infty$, we consider the time-domain behavior of a finite sum of sinusoidal components with harmonically related frequencies. Thus consider a sum of N sinusoidal components whose spacing in frequency is Δf and whose mean frequency is $f_0 = M \Delta f$, where M is an integer. Their amplitudes are assumed to be equal and their phase angles zero. The line spectrum of such a group of sinusoidal components is shown in Figure 3-9, where it is noted that the total bandwidth of the group is defined as $B = N \Delta f$, and their common amplitude is taken as $1/N$. Mathematically, the signal represented by the spectral plot of Figure 3-9 can be expressed as

$$x(t) = \frac{1}{N} \sum_{n=-(N-1)/2}^{(N-1)/2} \cos [2 \pi (f_0 - n \Delta f)t] \qquad (3\text{-}55)$$

where N is assumed to be odd. Expressing the cosine in exponential form and factoring out the $\exp (\pm j2\pi f_0 t)$ after separating into two sums, we obtain the form

$$x(t) = \frac{\exp (j2\pi f_0 t)}{2N} \sum \exp (-j2\pi n \Delta f t)$$

$$+ \frac{\exp (-j2\pi f_0 t)}{2N} \sum \exp (j2\pi n \Delta f t)$$

$$= \frac{1}{N} \cos 2\pi f_0 t \sum \exp (j2\pi n \Delta f t) \qquad (3\text{-}56)$$

where the indices have been omitted from the sums to simplify notation and we note that the sums of $\exp (-j2\pi n \Delta f t)$ and $\exp (j2\pi n \Delta f t)$ are equal because they are over symmetrical limits about $n = 0$. Letting $m = n + (N - 1)/2$ in the sum, we may express the sum as

$$A(t) \triangleq \frac{1}{N} \sum_{m=0}^{N-1} \exp \left[j2\pi \left(m - \frac{N - 1}{2} \right) \Delta f t \right]$$

$$= \frac{1}{N} \exp \left(-j2\pi \frac{N - 1}{2} \Delta f t \right) \sum_{m=0}^{N-1} \exp (j2\pi m \Delta f t)$$

$$= \frac{1}{N} \exp \left(-j2\pi \frac{N - 1}{2} \Delta f t \right) \frac{1 - \exp (j2\pi N \Delta f t)}{1 - \exp (j2\pi \Delta f t)} \qquad (3\text{-}57)$$

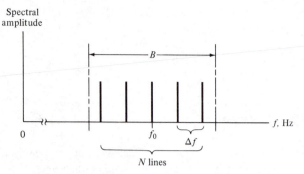

FIGURE 3-9. Group of N closely spaced spectral lines to illustrate constructive and destructive interference phenomena.

To arrive at (3-57), the summation formula for a geometric series, namely,

$$S_N = 1 + x + x^2 + \cdots + x^{N-1} = \frac{1 - x^N}{1 - x} \qquad (3\text{-}58)$$

has been used. It is left as an exercise for the student to show that (3-57) can be reduced to the form

$$A(t) = \frac{\sin \pi N \, \Delta f t}{N \sin \pi \, \Delta f t} \qquad (3\text{-}59)$$

so that

$$x(t) = \frac{\sin \pi N \, \Delta f t}{N \sin \pi \, \Delta f t} \cos 2\pi f_0 t \qquad (3\text{-}60)$$

The form of (3-60) is that of a slowly varying envelope or amplitude function, $A(t)$, multiplying a higher-frequency sinusoidal signal $\cos 2\pi f_0 t$. To the ear, a sound wave of the form (3-60) would alternately sound loud and soft due to the constructive and destructive interference of the component sinusoidal frequencies that make up (3-55) or, equivalently, (3-60).

To study this interference phenomenon further, it is instructive to graph the envelope (3-59). This is done in Figure 3-10 for $N = 7$, where we note that the reinforced portions are separated in time by $(\Delta f)^{-1}$ and that its zero crossings occur at multiples of $(N \, \Delta f)^{-1} = B^{-1}$.† The width of each reinforcement lobe is $2B^{-1}$.

If B is held constant and $\Delta f \to 0$, it is apparent from Figure 3-10 that the period between reinforcements grows without limit until finally the envelope has only one region of constructive interference or reinforcement in the entire interval $-\infty < t < \infty$. The limiting form of $A(t)$ is found by letting $\Delta f \to 0$ and $N \to \infty$ such that $B = N \, \Delta f$ remains constant. Replacing $\sin \pi \, \Delta f t$ in (3-59) by

---

†The reinforced envelope portions are often referred to as "beats" in music since it is the familiar phenomenon of the tones of two out-of-tune band instruments alternately reinforcing and destructively interfering with each other.

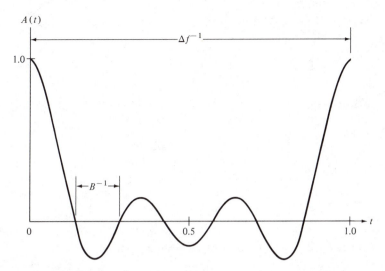

FIGURE 3-10. Amplitude function for the sum of **N** cosine waves spaced Δf hertz apart in frequency. **N** = 7, Δf = 1 Hz, and **B** = **N** Δf = 7 Hz.

its argument, we obtain

$$\lim_{\substack{\Delta f \to 0 \\ N \to \infty}} A(t) = \frac{\sin \pi B t}{\pi B t} = \operatorname{sinc} B t \tag{3-61}$$

which we have plotted in Figure 3-5.

3-11

THE FOURIER INTEGRAL

Our considerations in Section 3-10 have led us to the conclusion that whereas the linear superposition of a discrete set of uniformly spaced sinusoidal (or complex exponential) signal components yields a resultant signal that is periodic in nature, such superpositions have an aperiodic character when the components in the set are continuously distributed in frequency. We now use the insight gained in that exercise to develop a spectral representation for nonperiodic signals. Specifically, in the complex Fourier series sum and the integral for the coefficients of the Fourier series, we let the frequency spacing $f_0 = T_0^{-1}$ approach zeo and the index n approach infinity such that the product $n f_0$ approaches a continuous frequency variable f. Repeating the exponential form of the Fourier series below for convenience,

$$x(t) = \sum_{n=-\infty}^{\infty} X_n e^{j2\pi n f_0 t} \tag{3-62}$$

$$X_n = \frac{1}{T_0} \int_{-T/2}^{T_0/2} x(t) e^{-j2\pi n f_0 t} \, dt \tag{3-63}$$

we note that the limiting process $f_0 = T_0^{-1} \to 0$ evidently implies that $X_n \to 0$ provided that the integral

$$\left| \int_{-\infty}^{\infty} x(t)e^{-j2\pi nf_0 t}\,dt \right| \leq \int_{-\infty}^{\infty} |x(t)|\,dt \tag{3-64}$$

exists. Thus, before carrying out the limiting process, it is expedient to write (3-62) and (3-63) in the forms

$$x(t) = \sum_{nf_0=-\infty}^{\infty} \frac{X_n}{f_0} e^{j2\pi nf_0 t} \Delta(nf_0) \tag{3-65}$$

$$\tilde{X}(nf_0) \triangleq \frac{X_n}{f_0} = \int_{-1/(2f_0)}^{1/(2f_0)} x(t)e^{-j2\pi nf_0 t}\,dt \tag{3-66}$$

respectively, where the symbol $\Delta(nf_0)$ means the increment in the variable nf_0 which is f_0. The limiting process then formally becomes

$$T_0 \to \infty$$

$$f_0 \to 0 \text{ and } n \to \infty \text{ such that } nf_0 \to f$$

$$\Delta(nf_0) \to df$$

$$\frac{X_n}{f_0} \to \tilde{X}(f)$$

which results in the sum in (3-65) becoming an integral over the continuous variable f, and the integral in (3-66) extending over the entire t-axis, $-\infty < t < \infty$. The resulting expressions are

$$x(t) = \int_{-\infty}^{\infty} X(f)e^{j2\pi ft}\,df \tag{3-67}$$

and

$$X(f) = \int_{-\infty}^{\infty} x(t)e^{-j2\pi ft}\,dt \tag{3-68}$$

where the tilde has been dropped from $X(f)$ to simplify notation.

These expressions define a *Fourier transform pair* for the signal $x(t)$, and are sometimes denoted by the notation $x(t) \leftrightarrow X(f)$, where the double-headed arrow means that $X(f)$ is obtained from $x(t)$ by applying (3-68) and $x(t)$ is obtained from $X(f)$ by applying (3-67). The integral (3-68) is referred to as the *Fourier transform* of $x(t)$, and is frequently denoted symbolically as $X(f) = \mathcal{F}[x(t)]$. Conversion back to the time domain by means of (3-67) is referred to as the *inverse Fourier transform* and is often denoted symbolically as $x(t) = \mathcal{F}^{-1}[X(f)]$.†

†Sufficient conditions for the convergence of the Fourier transform integral are that (1) the integral of $|x(t)|$ from $-\infty$ to ∞ exist, and (2) any discontinuities in $x(t)$ be finite. These being sufficient conditions mean that signals exist which violate at least one of them and yet have Fourier transforms. An example is $\sin \alpha t/(\alpha t)$. For a discussion of other criteria that ensure convergence of the Fourier transform integral, see Bracewell (1978).

Later, we will include functions that do not possess Fourier transforms in the ordinary sense by generalization to Fourier transforms in the limit. Examples are the signals $x(t) = $ constant, all t, and $x(t) = \delta(t)$, the former violating condition (1) and the latter, condition (2).

Several symmetry properties of the Fourier transform may be shown by writing $X(f)$ in terms of magnitude and phase as

$$X(f) = |X(f)|e^{j\theta(f)} \qquad (3\text{-}69)$$

For example, assuming that $x(t)$ is real, it can be shown that

$$|X(f)| = |X(-f)| \qquad \text{and} \qquad \theta(f) = -\theta(-f) \qquad (3\text{-}70)$$

That is, the magnitude is an even function of frequency and the argument, or phase, is an odd function of frequency. Plots of $|X(f)|$ and $\theta(f) = \underline{/X(f)}$ with f chosen as the abscissa are referred to as the amplitude and phase spectra of $x(t)$, respectively.[†] Thus the amplitude spectrum of a real, Fourier transformable signal is even, and its phase spectrum is odd.

Further properties which can be derived for $X(f)$ if $x(t)$ is real is that the *real part* of $X(f)$ is an *even* function of frequency and the *imaginary part* of $X(f)$ is an *odd* function of frequency.

The foregoing properties are proved by expanding $\exp(-j2\pi ft)$ in (3-68) as $\cos 2\pi ft - j \sin 2\pi ft$ by employing Euler's theorem and writing $X(f)$ as the sum of two integrals which are the real and imaginary parts of $X(f)$. Since the frequency dependence of the real part of $X(f)$ is through $\cos 2\pi ft$, it is therefore an even function of f. Similarly, since the frequency dependence of the imaginary part of $X(f)$ is through $\sin 2\pi ft$, we conclude that it is odd.

Further properties that can be proved (see Problem 3-20) about $X(f)$ are:

1. If $x(t)$ is even [i.e., $x(t) = x(-t)$], then $X(f)$ is a *real, even function of f*.
2. If $x(t)$ is odd [i.e., $x(t) = -x(-t)$], then $X(f)$ is an *imaginary, odd function of f*.

To illustrate the computation of Fourier transforms as well as their symmetry properties discussed above, we consider the following example.

EXAMPLE 3-9

The Fourier transform of $x_a(t) = \Pi(t/2)$, which is sketched in Figure 3-11a, is real and even. On the other hand, the Fourier transform of $x_b(t) = \Pi(t + \frac{1}{2}) - \Pi(t - \frac{1}{2})$, which is the odd function sketched in Figure 3-11b, is imaginary and odd. To prove these statements, we derive their Fourier transforms in this example.

Since $\Pi(t/2) = 1$, $-1 < t < 1$, and is zero otherwise, the Fourier transform of $x_a(t)$ is given by

---

[†]More properly, the amplitude spectrum should be referred to as the magnitude spectrum, but the former term is customary. Note that $|X(f)|$ is an amplitude, or magnitude, *density*. Its units are volts (or whatever the dimensions of $x(t)$ are) per hertz. This differs from the Fourier coefficients, which have the same dimensions as the original signal, $x(t)$.

$$X_a(f) = \int_{-1}^{1} e^{-j2\pi ft}\, dt$$

$$= -\left. \frac{e^{-j2\pi ft}}{j2\pi f}\, \right|_{-1}^{1}$$

$$= 2\,\frac{\sin 2\pi f}{2\pi f}$$

$$= 2\operatorname{sinc} 2f \qquad\qquad (3\text{-}71)$$

which is clearly a real, even function of f.

From the sketch of $x_b(t)$ shown in Figure 3-11 we may write the integral for its Fourier transform as

$$X_b(f) = \int_{-1}^{0} e^{-j2\pi ft}\, dt - \int_{0}^{1} e^{-j2\pi ft}\, dt$$

$$= -\frac{1}{j2\pi f}\left[\left. e^{-j2\pi ft}\, \right|_{-1}^{0} - \left. e^{-j2\pi ft}\, \right|_{0}^{1} \right]$$

$$= \frac{j}{2\pi f}\,(2 - 2\cos 2\pi f)$$

$$= j2\,\frac{\sin^2 \pi f}{\pi f}$$

$$= j2\pi f \operatorname{sinc}^2 f \qquad\qquad (3\text{-}72)$$

This is clearly an imaginary, odd function of f.

The amplitude and phase spectra of $x_a(t)$ and $x_b(t)$ are shown in Figure 3-12. The student should convince himself that these are correct, particularly the phase spectrum for $x_b(t)$. ∎

3-12

ENERGY SPECTRAL DENSITY

The energy of a signal can be expressed in the frequency domain by proceeding as follows. By definition of the normalized energy for a signal, given by (1-27), we write

$$E \triangleq \int_{-\infty}^{\infty} |x(t)|^2\, dt$$

$$= \int_{-\infty}^{\infty} x^*(t)\left[\int_{-\infty}^{\infty} X(f)e^{j2\pi ft}\, df \right] dt \qquad\qquad (3\text{-}73)$$

where $x(t)$ has been written in terms of its Fourier transform. Reversal of the orders of integration results in

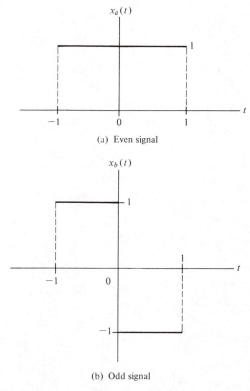

(a) Even signal

(b) Odd signal

FIGURE 3-11. Even and odd pulse-type signals for illustration of the symmetry properties of Fourier transforms.

$$E = \int_{-\infty}^{\infty} X(f) \left[\int_{-\infty}^{\infty} x^*(t) e^{j2\pi ft} \, dt \right] df$$

$$= \int_{-\infty}^{\infty} X(f) \left[\int_{-\infty}^{\infty} x(t) e^{-j2\pi ft} \, dt \right]^* df$$

$$= \int_{-\infty}^{\infty} X(f) X^*(f) \, df$$

or

$$E = \int_{-\infty}^{\infty} |x(t)|^2 \, dt = \int_{-\infty}^{\infty} |X(f)|^2 \, df \tag{3-74}$$

This is referred to as *Parseval's theorem for Fourier transforms*.

The units of $|X(f)|^2$, assuming that $x(t)$ is a voltage, are $(V\text{-}s)^2$ or, since E represents energy on a per ohm basis, $(W\text{-}s)/Hz = J/Hz$. Thus $|X(f)|^2$ has the units of energy density with frequency, and we define the *energy spectral density* of a signal as

$$G(f) \triangleq |X(f)|^2 \tag{3-75}$$

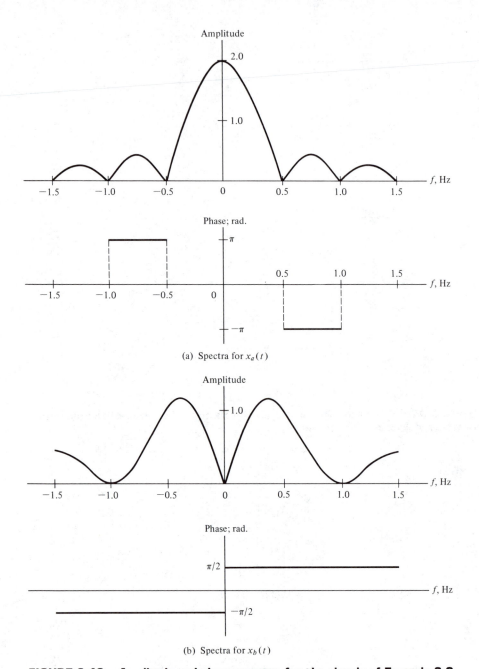

(a) Spectra for $x_a(t)$

(b) Spectra for $x_b(t)$

FIGURE 3-12. Amplitude and phase spectra for the signals of Example 3-9.

Integration of $G(f)$ over all frequencies from $-\infty$ to ∞ yields the total (normalized) energy contained in a signal. Similarly, integration of $G(f)$ over a finite range of frequencies gives the energy contained in the signal within the range of frequencies represented by the limits of integration.

EXAMPLE 3-10

The energy in the signal $x(t) = \exp(-\alpha t)u(t)$, $\alpha > 0$, has been shown to be $1/2\alpha$ in Example 1-6. The Fourier transform of this signal is (Problem 3-17)

$$X(f) = \frac{1}{\alpha + j2\pi f}$$

and its energy spectral density is

$$G(f) = \frac{1}{\alpha^2 + (2\pi f)^2}$$

The energy contained in this signal in the frequency range $-B < f < B$ is

$$E_B = \int_{-B}^{B} \frac{df}{\alpha^2 + (2\pi f)^2}$$

$$= \frac{1}{\pi\alpha} \int_{0}^{2\pi B/\alpha} \frac{dv}{1 + v^2}$$

$$= \frac{1}{\pi\alpha} \tan^{-1} \frac{2\pi B}{\alpha}$$

where the fact that the integral is even has been used together with the substitution $v = 2\pi f/\alpha$. The student should plot E_B as a function of B and show that $\lim_{B\to\infty} E_B = 1/2\alpha$, in agreement with the result calculated in Example 1-6. ∎

3-13

FOURIER TRANSFORMS IN THE LIMIT

We now wish to add to our repertoire of Fourier transform pairs. In doing so, we will soon be faced with transforms of signals that are not absolutely integrable, as expressed by (3-64), or which have infinite discontinuities. An example of a signal that is not absolutely integrable is a sinusoid extending over all time from $t = -\infty$ to $t = \infty$. This signal we know has a discrete spectrum and yet cannot be Fourier transformed through direct application of the defining integral (3-68). An example of a signal that has infinite discontinuities is the impulse function, which in the ordinary sense does not possess a Fourier transform. Rather than exclude such signals, which we know to be useful models, we broaden the definition of the Fourier transform to include Fourier transforms in the limit.

In order to define the Fourier transform of $\cos \omega_0 t$, $-\infty < t < \infty$, for example, we may first obtain the Fourier transform of the signal $\exp(-\alpha|t|) \cos \omega_0 t$, $\alpha > 0$, which is absolutely integrable. The Fourier transform of $\cos \omega_0 t$ then is defined as the limit of the Fourier transform of $\exp(-\alpha|t|) \cos \omega_0 t$ as $\alpha \to 0$ since $\lim_{\alpha\to 0} \exp(-\alpha|t|) = 1$, $-\infty < t < \infty$. The Fourier transform of $\cos \omega_0 t$ will be given later.

As a second example of obtaining a Fourier transform in the limit, consider the approximation for the unit impulse function $\delta_\epsilon(t) = \Pi(t/2\epsilon)/2\epsilon$ discussed in Chapter 1. The Fourier transform of $\delta_\epsilon(t)$ exists for each finite value of ϵ. In the limit as $\epsilon \to 0$, $\delta_\epsilon(t)$ approaches a unit impulse and its Fourier transform in the limit as $\epsilon \to 0$ is then defined to be the Fourier transform of the unit impulse function.

As a final, somewhat more subtle, example of a Fourier transform in the limit, consider $x(t) = A$, a constant. It should be clear that the Fourier integral in this case does not exist. (Write it down and try to evaluate the limits after integration!) On the other hand, the Fourier transform of $A\Pi(t/T)$ does exist and is similar to $x_a(f)$ derived in Example 3.9. It is, in fact, $\mathcal{F}[A\Pi(t/T)] = AT$ sinc fT. If we now let $T \to \infty$ in order to obtain the Fourier transform of a constant, it is not clear what the limit of AT sinc fT is. You should sketch this Fourier transform to show that it is a damped oscillatory function with a main lobe centered at $f = 0$ which becomes very narrow and high for T large. Furthermore, $\int_{-\infty}^{\infty} AT$ sinc $fT\, df = A$ for any T. Thus we formally write that $\mathcal{F}[A] = A\delta(f)$.

Easier ways of deriving these transforms will be shown after some basic transforms and theorems are considered.

3-14

FOURIER TRANSFORM THEOREMS

Several theorems involving operations on signals and the corresponding operations on their Fourier transforms are summarized in Table 3-3. These theorems are useful for obtaining additional Fourier transform pairs as well as in systems analysis. We now discuss these theorems, outlining proofs in some cases, and working examples to illustrate their application. To denote a Fourier transform pair we will sometimes use the notation $x(t) \leftrightarrow X(f)$, which means that $\mathcal{F}[x(t)] = X(f)$ and $\mathcal{F}^{-1}[X(f)] = x(t)$.

Linearity (Superposition) Theorem   Because the Fourier transform is an integral operation on a signal $x(t)$, with $x(t)$ appearing in the defining integral linearly, superposition is an obvious but extremely important property of the Fourier transform.

Time Delay Theorem   The time delay theorem follows by replacing $x(t)$ in (3-68) with $x(t - t_0)$ and making the change in variables $t = t' + t_0$, whereupon the factor exp $(-j2\pi f t_0)$ is taken outside the integral. The remaining integral is then the Fourier transform of $x(t)$ (note that what we call the variable of integration is immaterial).

Scale Change Theorem   To prove this theorem, we first suppose that $a > 0$, which gives

TABLE 3-3. Fourier Transform Theorems[a]

Name of Theorem		
1. Superposition (a_1 and a_2 arbitrary constants)	$a_1 x_1(t) + a_2 x_2(t)$	$a_1 X_1(f) + a_2 X_2(f)$
2. Time delay	$x(t - t_0)$	$X(f)e^{-j2\pi f t_0}$
3a. Scale change	$x(at)$	$\lvert a \rvert^{-1} X\left(\dfrac{f}{a}\right)$
b. Time reversal	$x(-t)$	$X(-f) = X^*(f)$
4. Duality	$X(t)$	$x(-f)$
5a. Frequency translation	$x(t)e^{j\omega_0 t}$	$X(f - f_0)$
b. Modulation	$x(t)\cos\omega_0 t$	$\tfrac{1}{2}X(f - f_0) + \tfrac{1}{2}X(f + f_0)$
6. Differentiation	$\dfrac{d^n x(t)}{dt^n}$	$(j2\pi f)^n X(f)$
7. Integration	$\displaystyle\int_{-\infty}^{t} x(t')\,dt'$	$(j2\pi f)^{-1} X(f) + \tfrac{1}{2}X(0)\delta(f)$
8. Convolution	$\displaystyle\int_{-\infty}^{\infty} x_1(t - t')x_2(t')\,dt'$ $\displaystyle= \int_{-\infty}^{\infty} x_1(t')x_2(t - t')\,dt'$	$X_1(f)X_2(f)$
9. Multiplication	$x_1(t)x_2(t)$	$\displaystyle\int_{-\infty}^{\infty} X_1(f - f')X_2(f')\,df'$ $\displaystyle= \int_{-\infty}^{\infty} X_1(f')X_2(f - f')\,df$

[a] $\omega_0 = 2\pi f_0$; $x(t)$ is assumed to be real in 3b.

$$\mathscr{F}[x(at)] = \int_{-\infty}^{\infty} x(at)e^{-j2\pi f t}\,dt$$

$$= \int_{-\infty}^{\infty} x(t')e^{-j2\pi f t'/a}\,\frac{dt'}{a}$$

$$= \frac{1}{a} X\left(\frac{f}{a}\right) \tag{3-76}$$

where the substitution $t' = at$ has been made. Now consider $a < 0$ with $at = -\lvert a \rvert t$ and the substitution $t' = -\lvert a \rvert t = at$ made in the integral for $\mathscr{F}[x(at)]$. This yields

$$\mathcal{F}[x(at)] = \int_{-\infty}^{\infty} x(-|a|t)e^{-j2\pi ft} \, dt$$

$$= \int_{-\infty}^{\infty} x(t')e^{j2\pi ft'/|a|} \frac{dt'}{|a|}$$

$$= \frac{1}{|a|} X\left(\frac{-f}{|a|}\right)$$

$$= \frac{1}{|a|} X\left(\frac{f}{a}\right) \tag{3-77}$$

where the last step results because $-|a| = a$.

As a special case, we let $a = -1$, which results in the *time reversal theorem*.

EXAMPLE 3-11

An application of the scale change theorem is provided by considering the recording on magnetic tape of a signal at, say, 15 inches per second (ips) and playing it back at $7\frac{1}{2}$ ips. Assuming a flat frequency response for the recorder which is independent of playback speed and an original signal spectrum of

$$X_r(f) = \frac{1}{1 + (2\pi f/1000)^2} \tag{3-78}$$

we find the spectrum of the played back signal to be

$$X_{pb}(f) = \frac{2}{1 + (2\pi f/500)^2} \tag{3-79}$$

where $a = \frac{1}{2}$ in the scale change theorem. To see that $a = \frac{1}{2}$, consider an easily visualized signal such as a pulse 1 s in duration recorded at 15 ips. When played back at $7\frac{1}{2}$ ips it will be 2 s in duration. Thus $x_{pb}(t) = x_r(at)$ or $x_{pb}(2) = x_r(2a)$, where $t = 2$ corresponds to the trailing edge of the pulse on playback, which must correspond to the trailing edge on record. This requires that $2a = 1$ or $a = \frac{1}{2}$. The spectrum of the played-back signal must be narrower than that of the recorded signal (time is stretched out on playback, thus lowering the frequency content). The student should sketch $X_r(f)$ and $X_{pb}(f)$ and convince himself that they are indeed reasonable. ∎

EXAMPLE 3-12

As another application of the scale change theorem, consider the signal $x_a(t)$ of Example 3-9, for which we obtained the transform pair

$$x_a(t) = \Pi\left(\frac{t}{2}\right) \leftrightarrow 2 \text{ sinc } 2f$$

Instead of the Fourier transforms of a pulse of width 2, we wish to generalize this to a pulse of arbitrary width, say τ. Thus we let $a = 2/\tau$ to obtain the transform pair

$$x_a(at) = \Pi\left(\frac{t}{\tau}\right) \leftrightarrow \tau \, \text{sinc} \, \tau f \qquad (3\text{-}80)$$

■

Duality Theorem

The duality theorem follows by virtue of the similarity of the direct and inverse Fourier transform relationships, the only difference in addition to the variable of integration being the sign in the exponent. The use of this theorem is illustrated in the following examples.

EXAMPLE 3-13

The duality theorem can be used to derive the Fourier transform pair

$$2W \, \text{sinc} \, 2Wt \leftrightarrow \Pi\left(\frac{f}{2W}\right)$$

given the transform pair

$$x(t) = \Pi\left(\frac{t}{\tau}\right) \leftrightarrow \tau \, \text{sinc} \, \tau f = X(f)$$

We do this by replacing f in $X(f)$ by t to get a new function of time, which is

$$X(t) = \tau \, \text{sinc} \, \tau t$$

According to the duality theorem, this new time function has the Fourier transform

$$x(f) = \Pi\left(\frac{t}{\tau}\right)\Big|_{t \to -f} = \Pi\left(\frac{-f}{\tau}\right)$$

Defining the new parameter $W = \tau/2$ to make the notation clearer, and noting that $\Pi(-f/\tau) = \Pi(f/\tau)$, we obtain the desired transform pair. ■

Frequency Translation Theorem

This theorem is the dual of the time delay theorem. It is proved by writing down the expression for the Fourier transform of $x(t)e^{j2\pi f_0 t}$, which is

$$\int_{-\infty}^{\infty} x(t)e^{2\pi f_0 t} e^{-j2\pi f t} \, dt = \int_{-\infty}^{\infty} x(t)e^{-j2\pi (f - f_0)t} \, dt$$

We recognize the right-hand side as the Fourier transform of $x(t)$, with f replaced by $f - f_0$ and the theorem is proved. The modulation theorem follows from the frequency translation theorem by using Euler's theorem to write $\cos \omega_0 t = \frac{1}{2} e^{j\omega_0 t} + \frac{1}{2} e^{-j\omega_0 t}$ and then applying superposition.

EXAMPLE 3-14

The Fourier transform of the signal

$$x_1(t) = \Pi\left(\frac{t}{2}\right) \exp(\pm j20\pi t)$$

is

$$X_1(f) = 2 \text{ sinc } [2(f \mp 10)]$$

where the transform pair obtained in Example 3-9 for $x_a(t)$ has been used. From the modulation theorem, the Fourier transform of

$$x_2(t) = \Pi\left(\frac{t}{2}\right) \cos 20\pi t$$

is

$$X_2(f) = \text{sinc } [2(f - 10)] + \text{sinc } [2(f + 10)]$$

The student should sketch $X_2(f)$. Note that the modulation process shifts the spectrum of a signal to a new frequency. It is an extremely useful property in communication system design and analysis. ■

EXAMPLE 3-15
Obtain the following transform pairs:

1. $A\delta(t) \leftrightarrow A$
2. $A\delta(t - t_0) \leftrightarrow Ae^{-j2\pi f t_0}$
3. $A \leftrightarrow A\delta(f)$
4. $Ae^{j2\pi f_0 t} \leftrightarrow A\delta(f - f_0)$
5. $A \cos 2\pi f_0 t \leftrightarrow A/2 \ [\delta(f - f_0) + \delta(f + f_0)]$

The impulse function is not a properly Fourier transformable signal. However, its transform is nevertheless useful. The same holds true for a constant. We could find their transforms using limiting operations, as discussed in Section 3-13. It is easier to prove the first transform pair by using the sifting property of the unit impulse:

$$\mathcal{F}[A\delta(t)] = A \int_{-\infty}^{\infty} \delta(t)e^{-j2\pi f t} \, dt = A$$

The second pair is obtained by applying the time delay theorem to pair 1.

The third pair can be shown by using duality in conjunction with the first transform pair. Thus

$$X(t) = A \leftrightarrow A\delta(-f) = A\delta(f) = x(-f)$$

where we recall that the impulse function was defined as even in Chapter 1.

Transform pair 4 follows by applying the frequency translation theorem to pair 3, and pair 5 is obtained by applying the modulation theorem to pair 3. ■

EXAMPLE 3-16
A convenient signal for use in obtaining transforms of periodic signals is the signal

$$y_s(t) = \sum_{m=-\infty}^{\infty} \delta(t - mT_s) \tag{3-81}$$

It is a periodic train of impulses referred to as the *ideal sampling waveform*.

The Fourier transform of $y_s(t)$ can be obtained by noting that it is periodic and, in a formal sense, can be represented by the Fourier series

$$y_s(t) = \sum_{n=-\infty}^{\infty} Y_n e^{j2\pi n f_s t}, \qquad f_s = \frac{1}{T_s},$$

The Fourier coefficients are given by

$$Y_n = \frac{1}{T_s} \int_{-T_s/2}^{T_s/2} \delta(t) e^{-j2\pi n f_s t} \, dt = f_s$$

where the sifting property of the unit impulse function has been used. Therefore, $y_s(t)$ can be represented by the Fourier series

$$y_s(t) = f_s \sum_{n=-\infty}^{\infty} e^{j2\pi n f_s t}$$

Using the Fourier transform pair $e^{j2\pi f_0 t} \leftrightarrow \delta(f - f_0)$ we may take the Fourier transform of this Fourier series term by term, to obtain

$$Y_s(f) = f_s \sum_{n=-\infty}^{\infty} \mathscr{F}[e^{j2\pi n f_s t}]$$

$$= f_s \sum_{n=-\infty}^{\infty} \delta(f - nf_s)$$

Summarizing, we have obtained the transform pair

$$\sum_{m=-\infty}^{\infty} \delta(t - mT_s) \leftrightarrow f_s \sum_{n=-\infty}^{\infty} \delta(f - nf_s) \qquad (3\text{-}82)$$

The transform pair (3-82) is useful in spectral representation of periodic signals by the Fourier transform, to be considered in Sections 3-17 and 3-21. ∎

Differentiation and Integration Theorems To prove the differentiation theorem, consider

$$\mathscr{F}\left[\frac{dx}{dt}\right] = \int_{-\infty}^{\infty} \frac{dx(t)}{dt} e^{-j2\pi f t} \, dt$$

Using integration by parts with $u = e^{-j2\pi f t}$ and $dv = (dx/dt)\,dt$, we obtain

$$\mathscr{F}\left[\frac{dx}{dt}\right] = x(t) e^{-j2\pi f t} \Big|_{-\infty}^{\infty} + j2\pi f \int_{-\infty}^{\infty} x(t) e^{-j2\pi f t} \, dt$$

But if $x(t)$ is absolutely integrable, $\lim_{t \to \pm\infty} |x(t)| = 0$. Therefore, $dx/dt \leftrightarrow j2\pi f X(f)$. Repeated application of integration by parts can be used to prove the theorem for n differentiations.

EXAMPLE 3-17

A useful application of the differentiation theorem is that of obtaining Fourier transforms of piecewise linear signals such as the triangle signal, defined as

$$\Lambda\left(\frac{t}{\tau}\right) \triangleq \begin{cases} 1 - \dfrac{|t|}{\tau}, & |t| < \tau \\ 0, & \text{otherwise} \end{cases} \tag{3-83}$$

Differentiating $\Lambda\,(t/\tau)$ twice, as shown graphically in Figure 3-13, we obtain

$$\frac{d^2\Lambda(t/\tau)}{dt^2} = \frac{1}{\tau}\delta(t + \tau) - \frac{2}{\tau}\delta(t) + \frac{1}{\tau}\delta(t - \tau)$$

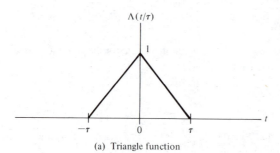

(a) Triangle function

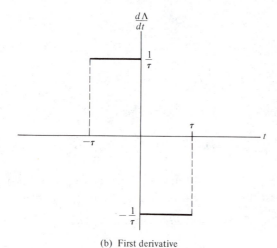

(b) First derivative

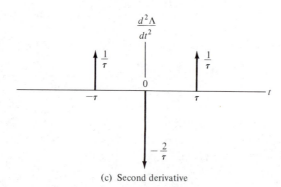

(c) Second derivative

FIGURE 3-13. Derivatives of triangle function used in obtaining its Fourier transform.

Use of the differentiation, superposition, and time shift theorems, and the results of Example 3-15 gives the Fourier transform

$$\mathcal{F}\left[\frac{d^2\Lambda(t/\tau)}{dt^2}\right] = (j2\pi f)^2 \mathcal{F}\left[\Lambda\left(\frac{t}{\tau}\right)\right] = \frac{1}{\tau}(e^{j2\pi f\tau} - 2 + e^{-j2\pi f\tau})$$

or

$$\mathcal{F}\left[\Lambda\left(\frac{t}{\tau}\right)\right] = \frac{2\cos 2\pi f\tau - 2}{\tau(j2\pi f)^2} = \tau \frac{\sin^2 \pi f\tau}{(\pi f\tau)^2}$$

where $\frac{1}{2}(1 - \cos 2\pi f\tau) = \sin^2 \pi f\tau$ has been used. That is, the transform pair

$$\Lambda\left(\frac{t}{\tau}\right) \leftrightarrow \tau \operatorname{sinc}^2 f\tau$$

has been derived. ■

The integration theorem is proved in a manner similar to that for the differentiation theorem if $X(0) = \int_{-\infty}^{\infty} x(t)\, dt = 0$. If $X(0) \neq 0$, the proof of the integration theorem must be carried out using a transform-in-the-limit approach. The second term of the transform of pair 7, $\frac{1}{2}X(0)\delta(f)$, simply represents the spectrum of the constant $\int_{-\infty}^{\infty} x(t)\, dt$, which is the net area of $x(t)$.

Note that differentiation enhances the high-frequency content of a signal, which is reflected by the factor $j2\pi f$ in the transform, while integration smooths out time fluctuations or suppresses the high-frequency content of a signal, as indicated by the $(j2\pi f)^{-1}$ in the transform.

EXAMPLE 3-18

As an application of the integration theorem, consider the Fourier transform of a unit step. Since $u(t) = \int_{-\infty}^{t} \delta(\lambda)\, d\lambda$ and $\int_{-\infty}^{\infty} \delta(t)\, dt = 1$, we obtain

$$\mathcal{F}[u(t)] = (j2\pi f)^{-1} + \frac{1}{2}\delta(f)$$

where $x(t)$ in the integration theorem of Table 3-2 has been taken as $\delta(t)$. A signal that is related to the unit step is the signum function, defined as

$$\operatorname{sgn} t = 2u(t) - 1 = \begin{cases} 1, & t > 0 \\ -1, & t < 0 \end{cases}$$

Using the result obtained above for the Fourier transform of a step, we find the Fourier transform of sgn t to be

$$\mathcal{F}[\operatorname{sgn} t] = (j\pi f)^{-1}$$

By the duality theorem, it follows that

$$\mathcal{F}\left[\frac{1}{j\pi t}\right] = \text{sgn}\,(-f) = -\text{sgn}\,f$$

or

$$\mathcal{F}\left[\frac{1}{\pi t}\right] = -j\,\text{sgn}\,f$$

which results because sgn f is odd. The last transform pair is useful in defining the Hilbert transform of a signal, which will be discussed in Example 3-20. ■

The Convolution Theorem

The proof of the convolution theorem follows by representing $x_2(t - t')$ in the second integral under Theorem 8 of Table 3-2 in terms of its inverse Fourier transform, which is

$$x_2(t - t') = \int_{-\infty}^{\infty} X_2(f)e^{j2\pi f(t - t')}\,df$$

Substitution into the convolution integral gives

$$x_1 * x_2 = \int_{-\infty}^{\infty} x_1(t')\left[\int_{-\infty}^{\infty} X_2(f)e^{j2\pi f(t - t')}\,df\right]dt'$$

$$= \int_{-\infty}^{\infty} X_2(f)\left[\int_{-\infty}^{\infty} x_1(t')e^{-j2\pi ft'}\,dt'\right]e^{j2\pi ft}\,df$$

The last step results from reversing the order of the two integrations. Recognizing the bracketed term inside the integral as $X_1(f)$, the Fourier transform of $x_1(t)$, we have obtained

$$x_1 * x_2 = \int_{-\infty}^{\infty} X_1(f)X_2(f)e^{j2\pi ft}\,df$$

which is the inverse Fourier transform of $X_1(f)X_2(f)$.

This theorem is useful in systems analysis applications as well as for deriving transform pairs. The latter application is illustrated by the following examples.

EXAMPLE 3-19

Consider the transform of the signal that is the convolution of two rectangular pulses. The student can verify that

$$\Pi\left(\frac{t}{\tau}\right) * \Pi\left(\frac{t}{\tau}\right) = \tau\Lambda\left(\frac{t}{\tau}\right)$$

where $\Lambda(t/\tau)$ is the triangular signal defined by (3-83). Therefore, according to the convolution theorem,

$$\mathscr{F}\left[\tau\Lambda\left(\frac{t}{\tau}\right)\right] = \mathscr{F}\left[\Pi\left(\frac{t}{\tau}\right)\right]^2$$

$$= \tau^2 \text{ sinc}^2 \, f\tau$$

Dividing through by τ, we obtain the same result as in Example 3-17 for the Fourier transform of $\Lambda(t/\tau)$. ∎

EXAMPLE 3-20

The Hilbert transform, $\hat{x}(t)$, of a signal, $x(t)$, is obtained by convolving $x(t)$ with $1/\pi t$. That is,

$$\hat{x}(t) = x(t) * \frac{1}{\pi t} = \frac{1}{\pi} \int_{-\infty}^{\infty} \frac{x(\lambda)}{t - \lambda} \, d\lambda$$

From Example 3-18 and the convolution theorem it follows that

$$\mathscr{F}[\hat{x}(t)] = \mathscr{F}\left[\frac{1}{\pi t}\right]\mathscr{F}[x(t)] = -j \text{ sgn } f \, X(f)$$

A Hilbert transform operation therefore multiplies all positive-frequency spectral components of a signal by $-j$, and all negative-frequency spectral components by j. The amplitude spectrum of the signal is left unchanged by the Hilbert transform operation. ∎

The Multiplication Theorem

The proof of the multiplication theorem proceeds in a manner analogous to the proof of the convolution theorem. It is left to the student as a problem. Its application will be illustrated by an example.

EXAMPLE 3-21

We will use the multiplication theorem to obtain the Fourier transform of the cosinusoidal pulse given by

$$x(t) = A\Pi\left(\frac{t}{\tau}\right) \cos 2\pi f_0 t$$

The makeup of this waveform is illustrated in Figure 3-14c on the left-hand side; that is, the product of the waveforms shown in Figure 3-14a and b is the oscillatory pulse shown in Figure 3-14c. Shown on the right-hand side in each part of the figure are the Fourier transforms of the corresponding waveforms on the left. From the multiplication theorem, we know that the convolution of the spectra shown in Figure 3-14a and b is the spectrum of the cosinusoidal pulse of Figure 3-14c. In terms of equations, we have the transform pair

$$A\Pi\left(\frac{t}{\tau}\right) \leftrightarrow A\tau \text{ sinc } f\tau$$

which was obtained previously. The Fourier transform of $\cos 2\pi f_0 t$ follows by using Euler's theorem to write

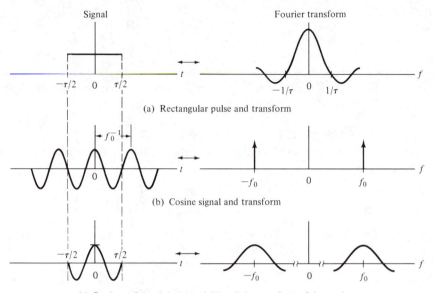

Signal   Fourier transform

$-\tau/2$ 0 $\tau/2$

t

$-1/\tau$ 0 $1/\tau$

f

(a) Rectangular pulse and transform

f_0^{-1}

0

t

$-f_0$ 0 f_0

f

(b) Cosine signal and transform

$-\tau/2$ $\tau/2$

0

t

$-f_0$ 0 f_0

f

(c) Product of signals in (a) and (b) and the transform of the product is the convolution of the transforms in (a) and (b)

FIGURE 3-14. Application of the multiplication theorem in obtaining the Fourier transform of a finite-duration cosine signal.

$$\mathcal{F}[\cos 2\pi f_0 t] = \mathcal{F}[\tfrac{1}{2}e^{j2\pi f_0 t} + \tfrac{1}{2}e^{-j2\pi f_0 t}]$$

$$= \tfrac{1}{2}\delta(f - f_0) + \tfrac{1}{2}\delta(f + f_0)$$

where the superposition theorem has been used together with the transform pair $e^{j2\pi f_0 t} \leftrightarrow \delta(f - f_0)$. The multiplication theorem for this example then states that

$$X(f) = \mathcal{F}\left[A\Pi\left(\frac{t}{\tau}\right) \cos 2\pi f_0 t \right]$$

$$= (A\tau \ \text{sinc} \ f\tau) * [\tfrac{1}{2}\delta(f - f_0) + \tfrac{1}{2}\delta(f + f_0)]$$

$$= \frac{A\tau}{2} [\text{sinc} \ (f - f_0)\tau + \text{sinc} \ (f + f_0)\tau]$$

where use has been made of

$$\text{sinc} \ f\tau * \delta(f \pm f_0) = \int_{-\infty}^{\infty} \text{sinc} \ \lambda\tau \ \delta(\lambda - f \pm f_0) \ d\lambda$$

$$= \text{sinc} \ (f \pm f_0)\tau$$

Note that the modulation theorem could have equally well been used to obtain this result. ∎

TABLE OF FOURIER TRANSFORM PAIRS

The Fourier transform pairs derived in the preceding section are collected in Table 3-4 together with other pairs that are often useful. A column is provided that gives suggestions on the derivation of each pair.

TABLE 3-4. Fourier Transform Pairs

Pair Number	$x(t)$	$X(f)$	Comments on Derivation
1.	$\Pi\left(\dfrac{t}{\tau}\right)$	$\tau \operatorname{sinc} \tau f$	Direct evaluation
2.	$2W \operatorname{sinc} 2Wt$	$\Pi\left(\dfrac{f}{2W}\right)$	Duality with pair 1, Example 3-13
3.	$\Lambda\left(\dfrac{t}{\tau}\right)$	$\tau \operatorname{sinc}^2 \tau f$	Convolution with pair 1
4.	$\exp\left(-\alpha t\right)u(t), \quad \alpha > 0$	$\dfrac{1}{\alpha + j2\pi f}$	Direct evaluation
5.	$t \exp\left(-\alpha t\right)u(t), \quad \alpha > 0$	$\dfrac{1}{(\alpha + j2\pi f)^2}$	Differentiation of pair 4 with respect to α
6.	$\exp\left(-\alpha|t|\right), \quad \alpha > 0$	$\dfrac{2\alpha}{\alpha^2 + (2\pi f)^2}$	Direct evaluation
7.	$\delta(t)$	1	Example 3-15
8.	1	$\delta(f)$	Duality with pair 7
9.	$\delta(t - t_0)$	$\exp\left(-j2\pi f t_0\right)$	Shift and pair 7
10.	$\exp\left(j2\pi f_0 t\right)$	$\delta(f - f_0)$	Duality with pair 9
11.	$\cos 2\pi f_0 t$	$\frac{1}{2}\delta(f - f_0) + \frac{1}{2}\delta(f + f_0)$	Exponential representation of cos and sin and pair 10
12.	$\sin 2\pi f_0 t$	$\dfrac{1}{2j}\delta(f - f_0) - \dfrac{1}{2j}\delta(f + f_0)$	
13.	$u(t)$	$(j2\pi f)^{-1} + \frac{1}{2}\delta(f)$	Integration and pair 7
14.	$\operatorname{sgn} t$	$(j\pi f)^{-1}$	Pair 8 and pair 13 with superposition
15.	$\dfrac{1}{\pi t}$	$j \operatorname{sgn} f$	Duality with pair 14
16.	$\hat{x}(t) = \dfrac{1}{\pi}\displaystyle\int_{-\infty}^{\infty} \dfrac{x(\lambda)}{t - \lambda}\,d\lambda$	$j \operatorname{sgn} f\, X(f)$	Convolution and pair 15
17.	$\displaystyle\sum_{m=-\infty}^{\infty} \delta(t - mT_s)$	$f_s \displaystyle\sum_{m=-\infty}^{\infty} \delta(f - mf_s),$ $f_s = T_s^{-1}$	Example 3-16

SYSTEM ANALYSIS WITH THE FOURIER TRANSFORM

In Chapter 2 the superposition integrals (2-13) and (2-14) were developed as systems analysis tools. Application of the convolution theorem of Fourier transforms, pair 8 of Table 3-2, to either (2-17) or (2-18) gives the result

$$Y(f) = H(f)X(f) \qquad (3\text{-}84)$$

where $X(f) = \mathcal{F}[x(t)]$, $Y(f) = \mathcal{F}[y(t)]$, and $H(f) = \mathcal{F}[h(t)]$. The latter is referred to as the *transfer function* of the system and is identical to the frequency-response function $\tilde{H}(\omega)$ defined by (3-48) but with $f = \omega/2\pi$ as the independent variable. Either $h(t)$ or $H(f)$ are equally good characterizations of the system, since its output can be found in terms of either one as illustrated in Figure 3-15. In the frequency domain, the output of the system is the inverse Fourier transform of (3-84), which is

$$y(t) = \int_{-\infty}^{\infty} X(f)H(f)e^{j2\pi ft}\, df \qquad (3\text{-}85)$$

Since $H(f)$ is, in general, a complex quantity, we write it as

$$H(f) = |H(f)|e^{j\underline{/H(f)}} \qquad (3\text{-}86)$$

where $|H(f)|$ is the *amplitude-response* function and $\underline{/H(f)}$ the *phase-response* function of the network. If $H(f)$ is the Fourier transform of a real time function $h(t)$, which it is in all cases being considered in this book, it follows that[†]

$$|H(f)| = |H(-f)| \qquad (3\text{-}87)$$

and

$$\underline{/H(f)} = -\underline{/H(-f)} \qquad (3\text{-}88)$$

Since $H(f)$ is such an important system function, we illustrate its computation by the following example.

EXAMPLE 3-22

To obtain the transfer function of the RC filter shown in Figure 3-16, any of the methods illustrated in Figure 3-16 may be used. The first method involves Fourier transformation of the governing differential equation (integrodifferential equation, in general), which for this system is

$$RC\frac{dy}{dt} + y(t) = x(t), \qquad -\infty < t < \infty \qquad (3\text{-}89)$$

Assuming that the Fourier transform of each term exists, we may apply the superposition and differentiation theorems to obtain

$$(j2\pi fRC + 1)Y(f) = X(f)$$

[†]It is useful to model some fixed, linear systems as having complex impulse responses. In such cases (3-87) and (3-88) do not hold.

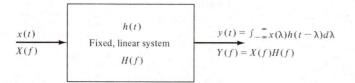

$$x(t) \qquad \boxed{\begin{array}{c} h(t) \\ \text{Fixed, linear system} \\ H(f) \end{array}} \qquad y(t) = \int_{-\infty}^{\infty} x(\lambda) h(t-\lambda) d\lambda$$

$$X(f) \qquad\qquad\qquad\qquad\qquad\qquad\qquad Y(f) = X(f)H(f)$$

FIGURE 3-15. Representation of a fixed, linear system in terms of its impulse response or, alternatively, its transfer function.

or

$$H(f) = \frac{Y(f)}{X(f)} = \frac{1}{1 + j(f/f_3)} = \frac{1}{\sqrt{1 + (f/f_3)^2}} \, e^{-j\tan^{-1}(f/f_3)} \qquad (3\text{-}90)$$

The frequency $f_3 = 1/2\pi RC$ is called the 3-dB or half-power frequency.†

Equivalently, we can use Laplace transform theory as discussed in Chapter 5 and replace s by $j2\pi f$.

A second method is to obtain the impulse response, as in Example 2-5, and Fourier transform it to obtain $H(f)$. This method is illustrated in Figure 3-16b.

A third alternative is to use ac sinusoidal steady-state analysis as shown in Figure 3-16c, and find the ratio of output to input phasors.

The amplitude and phase responses of this system are obtained from (3-90). They are

$$|H(f)| = \left[1 + \left(\frac{f}{f_3}\right)^2\right]^{-1/2} \qquad \text{and} \qquad \underline{/H(f)} = -\tan^{-1}\frac{f}{f_3}$$

respectively. They are illustrated in Figure 3-17.

To end this example, we find the response of the system to the input

$$x(t) = Ae^{-\alpha t}u(t), \qquad \alpha > 0$$

by using Fourier transform techniques. The Fourier transform of this input, from Table 3-3, is

$$X(f) = \frac{A}{\alpha + j2\pi f}$$

The Fourier transform of the output due to this input is

$$Y(f) = \frac{A/RC}{(\alpha + j2\pi f)(1/RC + j2\pi f)}$$

For $\alpha RC \neq 1$, this can be written as‡

---

†The half-power frequency of a two-port system is defined as the frequency of an input sine wave which results in a steady-state sinusoidal output of amplitude $1/\sqrt{2}$ of the maximum possible amplitude of the output sinusoid. The power of the output sinusoid is then one-half of the maximum possible output power. The ratio of actual to maximum powers in decibels (dB) is therefore (P_{out}/P_{max}) dB $= 10 \log_{10} \frac{1}{2} = -3$ dB.

‡Ways of obtaining partial fraction expansions such as this one will be reviewed in Chapter 4.

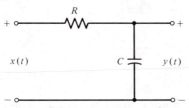

(a) Fourier transforming the differential equation relating input and output: $RC \dfrac{dy(t)}{dt} + y(t) = x(t)$

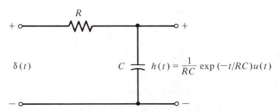

(b) Obtaining the impulse response of the system and finding its Fourier transform

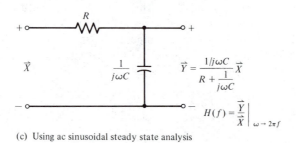

(c) Using ac sinusoidal steady state analysis

FIGURE 3-16. Illustration of methods that can be used to obtain the transfer function of a system.

$$Y(f) = \frac{A}{\alpha RC - 1} \left[\frac{1}{(1/RC) + j2\pi f} - \frac{1}{\alpha + j2\pi f} \right]$$

The inverse Fourier transform of $Y(f)$ can be found with the aid of Table 3-4 and the superposition theorem. It is given by

$$y(t) = \frac{A}{\alpha RC - 1} \left[\exp\left(\frac{-t}{RC}\right) - \exp\left(-\alpha t\right) \right] u(t)$$

We can find the response for $\alpha RC = 1$ by using pair 5 of Table 3-3 or by taking the limit of the foregoing result as $\alpha RC \rightarrow 1$. In either case, the result is

$$y(t) = A\left(\frac{t}{RC}\right) \exp\left(\frac{-t}{RC}\right) u(t), \qquad \alpha = \frac{1}{RC}$$

The student should plot $y(t)$ for various combinations of α and RC and consider what their relationship must be for the output to resemble the input except for a scale change. ∎

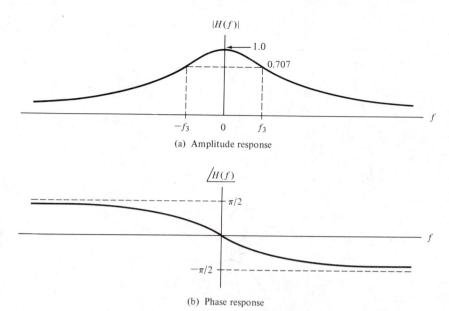

(a) Amplitude response

(b) Phase response

FIGURE 3-17. Amplitude and phase responses for a low-pass *RC* filter.

As an example of how system characteristics influence the response of the system to a given input, we consider the *RC*-circuit response to a rectangular-pulse input.

EXAMPLE 3-23

Again, we consider the system of Figure 3-16 but with the input

$$x(t) = A\Pi\left(\frac{t - T/2}{T}\right) = A[u(t) - u(t - T)]$$

Because the inverse Fourier transform relationship (3-85) is difficult to evaluate in this case, the output is found using time-domain techniques. From Example 2-3 the step response is

$$a_s(t) = (1 - e^{-t/RC})u(t)$$

Noting that the $x(t)$ consists of the difference of two steps and using superposition, we find the output to be

$$y(t) = \begin{cases} 0, & t < 0 \\ A(1 - e^{-t/RC}), & 0 \le t \le T \\ A[e^{-(t - T)/RC} - e^{-t/RC}], & t > T \end{cases}$$

This result is plotted in Figure 3-18 for several values of T/RC together with $|X(f)|$ and $|H(f)|$. The parameter $T/RC = 2\pi f_3/T^{-1}$ is proportional to the ratio of the 3-dB frequency of the filter to the spectral width (T^{-1}) of the pulse. When the filter bandwidth is large compared with the spectral width of the input pulse, the input is essentially passed undistorted by the system. On the other hand, for

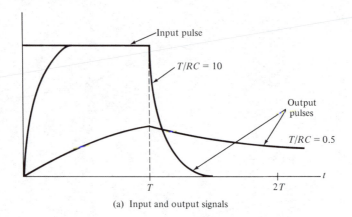

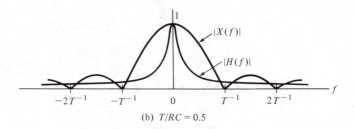

(b) $T/RC = 0.5$

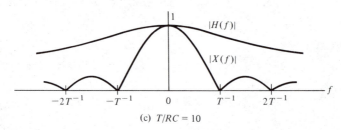

(c) $T/RC = 10$

FIGURE 3-18. Waveforms (a) and spectra (b, c) for a low-pass *RC* filter with pulse input.

$2\pi f_3/T^{-1} \ll 1$, the system distorts the input signal spectrum and the output does not resemble the input. ∎

3-17

STEADY-STATE SYSTEM RESPONSE TO SINUSOIDAL INPUTS BY MEANS OF THE FOURIER TRANSFORM

In Section 3-9 the transfer function of a fixed, linear system was defined in terms of the attenuation and phase shift suffered by a sinusoid in passing through the system. We may use (3-84) to relate the Fourier transforms of the input and output of a fixed, linear system in response to a periodic waveform. In particular, a periodic input signal $x(t)$ can be represented by its Fourier series as

$$x(t) = \sum_{n=-\infty}^{\infty} X_n e^{j2\pi n f_0 t}$$

Using the transform pair $e^{j2\pi n f_0 t} \leftrightarrow \delta(f - nf_0)$, this can be Fourier transformed term by term to yield

$$X(f) = \sum_{n=-\infty}^{\infty} X_n \delta(f - nf_0) \qquad (3\text{-}91)$$

When (3-91) is multiplied by the system transfer function $H(f)$, and use is made of the relationship $H(f)\delta(f - nf_0) = H(nf_0)\delta(f - nf_0)$, we may express the Fourier transform of the output as

$$Y(f) = \sum_{n=-\infty}^{\infty} X_n H(nf_0)\delta(f - nf_0) \qquad (3\text{-}92)$$

Writing X_n and $H(nf_0)$ in terms of magnitude and phase as $|X_n| \exp j\underline{/X_n}$ and $|H(nf_0)| \exp [j\underline{/H(nf_0)}]$, respectively, (3-92) may be inverse Fourier transformed term by term to give the output

$$y(t) = \sum_{n=-\infty}^{\infty} |X_n| |H(nf_0)| \exp [j\{2\pi f_0 t + \underline{/X_n} + \underline{/H(nf_0)} \}] \qquad (3\text{-}93)$$

which is equivalent to (3-51).

To emphasize the significance of (3-93), note that the nth spectral component of the input, X_n, appears at the output with amplitude attenuated (or amplified) by the amplitude response function $|H(nf_0)|$, and a phase which is the input phase shifted by the system phase response $\underline{/H(nf_0)}$, both of which are evaluated *at the frequency of the particular spectral component* under consideration.

EXAMPLE 3-24

Consider a system with amplitude- and phase-response functions given by

$$|H(f)| = K\Pi\left(\frac{f}{2B}\right) = \begin{cases} K, & |f| \leq B \\ 0, & \text{otherwise} \end{cases}$$

and

$$\underline{/H(f)} = -2\pi t_0 f$$

respectively. A filter with this transfer function is referred to as an *ideal low-pass filter*.

Its output in response to $x(t) = A \cos (2\pi f_0 t + \theta_0)$ is obtained from (3-93) by noting that

$$X_1 = \tfrac{1}{2}A e^{j\theta_0} = X_{-1}^*$$

with all other X_n's $= 0$. Thus the output is

$$y(t) = \begin{cases} 0, & f_0 > B \\ KA \cos [2\pi f_0(t - t_0) + \theta_0], & f_0 \leq B \end{cases}$$

Thus an ideal low-pass filter completely rejects all spectral components with frequencies greater than some cutoff frequency B, and passes all input spectral components below this cutoff frequency except that their amplitudes are multiplied by a constant K and they are phase shifted by an amount $-2\pi t_0 f_0$ or delayed in time by t_0. ■

Because ideal filters are used quite often in systems analysis, we consider the characteristics of three types of ideal filters in more detail in the following section.

3-18

IDEAL FILTERS

It is often connvenient to work with idealized filters having amplitude-response functions which are constant within the passband† and zero elsewhere. In general, three types of ideal filters, referred to as low-pass, high-pass, and bandpass filters, can be defined. A phase-shift function which is a linear function of frequency throughout the passband is assumed in each case. Figure 3-19 illustrates the frequency-response functions for each type of ideal filter.

The impulse response of any filter can be found by obtaining the inverse Fourier transform of its transfer function. For example, for the ideal low-pass filter, the impulse response can be found from

$$
h_{LP}(t) = \int_{-\infty}^{\infty} H_{LP}(f)e^{j2\pi ft}\, df
$$

$$
= \int_{-B}^{B} Ke^{-j2\pi ft_0}e^{j2\pi ft}\, df
$$

$$
= \int_{-B}^{B} Ke^{j2\pi f(t - t_0)}\, df
$$

$$
= 2BK \text{ sinc } [2B(t - t_0)] \tag{3-94}
$$

Since sinc $[2B(t - t_0)]$ is nonzero for all t, except when $2B(t - t_0)$ takes on an integer value, it follows that $h_{LP}(t)$ is nonzero for $t < 0$. In other words, an ideal low-pass filter is *noncausal*, as are all types of ideal filters.

In fact, it can be shown that the ideal bandpass filter has impulse response

$$
h_{BP}(t) = 2KB \text{ sinc } [B(t - t_0)] \cos [2\pi f_0(t - t_0)] \tag{3-95}
$$

and that the ideal high-pass filter has impulse response

$$
h_{HP}(t) = K\delta(t - t_0) - 2BK \text{ sinc } [2B(t - t_0)] \tag{3-96}
$$

Both of these impulse responses are nonzero for $t < 0$ if t_0 is finite. Figure 3-20 illustrates ideal filter impulse responses. The noncausal nature of ideal

†The passband of a filter is defined as the frequency range where its amplitude response is greater than some arbitrarily chosen minimum value. Quite often, this minimum value is chosen as $1/\sqrt{2}$ of the maximum value.

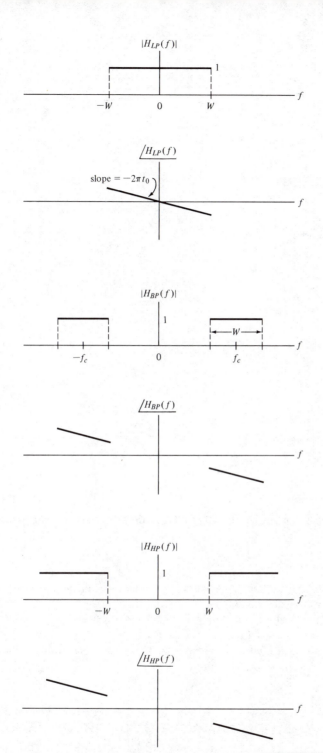

FIGURE 3-19. Ideal filter amplitude- and phase-response functions.

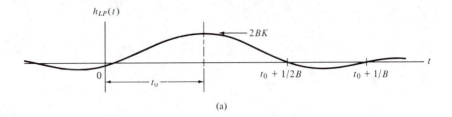

(a)

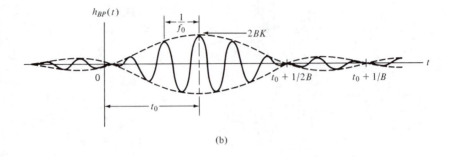

(b)

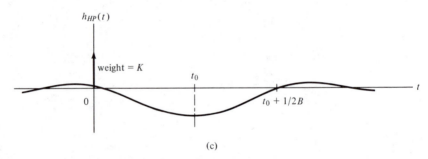

(c)

FIGURE 3-20. Impulse responses for ideal filters: (a) Low-pass; (b) bandpass; (c) high-pass.

filters is a consequence of their ideal attenuation characteristics in going from the passband to the stop band.

Several methods of approximating ideal filter frequency-response characteristics by means of causal filters are discussed in Appendix B. In general, the more closely a causal filter approximates a corresponding ideal filter, the more delay a signal suffers in passing through it.

WINDOW FUNCTIONS AND THE GIBBS PHENOMENON

In our consideration of trigonometric series at the beginning of this chapter, it was noted that the sum of such a series tended to overshoot the signal being approximated at a discontinuity. We can easily examine the reason for this behavior, referred to as the *Gibbs phenomenon,* with the aid of the Fourier transform. Consider a signal $x(t)$ with Fourier transform $X(f)$. We consider the effect of reconstructing $x(t)$ from only the low-pass part of its frequency spectrum. That is, we approximate the signal by

$$\tilde{x}(t) = \mathcal{F}^{-1}\left[X(f)\Pi\left(\frac{f}{2W}\right) \right] \tag{3-98}$$

where

$$\Pi\left(\frac{f}{2W}\right) = \begin{cases} 1, & |f| \le W \\ 0, & \text{otherwise} \end{cases}$$

According to the convolution theorem of Fourier transform theory,

$$\tilde{x}(t) = x(t) * \mathcal{F}^{-1}\left[\Pi\left(\frac{f}{2W}\right) \right] \tag{3-99}$$

$$= x(t) * (2W \text{ sinc } 2Wt)$$

where the last step follows by virtue of the transform pair developed in Example 3-13. Recalling that convolution is a folding-product, sliding-integration process, it is now apparent why the overshoot phenomenon occurs at a discontinuity. Figure 3-21 illustrates (3-99) for various values of W assuming that $x(t)$ is a square pulse. As W increases, more and more of the frequency content of the rectangular pulse is used to obtain the approximation $\tilde{x}(t)$. Nevertheless, we see that a finite value of W means that the signal $x(t)$ is viewed through the *window function* $2W$ sinc $2Wt$ in the time domain.† As a result, anticipatory wiggles are present in $\tilde{x}(t)$ before the onset of the main pulse, and echoing wiggles remain after the discontinuity of the leading edge of $x(t)$ passes.

---

†Actually, it is often customary to refer to the multiplying function $\Pi(f/2W)$ as the window function.

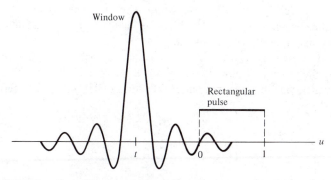

FIGURE 3-21. Convolution of the window function $2W$ sinc $2Wt$ with a rectangular pulse.

3-19 WINDOW FUNCTIONS AND THE GIBBS PHENOMENON

To put this discussion on a mathematical basis, we represent the square-pulse signal as

$$x(t) = u(t) - u(t - 1) \tag{3-100}$$

which results in

$$\tilde{x}(t) = [u(t) - u(t - 1)] * 2W \text{ sinc } 2wt \tag{3-101}$$

Considering, for the moment, only the first term of (3-101), we write it as

$$\tilde{u}(t) = u(t) * 2W \text{ sinc } 2Wt = 2W \text{ sinc } 2Wt * u(t)$$

$$= 2W \int_{-\infty}^{\infty} \text{ sinc } 2W\lambda \, u(t - \lambda) \, d\lambda$$

$$= 2W \int_{-\infty}^{t} \text{ sinc } 2W\lambda \, d\lambda \tag{3-102}$$

This integral cannot be evaluated in closed form but can be expressed in terms of the sine-integral function, Si (x), defined as

$$\text{Si } (x) = \int_0^x \frac{\sin u}{u} \, du \tag{3-103}$$

which is available in tabular form.† It is useful to note that Si(x) is an even function and that Si $(\infty) = \pi/2$. In terms of Si (x), $\tilde{u}(t)$ can be expressed as

$$\tilde{u}(t) = \begin{cases} \dfrac{1}{2} + \dfrac{1}{\pi} \text{ Si } (2\pi Wt), & t > 0 \\[2ex] \dfrac{1}{2} - \dfrac{1}{\pi} \text{ Si } (2\pi Wt), & t < 0 \end{cases} \tag{3-103}$$

A plot of $\tilde{u}(t)$ versus Wt is shown in Figure 3-22a.

From (3-101) it follows that we may obtain $\tilde{x}(t)$ from $\tilde{u}(t)$ as

$$\tilde{x}(t) = \tilde{u}(t) - \tilde{u}(t - 1) \tag{3-104}$$

an operation that is easily carried out graphically; $\tilde{x}(t)$ is shown in Figure 3-22b and c for $W = 2$ and $W = 5$. The approximation of $\tilde{x}(t)$ to $x(t)$ becomes better as W increases relative to the pulse width, except for the overshoot at the discontinuities at $t = 0$ and $t = 1$, which eventually approaches a value of 9% of the pulse height as W becomes large.

The question of what can be done to combat this behavior is answered by considering the possibility of reducing the sidelobes of the $2W$ sinc $2Wt$ window function, which was the inverse Fourier transform of $\Pi(f/2W)$. The function

$$A(f) = \begin{cases} 0.5 + 0.5 \cos \dfrac{\pi f}{W}, & |f| \leq W \\[2ex] 0, & \text{otherwise} \end{cases} \tag{3-105}$$

---

†M. Abramowitz and A. Stegun, *Handbook of Mathematical Functions with Formulas, Tables, and Graphs* (Washington, D.C.: National Bureau of Standards, 1964), p. 243.

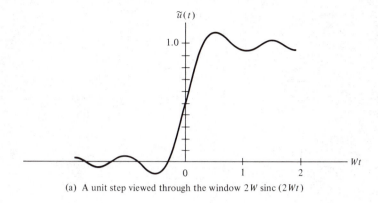

(a) A unit step viewed through the window $2W$ sinc $(2Wt)$

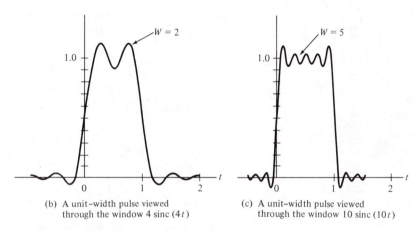

(b) A unit–width pulse viewed
through the window 4 sinc $(4t)$

(c) A unit–width pulse viewed
through the window 10 sinc $(10t)$

FIGURE 3-22.   Construction of a windowed rectangular pulse.

provides a smooth transition from zero at $f = \pm W$ to its maximum value of unity at $f = 0$.† Its inverse Fourier transform is

$$a(t) = W[\text{sinc } 2Wt + \tfrac{1}{2} \text{sinc } (2Wt - 1) + \tfrac{1}{2} \text{sinc } (2Wt + 1)] \quad (3\text{-}106)$$

This function and its transform are shown in Figure 3-23, where it is seen that its sidelobes are much lower than the function $2W$ sinc $2Wt$.

*3.20

RATE OF CONVERGENCE OF SPECTRA

It is often useful to have an idea of how rapidly the amplitude spectrum of a signal approaches zero with increasing frequency. In general, if the kth derivative of a periodic signal $x(t)$ is piecewise constant [implying that the $(k + 1)$st derivative contains impulses], the Fourier coefficients of $x(t)$ approach

†This function is called a *Hanning window*.

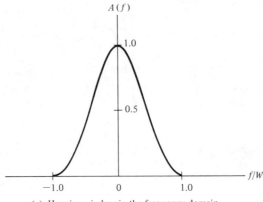

$A(f)$

1.0

0.5

−1.0   0   1.0 f/W

(a) Hanning window in the frequency domain

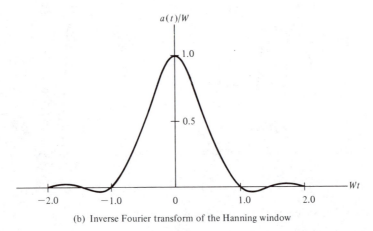

$a(t)/W$

1.0

0.5

−2.0   −1.0   0   1.0   2.0 Wt

(b) Inverse Fourier transform of the Hanning window

FIGURE 3-23.   Window with smoother transitions than the rectangular window.

zero as $1/n^{k+1}$. This can be shown as follows. In terms of its exponential Fourier series, $x(t)$ may be expressed as

$$x(t) = \sum_{n=-\infty}^{\infty} X_n e^{jn\omega_0 t} \qquad (3\text{-}107)$$

Differentiation of (3-107) $k + 1$ times with respect to t produces

$$\frac{d^{k+1}x(t)}{dt^{k+1}} = \sum_{n=-\infty}^{\infty} (jn\omega_0)^{k+1} X_n e^{jn\omega_0 t} \qquad (3\text{-}108)$$

which shows that the Fourier coefficients of $d^{k+1}x/dt^{k+1}$ are $(jn\omega_0)^{k+1}X_n$. Now $d^k x/dt^k$ is assumed to be piecewise continuous, and therefore $d^{k+1}x/dt^{k+1}$ possesses impulses. The Fourier coefficients of $d^{k+1}x/dt^{k+1}$ therefore have a term independent of $n\omega_0$ (i.e., a constant function of frequency). Thus $|X_n| = $ constant$/|n\omega_0|^{k+1}$ and it is seen that the amplitude spectrum of $x(t)$ approaches zero as $|n\omega_0|^{-k-1}$. A similar result can be shown for the Fourier transform.

FOURIER TRANSFORMS OF PERIODIC SIGNALS

The Fourier transform of a periodic signal, in a strict mathematical sense, does not exist since periodic signals are not Fourier transformable. However, using the transform pairs derived in Example 3-15 for a constant and a phasor signal we could, in a formal sense, write down the Fourier transform of a periodic signal by Fourier transforming its complex Fourier series term by term. Thus, for a periodic signal $x(t)$ with Fourier series $\sum_{n=-\infty}^{\infty} X_n e^{jn\omega_0 t}$, where $T_0 = 2\pi/\omega_0$ $= 1/f_0$ is the period, we have the transform pair

$$\sum_{n=-\infty}^{\infty} X_n e^{jn\omega_0 t} \leftrightarrow \sum_{n=-\infty}^{\infty} X_n \delta(f - nf_0) \qquad (3\text{-}109)$$

Either representation in (3-109) contains the same information about $x(t)$, and either result can be used to plot the two-sided spectra of a signal. If the Fourier transform representation is used [right-hand side of (3-109)], the amplitude spectrum consists of impulses rather than lines.

A somewhat more useful form for the Fourier transform of a periodic signal than (3-109) is obtained by applying the convolution theorem and pair 17 of Table 3-3 for the ideal sampling wave. To obtain it, consider the result of convolving the ideal sampling waveform with a pulse-type signal $p(t)$ to obtain a new signal $x(t)$. If $p(t)$ is an energy signal of limited time extent such that $p(t) = 0$, $|t| \geq T/2 \leq T_s/2$, then $x(t)$ is a periodic power signal. This is apparent when one carries out the convolution with the aid of the sifting property of the unit impulse:

$$x(t) = \left[\sum_{m=-\infty}^{\infty} \delta(t - mT_s) \right] * p(t) = \sum_{m=-\infty}^{\infty} p(t - mT_s) \qquad (3\text{-}110)$$

Applying the convolution theorem and the Fourier transform pair (3-82), the Fourier transform of $x(t)$ is

$$X(f) = \mathcal{F}\left[\sum_{m=-\infty}^{\infty} \delta(t - mT_s) \right] P(f)$$

$$= f_s P(f) \sum_{n=-\infty}^{\infty} \delta(f - nf_s)$$

$$= \sum_{n=-\infty}^{\infty} f_s P(nf_s) \delta(f - nf_s)$$

where $P(f) = \mathcal{F}[p(t)]$. Summarizing, we have obtained the Fourier transform pair

$$\sum_{m=-\infty}^{\infty} p(t - mT_s) \leftrightarrow \sum_{n=-\infty}^{\infty} f_s P(nf_s) \delta(f - nf_s) \qquad (3\text{-}111)$$

The usefulness of (3-111) will be illustrated with an example.

EXAMPLE 3-25

As an example of the use of (3-111), we obtain the spectrum of the signal shown in Figure 3-24 and of its periodic extension. We may write $x_1(t)$ as

$$x_1(t) = \Lambda(t) - \Pi(t - 1.5) \qquad (3\text{-}112)$$

Using previously derived Fourier transform pairs for the triangle and the pulse, we obtain

$$X_1(f) = \text{sinc}^2 f - \text{sinc } f \, e^{-j3\pi f} \qquad (3\text{-}113)$$

Recalling that sinc $u = (\sin \pi u)/\pi u$, we see that the spectrum of $x_1(t)$ goes to zero as $1/f$. Checking Figure 3-24a, we see that one derivative is required to produce impulses from the square-pulse portion of $x_1(t)$, which implies the decrease as n^{-1}.

We proceed to find the Fourier series of the periodic waveform $x(t)$ by using (3-111). Since $x_1(t)$ is repeated every 4 s to produce $x(t)$, $f_s = 0.25$ in (3-111), and the Fourier transform of $x(t)$, using (3-113) in (3-111), is

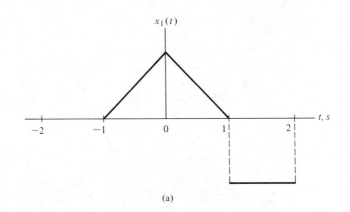

(a)

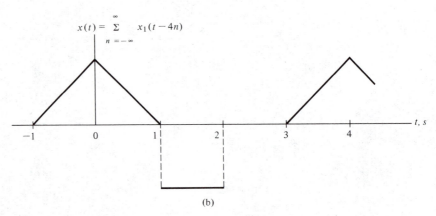

(b)

FIGURE 3-24. Pulse signal and its periodic extension.

$$X(f) = 0.25 \sum_{n=-\infty}^{\infty} [\text{sinc}^2 \, 0.25n - \text{sinc} \, 0.25n \, e^{-j0.75\pi n}]$$
$$\cdot \, \delta(f - 0.25n) \tag{3-114}$$

Recalling that $e^{j2\pi f_0 t} \leftrightarrow \delta(f - f_0)$, we may inverse Fourier transform (3-114) to obtain the exponential Fourier series of $x(t)$. The result is

$$x(t) = 0.25 \sum_{n=-\infty}^{\infty} (\text{sinc}^2 \, 0.25n - \text{sinc} \, 0.25n \, e^{-j0.75\pi n}) e^{j0.5\pi n t} \tag{3-115}$$

Again, it is seen that the predominant term as $n \to \infty$ is $\text{sinc} \, 0.25n \, \exp(-j0.75\pi n)$, which approaches zero as n^{-1}. The student should plot the amplitude and phase spectra of this signal. ∎

*3-22

FOURIER SERIES AND SIGNAL SPACES

It is often useful to visualize signals as analogous to vectors in a generalized vector space. The Fourier series and the Fourier transform that we have just studied are tools for resolving power and energy signals, respectively, into such generalized vector spaces. We briefly consider this approach to Fourier representations, referred to as *generalized Fourier series*.

We begin our consideration of generalized Fourier series with a review of some concepts about ordinary vectors in three-dimensional space. We were probably first introduced to vectors as physical quantities whose specification involves both magnitude and direction and that obey the parallelogram law of addition. Later, we were perhaps shown that it was convenient to represent a vector, $\mathbf{A}$ in a Cartesian coordinate system in terms of a mutually perpendicular triad of unit vectors $\hat{\mathbf{i}}$, $\hat{\mathbf{j}}$, and $\hat{\mathbf{k}}$ along the x, y, and z axes, respectively, as

$$\mathbf{A} = A_x \hat{\mathbf{i}} + A_y \hat{\mathbf{j}} + A_z \hat{\mathbf{k}} \tag{3-116}$$

The components A_x, A_y, and A_z can be expressed as

$$A_x = \mathbf{A} \cdot \hat{\mathbf{i}}, \qquad A_y = \mathbf{A} \cdot \hat{\mathbf{j}}, \qquad A_z = \mathbf{A} \cdot \hat{\mathbf{k}}$$

where the dot denotes the dot product of two vectors, which is the product of their magnitudes and the cosine of the angle between them. Since $\hat{\mathbf{i}}$, $\hat{\mathbf{j}}$, and $\hat{\mathbf{k}}$ are unit-length, mutually perpendicular vectors, the xyz components of $\mathbf{A}$ are easily obtained by simply projecting $\mathbf{A}$ onto the x, y, and z axes, respectively.

We now consider the possibility of representing a signal, $x(t)$, defined on a T-second interval $(t_0, t_0 + T)$ in terms of a set of preselected time functions, $\phi_1(t), \phi_2(t), \ldots, \phi_n(t)$. It is convenient to choose these functions with properties analogous to $\hat{\mathbf{i}}$, $\hat{\mathbf{j}}$, and $\hat{\mathbf{k}}$ of three-dimensional vector space. The mutually perpendicular property, referred to as *orthogonality*, is expressed as

$$\int_{t_0}^{t_0+T} \phi_m(t) \phi_n^*(t) \, dt = 0, \qquad m \neq n \tag{3-117}$$

where the conjugate suggests that complex-valued $\phi_n(t)$'s may be convenient in some cases. We further assume that the $\phi_n(t)$'s have been chosen such that

$$\int_{t_0}^{t_0+T} |\phi_n(t)|^2 \, dt = 1 \tag{3-118}$$

In view of (3-118), the $\phi_n(t)$'s are said to be *normalized*. This assumption is invoked to simplify future equations.

We now attempt to approximate $x(t)$ as best we can by a series of the form

$$y(t) = \sum_{n=1}^{N} d_n\phi_n(t) \tag{3-119}$$

where the d_n's are constants to be chosen such that $y(t)$ represents $x(t)$ as closely as possible according to some criterion. It is convenient to measure the error in the integral-square sense, which is defined as

$$\text{integral-square error} = \epsilon_N = \int_{t_0}^{t_0+T} |x(t) - y(t)|^2 \, dt \tag{3-120}$$

We wish to find $d_1, d_2, \ldots, d_N$ such that ϵ_N as expressed by (3-120) is a minimum. Substituting (3-119) into (3-120), we obtain

$$\epsilon_N = \int_T \left[x(t) - \sum_{n=1}^{N} d_n\phi_n(t) \right] \left[x^*(t) - \sum_{n=1}^{N} d_n^*\phi_n^*(t) \right] dt \tag{3-121}$$

where

$$\int_T (\cdot) \, dt \triangleq \int_{t_0}^{t_0+T} (\cdot) \, dt$$

This can be expanded to yield

$$\epsilon_N = \int_T |x(t)|^2 \, dt$$

$$- \sum_{n=1}^{N} \left[d_n^* \int_T x(t)\phi_n^*(t) \, dt + d_n \int_T x^*(t)\phi_n(t) \, dt \right] \tag{3-122}$$

$$+ \sum_{n=1}^{N} |d_n|^2$$

which was obtained by making use of (3-117) and (3-118) after interchanging the orders of summation and integration.

It is convenient to add and subtract the quantity

$$\sum_{n=1}^{N} \left| \int_T x(t)\phi_n^*(t) \, dt \right|^2$$

to (3-122) which, after rearrangement of terms, yields

$$\epsilon_n = \int_T |x(t)|^2 \, dt - \sum_{n=1}^{N} \left| \int_T x(t)\phi_n^*(t) \, dt \right|^2$$
(3-123)
$$+ \sum_{n=1}^{N} \left| d_n - \int_T x(t)\phi_n^*(t) \, dt \right|^2$$

To show the equivalence of (3-123) and (3-122) it is easiest to work backward from (3-123).

Now the first two terms on the right-hand side of (3-123) are independent of the coefficients d_n. The last summation of terms on the right-hand side is non-negative and is added to the first two terms. Therefore, to minimize ϵ_N through choice of the d_n's, the best we can do is make each term of the last sum zero. That is, we choose the nth coefficient, d_n, such that

$$d_n = \int_{t_0}^{t_0+T} x(t)\phi_n^*(t)dt, \qquad n = 1, 2, \ldots, N$$
(3-124)

This choice for $d_1, d_2, \ldots, d_N$ minimizes the integral-square error, ϵ_N. The resulting coefficients are called the *generalized Fourier coefficients*.

The minimum value for ϵ_N is

$$\epsilon_{N,\min} = \int_{t_0}^{t_0+T} |x(t)|^2 \, dt - \sum_{n=1}^{N} |d_n|^2$$
(3-125)

It is natural to inquire as to the possibility of $\epsilon_{N,\min}$ being zero. For special choices of the set of functions $\phi_1(t), \phi_2(t), \ldots, \phi_N(t)$, referred to as *complete sets* in the space of all integrable-square functions, it will be true that

$$\lim_{N \to \infty} \epsilon_{N,\min} = 0$$
(3-126)

for any signal that is square integrable, that is, for any signal for which

$$\int_{t_0}^{t_0+T} |x(t)|^2 \, dt < \infty$$

In the sense that the integral-square error is zero, we may then write

$$x(t) = \sum_{n=1}^{\infty} d_n \phi_n(t)$$
(3-127)

For a complete set of orthogonal functions, $\phi_1(t), \phi_2(t), \ldots$ we obtain from (3-125) that

$$\int_{t_0}^{t_0+T} |x(t)|^2 \, dt = \sum_{n=1}^{\infty} |d_n|^2$$
(3-128)

This is a generalized version of Parseval's theorem.

EXAMPLE 3-26

Given the set of functions shown in Figure 3-25. (a) Show that this is an orthogonal set and that each member of the set is normalized. (b) If $x(t) =$

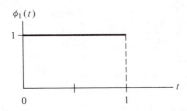

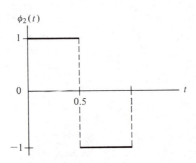

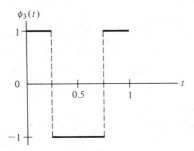

FIGURE 3-25. Orthogonal set of functions for Example 3-26.

$\cos 2\pi t$, $0 \le t \le 1$, find $y(t)$ as expressed by (3-119) such that the integral-square error is minimized. (c) Same question as in part (b) but $x(t) = \sin 2\pi t$, $0 \le t \le 1$.

Solution: (a) It is clear that each function is normalized according to (3-118). To show orthogonality, we calculate

$$\int_0^1 \phi_1(t)\phi_2(t) \, dt = \int_0^{0.5} (1)(1) \, dt + \int_{0.5}^1 (1)(-1) \, dt = 0$$

Similarly, it can be shown that $\int_0^1 \phi_1(t)\phi_3(t) = 0$ and $\int_0^1 \phi_2(t)\phi_3(t) \, dt = 0$.

(b) The generalized Fourier coefficients are found according to (3-124). The coefficients are

$$d_1 = \int_0^1 (1) \cos 2\pi t \, dt$$

$$= \left. \frac{\sin 2\pi t}{2\pi} \right|_0^1 = 0$$

$$d_2 = \int_0^{0.5} (1) \cos 2\pi t \, dt + \int_{0.5}^1 (-1) \cos 2\pi t \, dt$$

$$= \frac{1}{2\pi} \left[\left. \sin 2\pi t \right|_0^{0.5} - \left. \sin 2\pi t \right|_{0.5}^1 \right]$$

$$= 0$$

$$d_3 = \int_0^{0.25} (1) \cos 2\pi t \, dt + \int_{0.25}^{0.75} (-1) \cos 2\pi t \, dt + \int_{0.75}^1 (1) \cos 2\pi t \, dt$$

$$= \frac{1}{2\pi} \left[\left. \sin 2\pi t \right|_0^{0.25} - \left. \sin 2\pi t \right|_{0.25}^{0.75} + \left. \sin 2\pi t \right|_{0.75}^1 \right]$$

$$= \frac{2}{\pi}$$

Therefore,

$$y(t) = \frac{2}{\pi} \phi_3(t)$$

The minimum integral-square error is

$$\epsilon_3 = \int_0^1 \cos^2 2\pi t \, dt - \left(\frac{2}{\pi} \right)^2 = \frac{1}{2} - \left(\frac{2}{\pi} \right)^2$$

(c) In a similar manner to the calculations carried out for part (b), it follows that for $x(t) = \sin 2\pi t$,

$$d_1 = 0$$

$$d_2 = \frac{2}{\pi}$$

$$d_3 = 0$$

The minimum integral-square error is the same as for part (b).

The student is advised to sketch $y(t)$ for each case and compare it with $x(t)$. ∎

Note that the exponential Fourier series could have been derived as a generalized Fourier series with

$$\phi_n(t) = e^{jn\omega_0 t}, \qquad n = 0, \pm 1, \ldots \qquad (3\text{-}129)$$

where the interval under consideration is $(t_0, t_0 + T_0)$ and

$$\omega_0 = 2\pi f_0 = \frac{2\pi}{T_0}$$

Thus, *partial sums of exponential (and trigonometric) Fourier series minimize the integral-square error between the series and the signal under consideration.*

APPLICATIONS OF THE HILBERT TRANSFORM

The Hilbert transform, $\hat{x}(t)$ of a signal $x(t)$ was introduced in Example 3.20, and its Fourier transform in terms of $x(t)$ was given as pair 16 of Table 3-3. We discuss two uses of the Hilbert transform in this section.

Analytic Signals

An analytic signal $z(t)$ is a complex-valued signal whose spectrum is single-sided (i.e., is nonzero only for $f > 0$ or $f < 0$). Because of this property of its spectrum, it follows that the real and imaginary parts of an analytic signal cannot be specified independently. It turns out that one is the Hilbert transform of the other. To show this, suppose that $x(t)$ and $y(t)$ are the real and imaginary parts of $z(t)$, respectively. Let $Z(f)$, $X(f)$, and $Y(f)$ be the Fourier transforms of $z(t)$, $x(t)$, and $y(t)$, respectively. It follows that

$$Z(f) = X(f) + jY(f) \tag{3-130}$$

being zero for $f < 0$, say, requires that

$$Y(f) = j\, X(f), \qquad f < 0 \tag{3-131}$$

[note that $X(f)$ and $Y(f)$ may be complex]. If we double the positive-frequency portion of $Z(f)$, it then follows that

$$Y(f) = -jX(f), \qquad f > 0 \tag{3-132}$$

so that

$$Y(f) = -j\, \text{sgn}\, f\, X(f), \qquad \text{all } f \tag{3-133}$$

and

$$Z(f) = \begin{cases} 2X(f), & f > 0 \\ 0, & f < 0 \end{cases} \tag{3-134}$$

From (3-133) and pair 16 of Table 3-3, we see that

$$y(t) = -\hat{x}(t) \quad [Z(f) = 0, \quad f < 0] \tag{3-135}$$

where $\hat{x}(t)$ is the Hilbert transform of $x(t)$. Had we required that $Z(f)$ be zero for $f > 0$ and nonzero for $f < 0$, it follows that the signs in front of the j in

(3-131) and (3-132) would have been reversed, so that

$$y(t) = \hat{x}(t) \quad [Z(f) = 0, \quad f > 0] \tag{3-136}$$

Analytic signals are of use in modulation theory applications, and in particular to describe single-sideband modulated signals mathematically.

Causality

A causal system is defined as one whose output does not anticipate the input. If a causal system is also linear *and* fixed, so that it can be characterized by an impulse response, $h(t)$, the causality of the system requires that

$$h(t) = 0, \quad t < 0 \tag{3-137}$$

The Fourier transform of $h(t)$ is the transfer function $H(f)$ of the system and, from the discussion under analytic signals, it should be apparent that the real and imaginary parts of $H(f)$ cannot be independently specified, but are in fact Hilbert transforms of each other. To show this, consider $h(t)$ written in terms of its even and odd parts as

$$h(t) = h_e(t) + h_0(t) \tag{3-138}$$

where

$$h_e(t) = \tfrac{1}{2}[h(t) + h(-t)] \tag{3-139}$$

and

$$h_0(t) = \tfrac{1}{2}[h(t) - h(-t)] \tag{3-140}$$

In order that $h(t)$ be zero for $t < 0$, it follows that

$$h_0(t) = \begin{cases} h_e(t), & t > 0 \\ -h_e(t), & t < 0 \end{cases} \tag{3-141}$$
$$= \operatorname{sgn} t \, h_e(t)$$

Substitution of (3-141) into (3-138) results in

$$h(t) = h_e(t) + \operatorname{sgn} t \, h_e(t) \tag{3-142}$$

which, when Fourier transformed with the aid of pair 14 of Table 3-3 and the multiplication theorem, allows the transfer function to be written as

$$H(f) = H_e(f) + \frac{1}{j\pi f} * H_e(f)$$
$$= H_e(f) - j\hat{H}_e(f) \tag{3-143}$$

where $\hat{H}_e(f)$ is the Hilbert transform of $H_e(f) = \mathcal{F}[h_e(t)]$.

SUMMARY

In this chapter we have considered the representation of signals by trigonometric sums and as rotating phasor sums whose terms have frequencies which are harmonically related. The former representations are referred to as a trigonometric Fourier series and the latter as an exponential Fourier series. Both can be shown to be equivalent, although they provide different representations in terms of frequency spectra, trigonometric Fourier series resulting in single-sided and exponential Fourier series resulting in double-sided spectral representations, respectively. We naturally think of Fourier series as being useful representations of periodic signals, although a nonperiodic signal may be represented by a Fourier series over a finite interval. In order to obtain an analogous spectral representation for pulse-type signals, including energy signals, the Fourier transform was introduced as the limiting form of the exponential Fourier series as the interval of representation becomes large. In contrast to periodic signals whose spectra consist of lines at discrete frequencies, the spectra of energy signals exist over a continuum of frequencies. The rapidity with which the spectrum of a signal approaches zero as frequency increases can be shown to be directly related to the smoothness of the signal in the time domain (Section 3-20). Parseval's theorem and its generalization, known as Rayleigh's energy theorem, provide useful tools for examining the spectral distribution of power and energy for power and energy signals, respectively.

Several Fourier transform theorems provide means for obtaining Fourier transforms of more complex signals from those of simpler signals. An important Fourier transform theorem is the convolution theorem, which provides the basis for obtaining the output spectrum of a fixed, linear system given the input spectrum and the Fourier transform of the impulse response, which is referred to as the transfer function of the system.

In general, a system distorts the spectrum of any signal passing through it. The condition for a system to be distortionless for a given input signal is that it have linear phase and a constant amplitude response as a function of frequency over the bandwidth of the input signal spectrum (Section 3-9). Ideal filters (Section 3-18) are defined to have such ideal phase- and amplitude-response functions over suitable frequency bands, although they are noncausal since their impulse responses anticipate (or exist prior to) the impulsive input.

The Gibbs phenomenon, which refers to the overshoot of the partial Fourier series or inverse Fourier integral representation of a signal (i.e., the inclusion of only a portion of the spectrum to represent the signal) at a discontinuity was shown to exist precisely because the spectral limitation amounts to viewing the signal through a window via the convolution operation. Gibbs phenomenon effects may be reduced by employing weighting functions in the frequency domain whose inverse transforms have low sidelobes in the time domain.

An alternative way to develop the idea of a Fourier expansion of a signal is as a special case of the minimum integral-square-error expansion of a signal involving a complete orthonormal set of functions. When approached in this way, it is apparent that a Fourier series when truncated at any finite number of terms provides the best representation of the signal of any series in terms of minimum integral-squared error.

As a final note in this chapter, the implications of requiring that a signal be single-sided in the time or frequency domain were investigated. In particular, requiring that a spectrum be zero for frequencies less than zero implies that the real and imaginary parts of its inverse Fourier transform be Hilbert transforms. Requiring that a signal be causal, or exist only for $t > 0$, implies that the Fourier transforms of its even and odd parts be Hilbert transforms.

FURTHER READING

Fourier series and the Fourier transform are introduced in almost all beginning circuits analysis texts, but their use as systems analysis tools are probably not fully exploited until the student becomes involved in more systems-oriented areas such as communications and controls. Nevertheless, rather than cite references in these specialized areas, we will give a few that deal with these topics from a more mathematical viewpoint.

E. A. GUILLEMIN, *The Mathematics of Circuit Analysis*. New York: Wiley (original copyright 1949 with several printings). Chapter 7 presents an extensive and elegant treatment of Fourier series and the Fourier integral.

R. N. BRACEWELL, *The Fourier Transform and Its Applications*, 2nd ed. New York: McGraw-Hill, 1978.

A. PAPOULIS, *Signal Analysis*. New York: McGraw-Hill, 1977. Chapter 3 presents a brief but very readable treatment of Fourier analysis.

M. J. LIGHTHILL, *Introduction to Fourier Analysis and Generalized Functions*. New York: Cambridge University Press, 1962. This book presents the generalized function approach to singularity functions and their use in the treatment of properties of the Fourier series and integral.

PROBLEMS

SECTION 3-2

3-1. Plot the first through third partial sums of the series

$$x(t) = \frac{2}{\pi} (\cos \omega_0 t - \tfrac{1}{3} \cos 3\omega_0 t + \tfrac{1}{5} \cos 5\omega_0 t - \cdots)$$

Comment on its similarity to a square wave. Is it even or odd? Why is this series an alternating series whereas the series of Example 3-1 was not? (*Hint:* In calculating the data for making the plot, note that symmetry can be used to save work.)

SECTIONS 3-3 and 3-4

3-2. Prove the relationship (3-8).

3-3. Show all the steps that are necessary to obtain the result (3-16) for b_m.

3-4. Obtain the trigonometric Fourier series of the square wave

$$x(t) = \begin{cases} A, & \dfrac{-T_0}{4} < t \le \dfrac{T_0}{4} \\[2mm] -A, & \dfrac{-T_0}{2} < t \le \dfrac{-T_0}{4} \quad \text{and} \quad \dfrac{T_0}{4} < t \le \dfrac{T_0}{2} \end{cases}$$

with $x(t) = x(t + T_0)$, all t. Why is it composed of only cosine terms?

3-5. (a) Express the Fourier series for the odd-symmetrical square wave considered in Example 3-3 in the cosine–series form derived in Section 3-4 and expressed by (3-21).

(b) Plot A_n and θ_n versus $nf_0 = n\omega_0/2\pi$ to obtain the single-sided amplitude and phase spectra, respectively.

SECTIONS 3-5 to 3-15

3-6. Obtain the results for the exponential Fourier series examples given in Table 3-1 with the exception of number 3, which was worked in Example 3.5.

3-7. Suppose that differentiation of the periodic signal $x(t)$ results in a signal that has a Fourier series.

(a) Integrate by parts the expression

$$X_n' = \frac{1}{T_0} \int_{T_0} \frac{dx}{dt} e^{-jn\omega_0 t} \, dt$$

to show that

$$X_n' = (jn\omega_0)X_n$$

that is, the Fourier series coefficients of the signal dx/dt are related to the Fourier series coefficients of $x(t)$ through multiplication by $jn\omega_0$.

(b) From Table 3-1, waveform 5, obtain the Fourier series coefficients of an odd-symmetry square wave.

3-8. (a) Use Parseval's theorem expressed by (3-45) and (3-46) to find P_{av} for

$$x(t) = 2 \cos (10^4 \pi t) \sin^2 (2 \times 10^4 \pi t)$$

(b) If $x(t)$ is a signal that is transmitted through a telephone system which blocks dc and frequencies above 12 kHz, compute the ratio of received to transmitted power.

3-9. Consider the periodic pulse-train signal of Example 3.5. Find the ratio of the power in frequency components for which $|nf_0| \le \tau^{-1}$ to total power if:

(a) $T_0/\tau = 2$
(b) $T_0/\tau = 4$
(c) $T_0/\tau = 10$

3-10. Same question is Problem 3-9, but consider frequency components for which $|nf_0| \le 2\tau^{-1}$.

3-11. Plot the two-sided amplitude and phase spectra for the full-wave rectified sinewave (waveform 2 of Table 3-1). Convert the two-sided plots to single-sided plots.

3-12. Plot and compare amplitude spectra for the triangular and square waves (waveforms 4 and 5 of Table 3-1). Which requires the most bandwidth for a given fidelity of reproduction?

3-13. Plot and label accurately Figure 3-7 for specific values of τ and T_0 as follows:
(a) $\tau = 2$ ms; $T_0 = 8$ ms
(b) $\tau = 1$ ms; $T_0 = 8$ ms
(c) $\tau = 2$ ms; $T_0 = 16$ ms

3-14. (a) Prove the relationships (3-33a) and (3-33b), which relate the coefficients of the exponential Fourier series to the Fourier cosine series.
(b) Show that the trigonometric Fourier series of a real, even signal consists only of cosine terms, and that of an odd signal consists only of sine terms.

3-15. Show the following symmetry properties for the complex exponential Fourier coefficients:
(a) X_n is real and an even function of n if $x(t)$ is real and even.
(b) X_n is imaginary and odd if $x(t)$ is real and odd.
(c) $X_n = 0$ for even n if $x(t) = -x(t \pm T_0/2)$ (i.e., has odd half-wave symmetry).
Hint: (1) One way to do this is to write down the Fourier series for $x(t)$. Substitute $t + T_0/2$ for t to obtain a series for $x(t + T_0/2)$. What must be true of the coefficients in order for this new series to equal to original series for $x(t)$? (2) A second way to prove $X_n = 0$, n even, is to write down the integral for X_n and break the integral into one from $-T_0/2$ to 0 and 0 to $T_0/2$. Change variables in the one from $-T_0/2$ to 0 to change its limits to 0 to $T_0/2$. Use the fact that $x(t - T_0/2) = -x(t)$ in the integrand. Consider n even and n odd separately.

3-16. If the input to a system is $\cos 10\pi t + 2 \cos 20\pi t$ tell what kinds of distortion, if any, a system introduces if its output is:
(a) $\cos (10\pi t - \pi/4) + 5 \cos (20\pi t - \pi/2)$
(b) $\cos (10\pi t - \pi/4) + 2 \cos (20\pi t - \pi/4)$
(c) $\cos (10\pi t - \pi/4) + 2 \cos (20\pi t - \pi/2)$
(d) $2 \cos (10\pi t - \hbar/4) + 4 \cos (20\pi t - \pi/8)$

3-17. Obtain the Fourier transforms of the following signals ($\alpha > 0$):
(a) $x_a(t) = Ae^{-\alpha t} u(t)$;
(b) $x_b(t) = Ae^{\alpha t} u(-t)$;
(c) $x_c(t) = Ae^{-\alpha|t|}$;
(d) $x_d(t) = Ae^{-\alpha t} u(t) - Ae^{\alpha t} u(-t)$.

3-18. Plot and compare the amplitude and phase spectra of $x_a(t)$ and $x_b(t)$ in Prob. 3.17.

3-19. Plot and compare the amplitude and phase spectra of $x_c(t)$ and $x_d(t)$ in Prob. 3.17.

3-20. **(a)** Show that if $x(t)$ is real and even (i.e., $x(t) = x(-t)$), then the Fourier transform of $x(t)$ may be reduced to

$$X(f) = 2\int_0^\infty x(t) \cos(2\pi ft)\, dt.$$

Therefore, conclude that $X(f)$ is real and even in this case.
(b) Show that for $x(t)$ real and odd (i.e., $x(t) = -x(-t)$), then

$$X(f) = -2j\int_0^\infty x(t) \sin(2\pi ft)\, dt.$$

Therefore, conclude that $X(f)$ is imaginary and odd if $x(t)$ is real and odd.

3-21. If the Fourier transform of a real signal $x(t)$ is expressed in terms of its magnitude and phase as

$$X(f) = |X(f)| \exp[j\theta(f)]$$

show that

$$|X(f)| = |X(-f)|$$
$$\theta(f) = -\theta(-f)$$

(i.e., the magnitude is even and the phase is odd).

3-22. Using the transform pair of Example 3-12, plot the amplitude and phase spectra of the pulse signal $x(t) = A\Pi(t/\tau)$.

3-23. **(a)** Obtain the Fourier transform of the raised cosine pulse

$$x(t) = \tfrac{1}{2}A\left(1 + \cos\frac{2\pi t}{\tau}\right)\Pi\left(\frac{t}{\tau}\right)$$

(b) Plot its amplitude spectrum and compare with the amplitude spectrum of the square pulse (see Example 3-12).

3-24. Obtain and plot the energy spectral densities of the signals given in Problem 3-17.

3-25. **(a)** Find the energy contained in the signal $x(t) = A \exp(-\alpha t)u(t)$, $\alpha > 0$, for frequencies less than $|f| < \alpha$.
(b) Same question as part (a) for frequencies less than $|f| < 2\alpha$. What percent of the total energy is this?

3-26. **(a)** use the result of part (d), Problem 3.17, to find the Fourier transform of the signum function

$$\operatorname{sgn} t \triangleq \begin{cases} 1, & t > 0 \\ -1, & t < 0 \end{cases}$$

(b) Noting that the unit step $u(t)$ can be written in terms of the signum function as

$$u(t) = \tfrac{1}{2} (\operatorname{sgn} t + 1)$$

find its Fourier transform.

3-27. Consider the use of the transform pair

$$\tau^{-1} \Pi \left(\frac{t}{\tau} \right) \leftrightarrow \operatorname{sinc} f\tau$$

to obtain the Fourier transform of $\delta(t)$. In what sense does $\tau^{-1} \Pi(t/\tau)$ approximate a unit impulse? What is $\lim_{\tau \to 0} \operatorname{sinc} f\tau$?

3-28. **(a)** Use the superposition and time delay theorems together with the pair $\Pi(t/\tau)$ $\leftrightarrow \tau \operatorname{sinc} f\tau$ to obtain Fourier transforms for the signals shown.

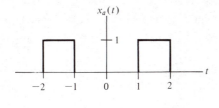

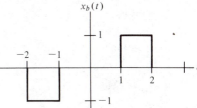

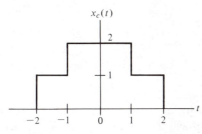

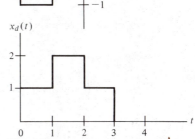

(b) Using symmetry, show that each transform is real, imaginary, or neither, as the case may be.

3-29. A signal with an approximately rectangular spectrum extending from 100 to 10000 Hz is recorded on a tape recorder at 15 inches per second (ips). It is played back at $7\tfrac{1}{2}$ ips. Sketch the spectra for the recorded and played-back signals.

3-30. Use the duality theorem to find the signal whose transform is shown.

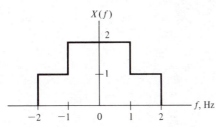

Hint: Work Problem 3-28 for $x_c(t)$ first.

3-31. If the signals of Problem 3-28 are each multiplied by the signal cos $20\pi t$, obtain and sketch the magnitudes of the Fourier transforms of the resulting signals.

3-32. Use the differentiation theorem and the transform pair $\delta(t - t_0) \leftrightarrow \exp(-j2\pi f t_0)$ to find the Fourier transforms of the signals shown. Which transform should be real and even and which one should be imaginary and odd?

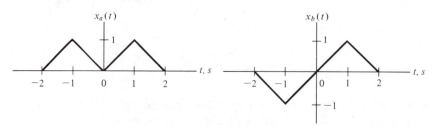

3-33. Obtain the Fourier transforms of the signals given below by applying (1) the time delay theorem and then the frequency translation theorem; (2) the frequency translation theorem and then the time delay theorem. Show that both procedures give the same Fourier transform.

(a) $\exp[j2\pi(t - 1) - (t - 1)]u(t - 1)$
(b) $\Pi[(t - 2)/4] \exp[j2\pi(t - 2)]$

3-34. Obtain the inverse Fourier transform of

$$X(f) = 10\,\frac{\text{sinc } 2f}{3 + j2\pi f}$$

Hint: Use the convolution theorem.

3-35. (a) Prove the multiplication theorem.
(b) Use the multiplication theorem and the Fourier transform pair derived in Example 3.13 to inverse Fourier transform

$$X(f) = 15\Lambda\left(\frac{f}{3}\right)$$

SECTION 3-16 AND 3-17

3-36. (a) Obtain the transfer function of the RC high-pass filter shown.

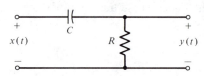

(b) Under what conditions will it perform as an ideal differentiator defined by the input–output relationship

$$y(t) = A\frac{dx}{dt}$$

3-37. Consider the RC high-pass filter of Problem 3-36. Obtain its output in response to an input of the form

$$x(t) = Ae^{-\alpha t}u(t)$$

by using Fourier transform techniques. Plot the output for $RC\alpha = 0.5, 1, 2$.

3-38. Consider an ideal integrator for which input and output are related by

$$y(t) = A \int^t x(t') \, dt'$$

Use the integration theorem to obtain its transfer function. Under what conditions will the RC low-pass filter considered in Example 3-22 perform as an integrator?

3-39. Obtain the impulse response and transfer function of the circuit shown. Plot $|H(f)|$ versus f for $a = 0.5$ and 1.0.

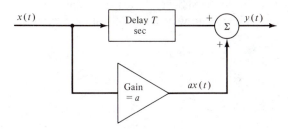

3-40. Consider a fixed, linear system with amplitude and phase response as shown.

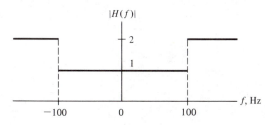

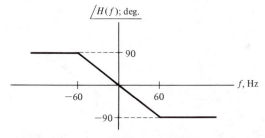

Obtain the output for the following inputs and tell what type of distortion, if any, results:

(a) $x(t) = \cos 20\pi t + \cos 60\pi t$
(b) $x(t) = \cos 20\pi t + \cos 140\pi t$
(c) $x(t) = \cos 20\pi t + \cos 220\pi t$
(d) $x(t) = \cos 130\pi t + \cos 220\pi t$

SECTION 3-18

3-41. Obtain the impulse response of the ideal high-pass filter with transfer function given by

$$H_{HP}(f) = H_0\left[1 - \Pi\left(\frac{f}{2B}\right)\right]e^{-j2\pi f t_0}$$

SECTION 3-19

3-42. Plot a figure similar to Figure 3-22b and c for $W = 2$ and $x(t) = u(t) - u(t - \frac{1}{2})$. What do you conclude about the relationship between pulse width and W so as to produce overshoot?

SECTIONS 3-20 AND 3-21

3-43. Given the transform pair $\Lambda(t/\tau) \leftrightarrow \tau \operatorname{sinc}^2 f\tau$, obtain the Fourier transforms of the periodic triangular waveforms shown. Sketch the amplitude spectrum for each. Comment on the rapidity with which $|X_n| \to 0$ with $n \to \infty$.

(a)

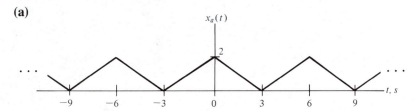

(b)

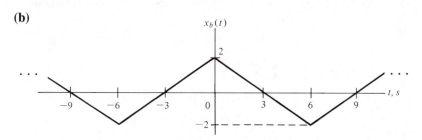

(c)

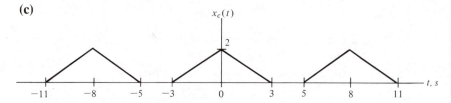

3-44. Obtain the Fourier transform of the periodic raised cosine pulse train

$$x(t) = \frac{1}{2}A \sum_{n=-\infty}^{\infty}\left[1 + \cos\left(\frac{2\pi(t - nT_0)}{\tau}\right)\Pi\left(\frac{t - nT_0}{\tau}\right)\right]$$

where $T_0 \geq \tau$. Sketch the waveform and the amplitude spectrum for the case $\tau = T_0$.

REVIEW PROBLEM

3-45. Consider the waveforms below and their complex exponential Fourier series. For which ones are the following true: real coefficients; purely imaginary coefficients; even-indexed coefficients zero; $X_0 = 0$?

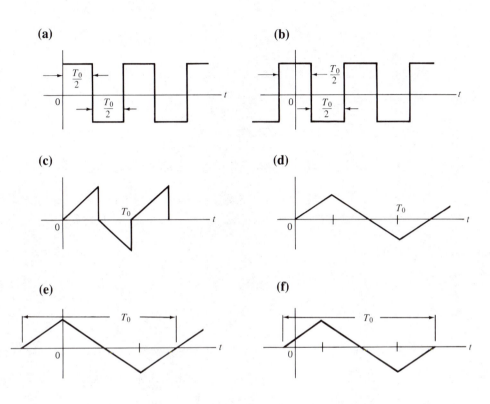

(a)

(b)

(c)

(d)

(e)

(f)

If true, check the appropriate box below.

Property	Waveform					
	a	**b**	**c**	**d**	**e**	**f**
Purely real coefficients						
Purely imaginary coefficients						
Complex coefficients						
Even-indexed coefficients zero						
$X_0 = 0$						

REVIEW PROBLEM

3-46. Given the periodic waveform shown:

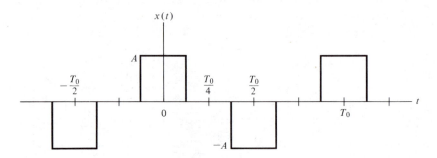

(a) What is the value of a_0 in the sine–cosine Fourier series? Why?

(b) What are the values of the b_m coefficients in the sine–cosine Fourier series? Why?

(c) Are the coefficients in the complex exponential Fourier series real, imaginary, or complex? Why?

(d) Does this waveform have half-wave odd symmetry?

CHAPTER **4**

The Laplace Transformation: Applications to Systems Analysis

4-1

INTRODUCTION

Systems analysis in the time domain involves the solution of differential equations or the evaluation of the superposition integral. Both techniques can result in tedious mathematical operations for relatively simple systems analysis problems. The Fourier transform provided an alternative approach wherein a differential equation relating the input and output of a system was transformed to an algebraic equation, the Fourier transform of the output was solved for, and the transform of the output then inverse Fourier transformed to provide the system output as a function of time. Unfortunately, there are many signals of interest that arise in systems analysis problems for which Fourier transforms do not exist.

The Fourier transform technique for systems analysis provides a hint for a more general transform analysis procedure, however. We seek a transform that applies to a wider class of signals than the Fourier transform does. This can be accomplished by multiplying a signal $x(t)$ by an exponential convergence factor and Fourier transforming the product. For signals that are zero for $t < 0$, an appropriate factor is $e^{-\sigma t}$, where σ is positive, resulting in the Fourier transform

$$X(\sigma + j\omega) = \int_0^\infty x(t)e^{-\sigma t}e^{-j\omega t}\, dt \qquad (4\text{-}1)$$

With $s = \sigma + j\omega$, (4-1) can be written as

$$\mathcal{L}[x(t)] = X(s) = \int_0^\infty x(t)e^{-st}\, dt \tag{4-2}$$

where the script letter $\mathcal{L}[\cdot]$ denotes the operation of obtaining the *Laplace transform of x(t)* defined by (4-2). Since the lower limit of the integral is zero, (4-2) is referred to as the *single-sided Laplace transformation*.

Use of the Laplace transform for systems analysis results in several advantages. Among these are:

1. The solution of differential equations progresses systematically and involves only algebraic manipulations.
2. The total solution—particular integral and complementary function (i.e., forced and transient responses)†—is obtained.
3. Initial conditions are automatically included in the solution of the equations.

Not every signal possesses a Laplace transform. An example is exp [exp (t)], $t > 0$, which grows faster than exp $(-\sigma t)$ decays. Thus the Laplace transform integral (4-2) does not converge for this signal. We will discuss convergence properties of Laplace transforms more fully in the following section.

The operation that changes $X(s)$ back to $x(t)$ is referred to as the *inverse Laplace transformation,* and is symbolized by $\mathcal{L}^{-1}[\cdot]$. To obtain $x(t)$ in terms of $X(s) = X(\sigma + j\omega)$, we observe from the inverse Fourier transform of (4-1) that

$$x(t)e^{-\sigma t} = \mathcal{F}^{-1}[X(\sigma + j\omega)] = \frac{1}{2\pi}\int_{-\infty}^\infty X(\sigma + j\omega)e^{j\omega t}\, d\omega \tag{4-3a}$$

Multiplying both sides by $e^{\sigma t}$ and assuming σ constant, we obtain the *inversion integral,*

$$x(t) = \frac{1}{2\pi j}\int_{\sigma - j\infty}^{\sigma + j\infty} X(s)e^{st}\, ds \tag{4-3b}$$

where the change of variable $s = \sigma + j\omega$ and $ds = jd\omega$ has been used. With $\omega = \pm\infty$ in (4-3a) the limits for (4-3b) are clearly $\sigma \pm j\infty$.

Although not easy to show rigorously, (4-3b) may be generalized to a *complex inversion integral* where $s = \sigma + j\omega$ is a complex variable with a real part that may vary along with its imaginary part ω. It is discussed in more detail in Section 4-5, where the technique of *contour integration* is briefly examined for the purpose of evaluating inverse Laplace transforms.

For the most part, contour integration can be avoided when finding inverse Laplace transforms in systems analysis problems simply by making use of a table of Laplace transform pairs. We now begin the construction of such a table and illustrate the evaluation of (4-2) for several simple signals.

---

†The particular integral, which emulates the forcing function or excitation of the system, is referred to as the *forced response*. That part of the total response which is influenced by characteristics of the system itself is referred to as the *transient response*. It is the nontrivial solution of the system differential equation with the excitation identically zero.

EXAMPLES OF EVALUATING LAPLACE TRANSFORMS

As an example of evaluating (4-2), let $x(t) = 1$. Then

$$X(s) = \int_0^\infty e^{-st}\, dt$$

$$= \frac{e^{-st}}{-s}\Big|_0^\infty = \frac{e^{-\sigma t} e^{-j\omega t}}{-s}\Big|_0^\infty$$

$$= -\frac{e^{-\sigma t}}{s}(\cos \omega t - j \sin \omega t)\Big|_0^\infty$$

Clearly, unless $\sigma = \text{Re}\,(s) > 0$, the limit as $t \to \infty$ does not exist since $\cos \omega t$ and $\sin \omega t$ oscillate with increasing t and $e^{-\sigma t}$ grows without bound if $\sigma < 0$. Requiring that $\text{Re}\,(s) > 0$, we obtain

$$\mathcal{L}[1] = \frac{1}{s}, \qquad \text{Re}\,(s) > 0 \qquad\qquad (4\text{-}4)$$

Because the lower limit of (4-2) is zero, values of $x(t)$ for $t < 0$ have no effect on $X(s)$. We see that $x(t) = 1$ and $x(t) = u(t)$ have the same single-sided Laplace transform. If we consider signals that are nonzero only for $t \geq 0$, this presents no problem.† This is reasonable for our purposes because all signals must start sometime. We may choose this starting time conveniently as $t = 0$.

Some discussion is in order about the convergence of the Laplace transform integral. In obtaining the Laplace transform of $x(t) = 1$, it was necessary to restrict $\text{Re}\,(s) > 0$ in order for the integral to exist. Recall that an integral with one or both of its limits unbounded is referred to as an *improper integral*. An improper integral of the form

$$\int_0^\infty x(t)\, dt \triangleq \lim_{L \to \infty} \int_0^L x(t)\, dt$$

is said to be convergent if the limit on the right-hand side exists. The integral *converges absolutely* if and only if

$$\int_0^\infty |x(t)|\, dt < \infty$$

The Laplace transform integral of a signal $x(t)$ can be shown to converge absolutely for all

$$\sigma = \text{Re}\,(s) > c \qquad \text{if} \qquad \int_0^L |x(t)|\, dt \leq K < \infty$$

for any positive real number L and some real constant K, and if

$$|x(t)| \leq Ae^{ct}, \qquad t > L,$$

---

†The *double-sided Laplace transform* can be used for signals that are nonzero for $t < 0$. An introduction to this transform is given in Section 4.6.

where A and c are appropriately chosen real constants. Such a signal is said to be of exponential order. The smallest possible value of c is called the *abscissa of absolute convergence*. A proof of this theorem is given in Appendix C.

Applying this theorem to $x(t) = u(t)$, we see that $A = 1$ and $c = 0$ are appropriate choices, and the Laplace transform of $u(t)$ therefore converges absolutely for $\text{Re}(s) > c = 0$.

By applying the theorem to $x(t) = t^{-1/2}$ with $L \geq 1$, $A = 1$, and $c = 0$, we know that this signal also has a Laplace transform which, in fact, can be shown to be $(\pi/s)^{1/2}$ for $\text{Re}(s) > 0$. Note the importance of the role of L in this example.

Consider next the Laplace transform of the signal

$$x(t) = e^{-\alpha t}u(t)$$

where α may be complex. By definition, the Laplace transform is

$$\mathcal{L}[e^{-\alpha t} u(t)] = \int_0^\infty e^{-(s+\alpha)t}\, dt$$

$$= -\frac{e^{-(\alpha+\sigma)t}e^{-j\omega t}}{(s + \alpha)} \Bigg|_0^\infty$$

or

$$X(s) = \frac{1}{s + \alpha}, \qquad \text{Re}(\alpha) + \sigma > 0 \text{ or } \text{Re}(s) > -\text{Re}(\alpha) \qquad (4\text{-}5)$$

The abscissa of absolute convergence, or simply abscissa of convergence, in this case is

$$c = -\text{Re}(\alpha)$$

As a final example, we obtain the Laplace transform of $\delta(t)$. By definition of the Laplace transform, it is

$$\mathcal{L}[\delta(t)] = \int_0^\infty \delta(t)e^{-st}\, dt \qquad (4\text{-}6)$$

whereupon we are faced with a dilemma. Since the unit impulse function occurs at $t = 0$, do we integrate through half of it, none of it, or all of it? We will assume in this book that the lower limit on the Laplace transform is $t = 0^-$. Thus (4-6) evaluates to $\mathcal{L}[\delta(t)] = 1$. The value of s makes no difference; that is, the region of convergence is the entire s plane.

A final word about convergence. The abscissas of convergence for the signals $u(t)$ and $e^{-\alpha t}u(t)$ are shown in Figure 4-1 together with the regions of absolute convergence of their respective Laplace transform integrals. For any value of s lying in the region of convergence, the respective Laplace transform is finite. It follows that any singularities of the Laplace transform of a signal must lie to the left of the abscissa of absolute convergence. The Laplace transforms of the unit step and decaying exponential signals, viewed as functions of the complex variable s, each have a single singularity which is referred to as a *first-order* pole.† They are shown as ×'s in Figure 4-1 and do indeed lie outside the regions of absolute convergence of the Laplace transform integral. One potential point of confusion is the distinction between the region of absolute convergence of the Laplace transform integral of a signal $x(t)$ and the region in the complex s-plane where $X(s) = \mathcal{L}[x(t)]$ is a well-behaved function of s.‡ The only place that $X(s)$ is undefined is at its singular points, which may be of the several different types defined in Appendix C. At every other point in the complex plane $X(s)$ is a well-behaved function of the complex variable s, both within and outside the region of absolute convergence of the Laplace transform integral.

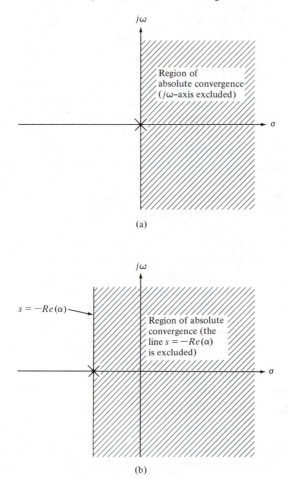

(a)

(b)

FIGURE 4-1. Regions of absolute convergence for the Laplace transforms of (a) $u(t)$ and (b) $e^{-\alpha t}u(t)$.

†Appendix C gives a short summary of several definitions and theorems pertaining to functions of a complex variable. For our purposes here, a pole can be defined as a value for s for which $X(s) \to \infty$.

‡In terms of the terminology of functions of a complex variable $X(s)$ is an *analytic function* of s. See Appendix C.

TABLE 4-1. Table of Laplace Transforms

Signal	Laplace Transform	Abscissa of Convergence
$\delta(t)$	1	$-\infty$
1	$\dfrac{1}{s}$	0
$e^{-\alpha t}$	$\dfrac{1}{s + \alpha}$	$-\alpha$

The integral of (4-3) is carried out along any line to the right of the singularities of $X(s)$. If this line can be chosen as the $j\omega$-axis, it then follows from (4-1) and (4-3) that the ordinary Fourier transform of $x(t)$ exists and can be obtained from $X(s)$ by substituting $s = j\omega$.

The three Laplace transform pairs just derived are summarized in Table 4-1 together with their abscissas of absolute convergence. We will dispense with specifying the abscissas of absolute convergence of single-sided Laplace transforms since it can be shown that there is no ambiguity in the inverse single-sided Laplace transform, even with the region of convergence unspecified. This is not true of the double-sided Laplace transform for which it is necessary to specify the region of convergence together with the Laplace transform in order to specify uniquely the inverse Laplace transform. Since we are not considering the two-sided Laplace transform, specifications of the abscissas of absolute convergence are not necessary.

We now will extend Table 4-1, but to avoid the labor of carrying out integrations, we will give several Laplace transform theorems and use them to extend it. Because the Fourier and Laplace transforms are both linear integral transforms, many of the theorems to be stated will be self-evident from our consideration of the Fourier transform.

4-3

SOME LAPLACE TRANSFORM THEOREMS

Theorem 1: Linearity: Since (4-2) is an integral in which $x(t)$ appears linearly, it follows, for two constants a_1 and a_2 which may be complex, that

$$\mathscr{L}[a_1 x_1(t) + a_2 x_2(t)] = a_1 X_1(s) + a_2 X_2(s) \qquad (4\text{-}7)$$

where $X_1(s) = \mathscr{L}[x_1(t)]$ and $X_2(t) = \mathscr{L}[x_2(t)]$. The linearity theorem will be made use of in the Laplace transformation of differential equations. We illustrate its use for extending Table 4-1 by an example.

EXAMPLE 4-1

The Laplace transform of $\cos \omega_0 t$ may be obtained by expressing it in exponential form, applying the linearity theorem, and using the transform pair

(4-5) with $\alpha = \pm j\omega_0$, which gives

$$\mathscr{L}[\cos \omega_0 t] = \mathscr{L}[\tfrac{1}{2}e^{j\omega_0 t} + \tfrac{1}{2}e^{-j\omega_0 t}]$$

$$= \tfrac{1}{2}\mathscr{L}[e^{j\omega_0 t}] + \tfrac{1}{2}\mathscr{L}[e^{-j\omega_0 t}]$$

$$= \frac{1}{2}\frac{1}{s - j\omega_0} + \frac{1}{2}\frac{1}{s + j\omega_0}$$

$$= \frac{s}{s^2 + \omega_0^2} \tag{4-8}$$

In a similar manner, the Laplace transform of $\sin \omega_0 t$ can be derived as follows:

$$\mathscr{L}[\sin \omega_0 t] = \mathscr{L}\left[\frac{1}{2j}e^{j\omega_0 t} - \frac{1}{2j}e^{-j\omega_0 t}\right]$$

$$= \frac{1}{2j}\mathscr{L}[e^{j\omega_0 t}] - \frac{1}{2j}\mathscr{L}[e^{-j\omega_0 t}]$$

$$= \frac{1}{2j}\frac{1}{s - j\omega_0} - \frac{1}{2j}\frac{1}{s + j\omega_0}$$

$$= \frac{\omega_0}{s^2 + \omega_0^2} \tag{4-9}$$

∎

Theorem 2: Transforms of Derivatives From (4-2) the Laplace transform of $dx(t)/dt$ is

$$\mathscr{L}\left[\frac{dx(t)}{dt}\right] = \int_0^\infty \frac{dx(t)}{dt}e^{-st}\,dt \tag{4-10}$$

Which may be integrated by parts by letting $u = e^{-st}$, and $dv = dx(t)$ in the equation

$$\int_a^b u\,dv = uv\bigg|_a^b - \int_a^b v\,du \tag{4-11}$$

Then $du = -se^{-st}\,dt$ and $v = x(t)$, so that (4-10) becomes

$$\mathscr{L}\left[\frac{dx(t)}{dt}\right] = e^{-st}x(t)\bigg|_0^\infty + s\int_0^\infty x(t)e^{-st}\,dt$$

$$= sX(s) - x(0^-) \tag{4-12}$$

provided that $\lim_{t\to\infty} x(t)e^{-st} = 0$, where we recall that we have agreed to use 0^- for the lower limit.

To find $\mathscr{L}\left[\dfrac{d^2x(t)}{dt^2}\right]$, we write

$$\frac{d^2x(t)}{dt^2} = \frac{d}{dt}\left[\frac{dx(t)}{dt}\right]$$

which results in

$$\mathcal{L}\left[\frac{d^2x(t)}{dt^2}\right] = s\mathcal{L}\left[\frac{dx(t)}{dt}\right] - \frac{dx}{dt}\bigg|_{t=0^-}$$

$$= s[sX(s) - x(0^-)] - x^{(1)}(0^-)$$

$$= s^2X(s) - sx(0^-) - x^{(1)}(0^-) \qquad (4\text{-}13)$$

Using induction, we may show that

$$\mathcal{L}\left[\frac{d^nx(t)}{dt^n}\right] = s^nX(s) - s^{n-1}x(0^-) - \cdots - x^{(n-1)}(0^-)$$

$$= s^n X(s) - \sum_{i=0}^{n-1} s^{n-1-i} X^{(i)}(0)$$

where $x^{(n)}(0^-)$ denotes the nth derivative of $x(t)$ evaluated at $t = 0^-$. The use of (4-12) will be illustrated with an example.

EXAMPLE 4-2

Consider the circuit shown in Figure 4-2, where the switch is switched from ① to ② at $t = 0$. Find the current through the inductor as a function of time.

Solution: The inductor current obeys the differential equation

$$\frac{di(t)}{dt} + 2i(t) = \begin{cases} 4, & t \le 0 \\ 0, & t > 0 \end{cases}$$

Taking the Laplace transform of both sides starting at $t = 0^-$, we obtain

$$sI(s) - i(0^-) + 2I(s) = 0$$

To proceed further, we require $i(0^-)$. Assuming that the circuit was in steady state for $t < 0$, we see from the circuit diagram that

$$i(0^-) = \frac{4}{2} = 2$$

Thus

$$I(s)(s + 2) - 2 = 0$$

or

$$I(s) = \frac{2}{s + 2}$$

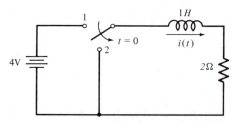

FIGURE 4-2. Circuit for the illustration of Laplace transform solution techniques.

Using (4-5) the current through the inductor is

$$i(t) = \begin{cases} 2e^{-2t}; & t > 0 \\ 2, & t \leq 0 \end{cases}$$

■

Theorem 3: Laplace Transform of an Integral   The Laplace transform of

$$y(t) = \int_{-\infty}^{t} x(\lambda)\, d\lambda \tag{4-14}$$

is

$$\mathcal{L}\left[\int_{-\infty}^{t} x(\lambda)\, d\lambda\right] = \frac{X(s)}{s} + \frac{x^{(-1)}(0^-)}{s} \tag{4-15}$$

where $x^{(-1)}(0^-)$ is shorthand notation for

$$\int_{-\infty}^{t} x(\lambda)\, d\lambda \Big|_{t=0^-}$$

This theorem is proved by using the formula for integration by parts. With

$$u = \int_{-\infty}^{t} x(\lambda)\, d\lambda, \qquad du = x(t)\, dt \tag{4-16a}$$

and

$$dv = e^{-st}\, dt, \qquad v = \frac{-e^{-st}}{s} \tag{4-16b}$$

the Laplace transform of (4-14) beomes

$$\mathcal{L}\left[\int_{-\infty}^{t} x(\lambda)\, d\lambda\right] = -\frac{e^{-st}}{s} \int_{-\infty}^{t} x(\lambda)\, d\lambda \Big|_{0^-}^{\infty}$$

$$+ \frac{1}{s} \int_{0^-}^{\infty} x(t)e^{-st}\, dt \tag{4-17}$$

which gives (4-15) when the limits are substituted provided that

$$\lim_{t \to \infty} e^{-st} \int_{-\infty}^{t} x(\lambda)\, d\lambda = 0$$

EXAMPLE 4-3

The use of the integration theorem will be illustrated by finding an expression for the Laplace transform of the current in the circuit of Figure 4-3. Kirchhoff's voltage law results in the loop equation

$$L\frac{di(t)}{dt} + Ri(t) + \frac{1}{C}\int_{-\infty}^{t} i(\lambda)\, d\lambda = x(t) \tag{4-18}$$

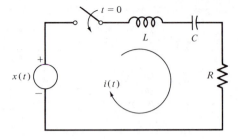

FIGURE 4-3. Circuit for illustration of the integration theorem.

where the voltage–current relationships for each element have been substituted. Application of the linearity, differentiation, and integration theorems yields

$$LsI(s) + RI(s) + \frac{I(s)}{sC} + \frac{v_c(0^-)}{s} = X(s) \tag{4-19}$$

where $i(0^-) = 0$ because the switch is open prior to $t = 0$ and

$$\frac{i^{(-1)}(0^-)}{C} = \frac{1}{C} \int_{-\infty}^{0^-} i(\lambda)\, d\lambda = \frac{q_c(0^-)}{C} = v_c(0^-)$$

is the voltage across the capacitor at $t = 0^-$. Solving for $I(s)$, we obtain

$$I(s) = \frac{sX(s) - v_c(0^-)}{L[s^2 + (R/L)s + 1/LC]} \tag{4-20}$$

The inversion of $I(s)$ in response to particular inputs will be postponed until the next section. ∎

Theorem 4: Complex Frequency Shift (S-Shift) Theorem
The Laplace transform of

$$y(t) = x(t)e^{-\alpha t} \tag{4-21}$$

is

$$Y(s) = X(s + \alpha) \tag{4-22}$$

where $X(s) = \mathcal{L}[x(t)]$.

The theorem is proved by substituting (4-21) into the definition of the Laplace transform and noting that the integral is the Laplace transform of $x(t)$ with the variable $s + \alpha$. The proof is left to the problems at the end of the chapter.

The s-shift theorem is useful for extending our table of Laplace transforms. For example, application of the s-shift theorem to (4-8) shows that

$$\mathcal{L}[e^{-\alpha t} \cos \omega_0 t] = \frac{s + \alpha}{(s + \alpha)^2 + \omega_0^2} \tag{4-23}$$

while application to (4-9) results in the Laplace transform pair

$$\mathcal{L}[e^{-\alpha t} \sin \omega_0 t] = \frac{\omega_0}{(s + \alpha)^2 + \omega_0^2} \tag{4-24}$$

EXAMPLE 4-4

Consider the function of s,

$$X(s) = \frac{s + 8}{s^2 + 6s + 13} \tag{4-25}$$

To find the corresponding function of time, we write it as

$$X(s) = \frac{s + 8}{(s + 3)^2 + 4}$$

$$= \frac{s + 3}{(s + 3)^2 + 2^2} + \frac{5}{(s + 3)^2 + 2^2}$$

Applying the transform pairs (4-23) and (4-24), we obtain

$$x(t) = e^{-3t}(\cos 2t + \tfrac{5}{2} \sin 2t), \qquad t > 0 \qquad \blacksquare$$

Theorem 5: Delay Theorem If the Laplace transform of $x(t)u(t)$ is $X(s)$, then

$$\mathcal{L}[x(t - t_0)u(t - t_0)] = e^{-st_0}X(s), \qquad t_0 > 0 \tag{4-26}$$

The proof follows easily by using the definition of the Laplace transform (4-2) with $x(t - t_0)u(t - t_0)$ substituted. Since $u(t - t_0) = 0$ for $t < t_0$, we obtain

$$\mathcal{L}[x(t - t_0)u(t - t_0)] = \int_{t_0}^{\infty} x(t - t_0)e^{-st}\, dt \tag{4-27}$$

Letting $t' = t - t_0$ in the integrand, this becomes

$$\mathcal{L}[x(t - t_0)u(t - t_0)] = \int_0^{\infty} x(t')e^{-s(t' + t_0)}\, dt' = X(s)e^{-st_0}$$

which proves the theorem.

We note that the step function $u(t - t_0)$ is necessary in (4-26) to give the proper lower limit on the Laplace transform. Equation (4-26) does not hold if $t_0 < 0$ since the single-sided Laplace transform will not include the portion of $x(t - t_0)u(t - t_0)$, that exists for $t < 0$.

EXAMPLE 4-5

The Laplace transform of a square wave beginning at $t = 0$ is

$$\mathcal{L}\left[u(t) - 2u\left(t - \frac{T_0}{2}\right) + 2u(t - T_0) - \cdots \right]$$

$$= \frac{1}{s}(1 - 2e^{-sT_0/2} + 2e^{-sT_0} - \cdots)$$

Using the series

$$\frac{1}{1 + x} = 1 - x + x^2 - x^3 + \cdots, \qquad |x| < 1,$$

we obtain

$$\mathcal{L}[x_{sq}(t)] = \frac{1}{s}\left(\frac{2}{1 + e^{-sT_0/2}} - 1\right)$$

$$= \frac{1}{s}\frac{1 - e^{-sT_0/2}}{1 + e^{-sT_0/2}}$$

where $x_{sq}(t)$ is a square wave with amplitudes ± 1 and period T_0. ∎

Theorem 6: Laplace Transform of the Convolution of Two Signals Given two signals, $x_1(t)$ and $x_2(t)$, which are zero for $t < 0$, their convolution is

$$y(t) \triangleq x_1(t) * x_2(t) = \int_0^t x_1(\lambda)x_2(t - \lambda)\,d\lambda \tag{4-28}$$

$$= \int_0^\infty x_1(\lambda)x_2(t - \lambda)\,d\lambda$$

The last equation follows by virtue of $x_2(t) = 0$, $t < 0$ or $x_2(t - \lambda) = 0$, $\lambda > t$. By definition of the single-sided Laplace transform, we have

$$Y(s) = \mathcal{L}[y(t)] = \int_0^\infty \left[\int_0^\infty x_1(\lambda)x_2(t - \lambda)\,d\lambda\right]e^{-st}\,dt$$

Integrating over t first and λ last, this becomes

$$Y(s) = \int_0^\infty x_1(\lambda)\left[\int_0^\infty x_2(\eta)e^{s\eta}\,d\eta\right]e^{-s\lambda}\,d\lambda$$

$$= X_1(s)X_2(s) \tag{4-29}$$

where the change of variable $\eta = t - \lambda$ has been made. We recognize the inner integral as the Laplace transform of $x_2(t)$ and the outer integral as the Laplace transform of $x_1(t)$. Thus, the Laplace transform of the convolution of two signals is the product of their respective Laplace transforms, and the region of absolute convergence consists of the intersection of the regions for $X_1(s)$ and $X_2(s)$.

Theorem 7: Laplace Transform of a Product The Laplace transform of a product of two signals can be expressed as an integral in the complex plane. We will not use it in this chapter, but it is given in the table of Laplace transform theorems at the end of this section for reference.

Theorem 8: Initial Value Theorem The Laplace transform of the derivative of a signal has been shown to be

$$\mathcal{L}\left[\frac{dx}{dt}\right] \triangleq \int_0^\infty \frac{dx}{dt}e^{-st}\,dt = sX(s) - x(0^-)$$

If $x(t)$ is continuous at $t = 0$, dx/dt does not contain an impulse at $t = 0$.

Therefore, as $s \to \infty$ with σ greater than the abscissa of convergence, the integral vanishes, giving

$$\lim_{s \to \infty} sX(s) = x(0^-) = x(0^+)$$

If $x(t)$ is discontinuous at $t = 0$, then dx/dt contains an impulse $[x(0^+) - x(0^-)]\delta(t)$, and

$$\lim_{s \to \infty} \int_{0^-}^{\infty} \frac{dx}{dt} e^{-st}\, dt = x(0^+) - x(0^-)$$

Thus

$$x(0^+) - x(0^-) = \lim_{s \to \infty} sX(s) - x(0^-)$$

or

$$\lim_{s \to \infty} sX(s) = x(0^+) \qquad\qquad (4\text{-}30)$$

provided that $x(0^+)$ exists. If the lower limit on the Laplace transform had been taken as 0^+, the impulse would not have been included and we would have again obtained (4-30). Thus $\lim_{s \to 0} sX(s)$, provided that it exists, always gives the initial value of $x(t)$ as $t \to 0^+$ regardless of the lower limit used on the Laplace transform integral.

EXAMPLE 4-6

The initial value of $\exp(-\alpha t) \cos \omega_0 t\, u(t)$ is given by

$$\lim_{s \to \infty} s \frac{s + \alpha}{(s + \alpha)^2 + \omega_0^2} = 1$$

where the transform pair given by (4-23) has been used.
    The initial value of $\exp(-\alpha t) \sin \omega_0 t\, u(t)$ is given by

$$\lim_{s \to \infty} s \frac{\omega_0}{(s + \alpha)^2 + \omega_0^2} = 0$$

where the transform pair (4-24) has been used.
    The initial value of the time function whose Laplace transform is $s/(s + 10)$ cannot be found since $\lim_{s \to \infty} s^2/(s + 10)$ does not exist. The reason is that

$$\frac{s}{s + 10} = 1 - \frac{10}{s + 10}$$

has the inverse Laplace transform $\delta(t) - 10e^{-10t}u(t)$, which has an impulse at $t = 0$. ∎

Theorem 9: Final Value Theorem   If $x(t)$ and $dx(t)/dt$ are Laplace transformable, then

$$\lim_{t \to \infty} x(t) = \lim_{s \to 0} sX(s)$$

provided that $\lim_{t\to\infty} x(t)$ exists, which is the case if $sX(s)$ has no poles on the $j\omega$-axis or in the right-half plane. The proof of this theorem is left to the problems.

EXAMPLE 4-7

The final value of a unit step is given by

$$\lim_{s\to 0} s \frac{1}{s} = 1$$

The final value of $\exp[(-\alpha t)u(t)]$ is obtained from

$$\lim_{s\to 0} s \left(\frac{1}{s + \alpha} \right) = 0$$

∎

Theorem 10: The Scaling Theorem The Laplace transform of $x(at)$, where a is a positive constant, is $a^{-1}X(s/a)$. This is the generalization of the scale change theorem of Fourier transforms to the complex s-domain. Note that since we are dealing with the single-sided Laplace transform, a must be positive. If $a = -1$, for example, a time function to be Laplace transformed would be reversed and exist for t negative. The resulting negative-time portion would then not be included in the Laplace transform integral.

The theorems and transform pairs that have been developed so far are collected for easy reference in Tables 4-2 and 4-3, respectively.

4-4

INVERSION OF RATIONAL FUNCTIONS

The examples of Section 4.3 illustrate that functions involving the ratio of two polynomials in s, called rational functions of s, are commonly occurring Laplace transforms. Indeed, this will be the form of the Laplace transform obtained when considering any fixed, linear, lumped system with a forcing function that is a power of t, exponential in t, sinusoidal, or a combination of these.

The techniques that we shall develop for inversion of rational functions of s are valid only for *proper rational functions*, that is, functions for which the numerator polynomial is of degree less than the degree of the denominator polynomial. For cases where this is not true, it is very simple to obtain a proper rational function through use of long division. For example, consider the nonproper rational function of s

$$Z(s) = \frac{s + 2}{s + 1}$$

This can be written as

$$Z(s) = 1 + \frac{1}{s + 1} \tag{4-31}$$

TABLE 4-2 Laplace Transform Theorems

Name	Operation in Time Domain	Operation in Frequency Domain
1. Linearity	$a_1 x_1(t) + a_2 x_2(t)$	$a_1 X_1(s) + a_2 X_2(s)$
2. Differentiation	$\dfrac{d^n x(t)}{dt^n}$	$s^n X(s) - s^{n-1}x(0^-) - \cdots - x^{(n-1)}(0^-)$
3. Integration	$\displaystyle\int_{-\infty}^{t} x(\lambda)\, d\lambda$	$\dfrac{X(s)}{s} + \dfrac{x^{(-1)}(0^-)}{s}$
4. s-shift	$x(t)\exp(-\alpha t)$	$X(s + \alpha)$
5. Delay	$x(t - t_0)u(t - t_0)$	$X(s)\exp(-s t_0)$
6. Convolution	$x_1(t) * x_2(t)$	$X_1(s)X_2(s)$
	$\displaystyle = \int_0^{\infty} x_1(\lambda)x_2(t - \lambda)\, d\lambda$	
7. Product	$x_1(t)x_2(t)$	$\dfrac{1}{2\pi j}\displaystyle\int_{c-j\infty}^{c+j\infty} X_1(s - \lambda)X_2(\lambda)\, d\lambda$
8. Initial value (provided limits exist)	$\displaystyle\lim_{t\to 0^+} x(t)$	$\displaystyle\lim_{s\to\infty} sX(s)$
9. Final value (provided limits exist)	$\displaystyle\lim_{t\to\infty} x(t)$	$\displaystyle\lim_{s\to 0} sX(s)$
10. Time scaling	$x(at),\quad a > 0$	$a^{-1}X\left(\dfrac{s}{a}\right)$

Handwritten annotation in row 2: $= s^n X(s) - \sum_{i=0}^{n-1} s^{n-1-i}\, x^{(i)}(0)$

by use of long division. We may easily apply the Laplace transform pairs of Table 4-3 to obtain its inverse Laplace transform. The important point is that the second term is a proper rational function.

To illustrate techniques for the inverse Laplace transformation of proper rational functions, we consider (4-20) for several special cases. To simplify the notation, we consider the voltage across the resistor in Figure 4-2 as the system output $y(t)$, and denote its Laplace transform as $Y(s)$. Also, let

$$2\zeta\omega_n = \frac{R}{L} \tag{4-32}$$

and

$$\omega_n^2 = \frac{1}{LC} \tag{4-33}$$

Thus (4-20) becomes

$$Y(s) = \frac{2\zeta\omega_n [sX(s) - v_c(0^-)]}{s^2 + 2\zeta\omega_n s + \omega_n^2} \tag{4-34}$$

TABLE 4-3   Extended Table of Laplace Transforms

Signal	Laplace Transform	Comments on Derivation
1. $\delta^{(n)}(t)$	s^n	Direct evaluation with aid of (1-25)
2. 1 or $u(t)$	$\dfrac{1}{s}$	Direct evaluation
3. $\dfrac{t^n \exp(-\alpha t)u(t)}{n!}$	$\dfrac{1}{(s+\alpha)^{n+1}}$	Differentiation applied to pair 3, Table 4-1
4. $\cos \omega_0 t\, u(t)$	$\dfrac{s}{s^2+\omega_0^2}$	Example 4-1
5. $\sin \omega_0 t\, u(t)$	$\dfrac{\omega_0}{s^2+\omega_0^2}$	Example 4-1
6. $\exp(-\alpha t)\cos \omega_0 t\, u(t)$	$\dfrac{s+\alpha}{(s+\alpha)^2+\omega_0^2}$	s-shift and pair 4
7. $\exp(-\alpha t)\sin \omega_0 t\, u(t)$	$\dfrac{\omega_0}{(s+\alpha)^2+\omega_0^2}$	s-shift and pair 5
8. Square wave: $u(t) - 2u\left(t - \dfrac{T_0}{2}\right) + 2u(t - T_0) - \cdots$	$\dfrac{1}{s}\dfrac{1-e^{-sT_0/2}}{1+e^{-sT_0/2}}$	Example 4-5
9. $(\sin \omega_0 t - \omega_0 t \cos \omega_0 t)u(t)$	$\dfrac{2\omega_0^3}{(s^2+\omega_0^2)^2}$	Example 4-12, pair 5, and convolution
10. $(\omega_0 t \sin \omega_0 t)u(t)$	$\dfrac{2\omega_0^2 s}{(s^2+\omega_0^2)^2}$	Pair 4 and convolution
11. $\omega_0 t \exp(-\alpha t)\sin \omega_0 t\, u(t)$	$\dfrac{2\omega_0^2(s+\alpha)}{[(s+\alpha)^2+\omega_0^2]^2}$	s-shift and pair 10
12. $\exp(-\alpha t)(\sin \omega_0 t - \omega_0 t \cos \omega_0 t)u(t)$	$\dfrac{2\omega_0^3}{[(s+\alpha)^2+\omega_0^2]^2}$	s-shift and pair 9

We refer to ζ and ω_n as the damping ratio and natural frequency, respectively. The roots of the denominator of (4-34) are given by

$$s_{1,2} = [-\zeta \pm \sqrt{\zeta^2 - 1}]\,\omega_n \qquad (4\text{-}35)$$

Thus, if $\zeta > 1$, the roots are real and distinct; if $\zeta < 1$, the roots are complex conjugates; and if $\zeta = 1$, they are real and equal.

EXAMPLE 4-8

Let the input to the circuit shown in Figure 4-3 be a unit step with Laplace transform $X(s) = 1/s$. Also, assume that $v_c(0^-) = 0$, $\omega_n^2 = 16$, and $2\zeta\omega_n = 10$. Thus (4-34) becomes

$$Y(s) = \frac{10}{s^2 + 10s + 16}$$

$$= \frac{10}{(s + 2)(s + 8)} \tag{4-36}$$

Checking Table 4-3, we see that the required form for (4-36) does not appear. However, (4-36) may be expanded in partial fractions as

$$\frac{10}{(s + 2)(s + 8)} = \frac{A}{s + 2} + \frac{B}{s + 8} \tag{4-37}$$

We have three routes we can follow in determining the unknown coefficients, A and B.

1. *Common Denominator*. Placing each factor on the right-hand side of (4-37) over a common demonimator results in

$$10 = (s + 8)A + (s + 2)B$$

or

$$10 = (A + B)s + (8A + 2B)$$

Setting coefficients of like powers of s equal on either side of this equation, we obtain the simultaneous equations

$$A + B = 0$$

$$8A + 2B = 10$$

The first of these equations gives $A = -B$, which, when substituted into the second, yields

$$-8B + 2B = 10$$

or $B = -\frac{5}{3}$, and $A = \frac{5}{3}$. Thus

$$Y(s) = \frac{5}{3}\left(\frac{1}{s + 2} - \frac{1}{s + 8}\right) \tag{4-38}$$

Using the linearity theorem and transform pair 3, we find $y(t)$ to be

$$y(t) = \frac{5}{3}(e^{-2t} - e^{-8t})\,u(t) \tag{4-39}$$

2. *Substituting Specific Values of s*. Since (4-37) is an identity for any s, we may obtain two simultaneous equations for A and B by substituting two convenient values of s (neither of which are equal to either root of the denominator). For example, substituting $s = 0$ and $s = 2$ in (4-37), we obtain the two equations

$$\frac{10}{(2)(8)} = \frac{A}{2} + \frac{B}{8}$$

and

$$\frac{10}{(4)(10)} = \frac{A}{4} + \frac{B}{10}$$

or

$$4A + B = 5$$

$$5A + 2B = 5$$

which give the same values of A and B as before.

3. *Heaviside's Expansion Theorem.* The coefficients in (4-37) may be found by yet another technique, known as Heaviside's expansion theorem.† This procedure can be justified, in the case of (4-37), by noting that multiplication of both sides by $s + 2$ results in the equation

$$\frac{10}{s + 8} = A + B\frac{s + 2}{s + 8} \qquad (4\text{-}40)$$

Since (4-40) holds for all values of s, we can obtain an equation for A by letting $s = -2$, which eliminates the second term on the right-hand side. The result is

$$A = \tfrac{10}{6} = \tfrac{5}{3}$$

as before. Similarly, B can be found by multiplying both sides of (4-37) by $s + 8$ and setting $s = -8$. This results in

$$\frac{10}{s + 2}\bigg|_{s=-8} = A\frac{s + 8}{s + 2}\bigg|_{s=-8} + B$$

or $B = 10/(-6) = -\tfrac{5}{3}$, as before. ∎

We have just seen how to expand a rational function of s, which contains only *simple factors* in the denominator, in partial fractions. That is, the procedures used in Example 4-8 can be used as long as the factors in the denominator are different or are not raised to a power. Furthermore, the degree of the numerator must be less than the degree of the denominator; that is, it must be a proper rational function of s. If not, long division is used to obtain a proper rational function.

We now wish to consider examples involving more complicated factors than simple ones in the denominator.

EXAMPLE 4-9

As our next example, suppose that the input to the circuit shown in Figure 4-3 is $x(t) = (-0.5 \cos t + 2.5 \sin t)u(t)$, with Laplace transform

$$X(s) = -\frac{0.5s}{s^2 + 1} + \frac{2.5}{s^2 + 1} = \frac{-0.5s + 2.5}{s^2 + 1}$$

†Named after Oliver Heaviside (1850–1925), an English engineer, who originated *operational calculus.*

and that $v_c(0^-) = -2$ v. All other parameters are assumed to be the same as in Example 4-7. Then the Laplace transform of the output is

$$Y(s) = \frac{10[(-0.5s^2 + 2.5s)/(s^2 + 1) + 2]}{s^2 + 10s + 16}$$

$$= \frac{15s^2 + 25s + 20}{(s^2 + 1)(s + 2)(s + 8)} \tag{4-41}$$

Again, we have several approaches at our disposal for obtaining the partial-fraction expansion of (4-41). For example, the factor $s^2 + 1$ in the denominator can be factored as $(s + j)(s - j)$ and Heaviside's expansion theorem used as in part (3) of Example 4-8. Using such a procedure, we would represent (4-41) as

$$Y(s) = \frac{A_1}{s + j} + \frac{A_2}{s - j} + \frac{A_3}{s + 2} + \frac{A_4}{s + 8} \tag{4-42}$$

Using Heaviside's expansion formula, A_3 and A_4 are found to be

$$A_3 = (s + 2)Y(s)\Big|_{s=-2}$$

$$= \frac{15s^2 + 25s + 20}{(s^2 + 1)(s + 8)}\Big|_{s=-2} = 1 \tag{4-43}$$

and

$$A_4 = (s + 8)Y(s)\Big|_{s=-8}$$

$$= \frac{15s^2 + 25s + 20}{(s^2 + 1)(s + 2)}\Big|_{s=-8} = -2 \tag{4-44}$$

respectively. A_1 and A_2 could be found similarly. However, by examining the first two terms in (4-42) further, we can simplify things to some extent. Putting the first two terms over a common denominator, we obtain

$$\frac{A_1}{s + j} + \frac{A_2}{s - j} = \frac{(A_1 + A_2)s + j(A_2 - A_1)}{s^2 + 1} \tag{4-45}$$

The numerator of (4-45) must be a real function of s. The only way that this can be true is for $A_1 = A_2^*$, so that $A_1 + A_2^* = 2\,\text{Re}\,(A_1) \triangleq B_1$ and $j(A_1 - A_2) = -2\,\text{Im}\,(A_1) \triangleq B_2$. Thus (4-42) can be written as

$$Y(s) = \frac{B_1 s + B_2}{s^2 + 1} + \frac{1}{s + 2} - \frac{2}{s + 8} \tag{4-46}$$

The unknown constants B_1 and B_2 in the first term on the right-hand side can be found in two ways.

1. Placing the right-hand side of (4-46) over a common denominator and setting the result equal to (4-41) results in the identity

$$(B_1s + B_2)(s + 2)(s + 8) + (s^2 + 1)(s + 8) - 2(s^2 + 1)(s + 2)$$

$$= 15s^2 + 25s + 20$$

Multiplying factors, collecting like powers in s, and setting coefficients of like powers of s equal, we obtain

$$B_1 + 1 - 2 = 0 \quad \text{or} \quad B_1 \qquad\qquad = 1$$

$$10B_1 + B_2 + 8 - 4 = 15 \quad \text{or} \quad 10B_1 + B_2 \quad = 11$$

$$16B_1 + 10B_2 + 1 - 2 = 25 \quad \text{or} \quad 16B_1 + 10B_2 = 26$$

$$16B_2 + 8 - 4 = 20 \quad \text{or} \quad B_2 \qquad\qquad = 1$$

We note that although there are only two unknowns and four equations, the four equations are consistent; the second and third provide checks on the first and the fourth. Substituting these results in (4-46), $Y(s)$ becomes

$$Y(s) = \frac{s + 1}{s^2 + 1} + \frac{1}{s + 2} - \frac{2}{s + 8} \qquad (4\text{-}47)$$

Using Table 4-3, the output time function is

$$y(t) = [\cos t + \sin t + \exp(-2t) - 2\exp(-8t)]u(t) \qquad (4\text{-}48)$$

The same result would have been obtained by finding A_1 and A_2 in (4-42) using Heaviside's expansion theorem. The student should show this.

2. Another approach that can be used to find the first term in (4-46) is to equate (4-46) and (4-41). Solving for $(B_1s + B_2)/(s^2 + 1)$ results in

$$\frac{B_1s + B_2}{s^2 + 1} = \frac{15s^2 + 25s + 20}{(s^2 + 1)(s + 2)(s + 8)} - \frac{1}{s + 2} + \frac{2}{s + 8}$$

$$= \frac{15s^2 + 25s + 20 - (s^2 + 1)(s + 8) + 2(s^2 + 1)(s + 2)}{(s^2 + 1)(s + 2)(s + 8)}$$

$$= \frac{s^3 + 11s^2 + 26s + 16}{(s^2 + 1)(s + 2)(s + 8)} \qquad (4\text{-}49)$$

This looks worse than our initial expression, but if (4-46) holds, and we know that it does, then $(s + 2)(s + 8) = s^2 + 10s + 16$ must be a factor of the numerator. Carrying out the long division on (4-49), we find this to be the case, and

$$\frac{B_1s + B_2}{s^2 + 1} = \frac{s + 1}{s^2 + 1}$$ ∎

as before.

EXAMPLE 4-10

In this example we consider the case of repeated linear factors. Let the forcing function for the circuit shown in Figure 4-3 be $x(t) = \exp(-2t)u(t)$, for which the Laplace transform is

$$X(s) = \frac{1}{s + 2}$$

Assume that $v_c(0^-) = 0$ and that the other parameters are the same as in Example 4-8. Thus

$$Y(s) = \frac{10s}{(s + 2)^2(s + 8)} \tag{4-50}$$

The partial-fraction expression of this rational function must be of the form

$$Y(s) = \frac{A_1}{s + 8} + \frac{A_2}{s + 2} + \frac{A_3}{(s + 2)^2} \tag{4-51}$$

where A_1 and A_3 can be found using the Heaviside technique. We obtain

$$A_1 = (s + 8)Y(s)\Big|_{s=-8} = -\tfrac{20}{9}$$

and

$$A_3 = (s + 2)^2 Y(s)\Big|_{s=-2} = -\tfrac{10}{3}$$

The same approach will not work for A_2 because multiplication by $(s + 2)$ removes only one of the $(s + 2)$ factors in the denominator of (4-50). Substitution of $s = -2$ then gives an undefined result for A_2. We could use the procedure employed in the preceding example and subtract $A_1/(s + 8)$ and $A_3/(s + 2)^2$ from $Y(s)$. However, it is easier to note that

$$\frac{d}{ds}[(s + 2)^2 Y(s)]\Big|_{s=-2}$$

$$= \frac{d}{ds}\frac{A_1(s + 2)^2}{s + 8}\Big|_{s=-2} + \frac{d}{ds}(s + 2)A_2\Big|_{s=-2} = A_2$$

or

$$A_2 = \frac{d}{ds}\left(\frac{10s}{s + 8}\right)\Big|_{s=-2} = \frac{10(s + 8) - 10s}{(s + 8)^2}\Big|_{s=-2} = \frac{20}{9}$$

Thus

$$Y(s) = \frac{20}{9}\left[-\frac{1}{s + 8} + \frac{1}{s + 2} - \frac{\tfrac{3}{2}}{(s + 2)^2}\right] \tag{4-52}$$

Using Table 4-3, we find the inverse Laplace transform to be

$$y(t) = \tfrac{20}{9} \left[-\exp(-8t) + \exp(-2t) - \tfrac{3}{2}t \exp(-2t) \right] u(t) \quad (4\text{-}53)$$

■

EXAMPLE 4-11

We consider next the generalization of the method for finding the partial-fraction expansion for repeated linear factors developed in Example 4-10. Suppose that the forcing function for the circuit of Figure 4-3 is now $x(t) = t \exp(-2t)u(t)$, with all other circuit parameters the same as in Example 4-10. Thus

$$X(s) = \frac{1}{(s+2)^2}$$

and

$$Y(s) = \frac{10s}{(s+2)^3(s+8)}$$

The partial-fraction expansion of $Y(s)$ is of the form

$$Y(s) = \frac{A_1}{s+8} + \frac{A_2}{s+2} + \frac{A_3}{(s+2)^2} + \frac{A_4}{(s+2)^3} \quad (4\text{-}54)$$

Using Heaviside's expansion formula, we obtain

$$A_1 = (s+8)Y(s) \Big|_{s=-8} = \tfrac{10}{27}$$

$$A_4 = (s+2)^3 Y(s) \Big|_{s=-2} = -\tfrac{10}{3}$$

Using the differentiation technique of Example 4-10, we find that

$$A_3 = \frac{d}{ds}[(s+2)^3 Y(s)] \Big|_{s=-2}$$

$$= \frac{d}{ds}\left(\frac{10s}{s+8}\right)_{s=-2}$$

$$= \frac{80}{(s+8)^2} \Big|_{s=-2}$$

$$= \tfrac{20}{9}$$

To find A_2, we note that multiplication of both sides of (4-54) by $(s+2)^3$ gives

$$(s+2)^3 Y(s) = \frac{(s+2)^3 A_1}{s+8} + (s+2)^2 A_2 + (s+2)A_3 + A_4$$

Differentiation twice with respect to s gives

$$\frac{d^2}{ds^2}[(s + 2)^3 Y(s)] = \frac{d^2}{ds^2}\left[\frac{(s + 2)^3}{s + 8}A_1\right] + 2A_2 + 0 + 0$$

Setting $s = -2$, we obtain

$$2A_2 = \frac{d^2}{ds^2}\left[\frac{10s}{s + 8}\right]_{s = -2}$$

$$= \frac{d}{ds}\left[\frac{80}{(s + 8)^2}\right]_{s = -2}$$

$$= -\frac{160}{(s + 8)^3}\Big|_{s = -2}$$

$$= -\tfrac{20}{27}$$

Thus $A_2 = -\tfrac{10}{27}$ and

$$Y(s) = \frac{10}{27}\left(\frac{1}{s + 8} - \frac{1}{s + 2}\right) + \frac{20}{9}\frac{1}{(s + 2)^2} - \frac{10}{3}\frac{1}{(s + 2)^3} \quad (4\text{-}55)$$

Using Table 4-3, we find $y(t)$ to be

$$y(t) = [\tfrac{10}{27}(e^{-8t} - e^{-2t}) - \tfrac{5}{3}t(t - \tfrac{4}{3})e^{-2t}]u(t) \quad (4\text{-}56)$$

■

Based on the preceding two examples, we may deduce a general result for expanding rational functions involving repeated linear factors. If the rational function is of the form

$$Y(s) = \frac{P(s)}{(s + \alpha)^n Q(s)} \quad (4\text{-}57)$$

then its partial-fraction expansion is of the form

$$Y(s) = \frac{A_1}{(s + \alpha)} + \frac{A_2}{(s + \alpha)^2} + \cdots + \frac{A_n}{(s + \alpha)^n} + \frac{R(s)}{Q(s)} \quad (4\text{-}58)$$

where $P(s)$, $Q(s)$, and $R(s)$ are polynomials in s and the factors of $P(s)$ and $Q(s)$ are distinct from $s + \alpha$. The mth coefficient, A_m, is given by

$$A_m = \frac{1}{(n - m)!}\frac{d^{(n-m)}}{ds^{(n-m)}}[(s + \alpha)^n Y(s)]_{s = -\alpha} \quad (4\text{-}59)$$

To continue with the use of partial-fraction expansions in obtaining inverse Laplace transforms, we consider two examples which are really special cases of previous examples, but which are conveniently carried out in special ways.

EXAMPLE 4-12

Complex-conjugate roots. Consider the rational function of s

$$Y(s) = \frac{2s^2 + 6s + 6}{(s + 2)(s^2 + 2s + 2)}$$

(4-60)

$$= \frac{2s^2 + 6s + 6}{(s + 2)[(s + 1)^2 + 1]}$$

The quadratic factor in the denominator can be factored as

$$(s + 1)^2 + 1 = (s + 1 + j)(s + 1 - j)$$

(4-61)

and Heaviside's expansion theorem used as in Example 4-8 since all factors are distinct. However, it is easier to keep both of the complex-conjugate factors expressed by (4-61) together and expand (4-60) as

$$Y(s) = \frac{A}{s + 2} + \frac{Bs + C}{s^2 + 2s + 2}$$

(4-62)

This allows the inverse Laplace transform to be found easily with the help of pairs 3, 6, and 7 in Table 4-3.

As before, we may calculate A using Heaviside's theorem:

$$A = (s + 2)Y(s) \Big|_{s = -2}$$

$$= \frac{2s^2 + 6s + 6}{s^2 + 2s + 2} \Big|_{s = -2} = 1$$

(4-63)

The coefficients B and C may be obtained by substituting specific values of s in (4-62), or by placing both terms on the right-hand side over a common denominator and equating coefficients of like powers of s in the numerator. Or, one could subtract $1/(s + 2)$ from $Y(s)$ to produce the second term on the right-hand side of (4-62). We will use the first technique. First, letting $s = 0$ in (4-62), we obtain

$$Y(0) = \frac{A}{2} + \frac{C}{2}$$

or

$$C = 2Y(0) - A$$

$$= \frac{2(6)}{(4)} - 1$$

$$= 2$$

To find B, we multiply both sides of (4-62) by s and let $s \to \infty$. This gives

$$\lim_{s \to \infty} sY(s) = A + B$$

or

$$B = \lim_{s \to \infty} sY(s) - A$$

$$= 2 - 1 = 1$$

Thus (4-62) becomes

$$Y(s) = \frac{1}{s + 2} + \frac{s + 2}{s^2 + 2s + 2} \tag{4-64}$$

$$= \frac{1}{s + 2} + \frac{s + 1}{(s + 1)^2 + 1} + \frac{1}{(s + 1)^2 + 1}$$

Application of transform pairs 3, 6, and 7 in Table 4-3 then yields

$$y(t) = [e^{-2t} + e^{-t}(\cos t + \sin t)]u(t) \tag{4-65}$$

■

EXAMPLE 4-13 Repeated Quadratic Factors

Consider the rational function

$$Y(s) = \frac{s^4 + 5s^3 + 12s^2 + 7s + 15}{(s + 2)(s^2 + 1)^2} \tag{4-66}$$

The second factor in the denominator could be expanded as $(s + j)^2 (s - j)^2$ and (4-59) used to treat the repeated roots. However, it is more convenient to keep the complex-conjugate factors in the denominator together and expand (4-66) as

$$Y(s) = \frac{A_1}{s + 2} + \frac{B_1 s + C_1}{s^2 + 1} + \frac{B_2 s + C_2}{(s^2 + 1)^2} \tag{4-67}$$

Using Heaviside's technique, we find A_1 to be

$$A_1 = (s + 2)Y(s) \big|_{s = -2} = 1$$

We can obtain the remaining coefficients by putting the right-hand side of (4-67) over a common denominator and equating the resulting numerator to the numerator of (4-66). This results in the identity

$$A_1 (s^2 + 1)^2 + (B_1 s + C_1) (s + 2) (s^2 + 1) + (B_2 s + C_2)(s + 2)$$

$$= s^4 + 5s^3 + 12s^2 + 7s + 15$$

Multiplying the factors on the left-hand side together and collecting like powers of s, we obtain

$$(A_1 + B_1)s^4 + (C_1 + 2B_1)s^3$$

$$+ (2A_1 + B_1 + 2C_1 + B_2)s^2 \tag{4-68}$$

$$+ (2B_1 + C_1 + 2B_2 + C_2)s$$

$$+ (A_1 + 2C_1 + 2C_2) = s^4$$

$$+ 5s^3 + 12s^2 + 7s + 15$$

The coefficients of like powers of s on either side of (4-68) must be equal since (4-68) is an identity. This results in the equations

$$A_1 + B_1 = 1 \tag{4-69a}$$

$$C_1 + 2B_1 = 5 \tag{4-69b}$$

$$2A_1 + B_1 + 2C_1 + B_2 = 12 \tag{4-69c}$$

$$A_1 + 2C_1 + 2C_2 = 15 \tag{4-69d}$$

Since $A_1 = 1$, the first of these equations results in

$$B_1 = 1 - A_1 = 0$$

This immediately gives

$$C_1 = 5$$

from (4-69b). Thus (4.69c) and (4.69d) become

$$2 + 10 + B_2 = 12 \quad \text{or} \quad B_2 = 0$$

and

$$1 + 10 + 2C_2 = 15 \quad \text{or} \quad C_2 = 2$$

respectively. The partial-fraction expansion for $Y(s)$, as given by (4-67), is

$$Y(s) = \frac{1}{s + 2} + \frac{5}{s^2 + 1} + \frac{2}{(s^2 + 1)^2} \tag{4-70}$$

This can be inverse Laplace transformed by using the theorems and transform pairs given earlier. The first two terms are immediately inverse transformed by using pairs 3 and 5, respectively, of Table 4-3. The last term can be handled by using the convolution theorem on pair 5 of Table 4-3. Thus

$$\mathscr{L}\{[\sin \omega_0 t \, u(t)] * [\sin \omega_0 t \, u(t)]\} = \frac{\omega_0^2}{(s^2 + \omega_0^2)^2}$$

The left-hand side, using trigonometric identities, is

$$\int_0^t \sin \omega_0 \lambda \sin \omega_0 (t - \lambda) \, d\lambda$$

$$= \tfrac{1}{2} \int_0^t \cos (2\omega_0 \lambda - \omega_0 t) \, d\lambda - \tfrac{1}{2} \int_0^t \cos \omega_0 t \, d\lambda$$

$$= \frac{1}{2\omega_0} \sin \omega_0 t - \tfrac{1}{2} t \cos \omega_0 t$$

Therefore we have the transform pair

$$\sin \omega_0 t - \omega_0 t \cos \omega_0 t \leftrightarrow \frac{2\omega_0^3}{(s^2 + \omega_0^2)^2} \tag{4-71}$$

which is pair 9 of Table 4-3. The inverse Laplace transform of (4-70) can be written as

$$y(t) = (e^{-2t} + 5 \sin t + \sin t - t \cos t)u(t)$$

$$= (e^{-2t} + 6 \sin t - t \cos t)u(t) \tag{4-72}$$

∎

*4-5

THE INVERSION INTEGRAL AND ITS USE IN OBTAINING INVERSE LAPLACE TRANSFORMS

In the introductory discussion on the Laplace transform given at the beginning of the chapter, we derived the inversion integral (4-3) in a logical way by making use of the inverse Fourier transform. In that derivation σ was assumed constant. It is much more useful to allow s to be a *complex variable* where both σ and ω may vary. The inversion integral in this case takes the form

$$\frac{1}{2\pi j} \int_{c-j\infty}^{c+j\infty} X(s)e^{st}\, ds = \begin{cases} 0, & t < 0 \\ \frac{1}{2}x\,(0^+), & t = 0 \\ \frac{1}{2}[x(t^+) + x(t^-)], & t > 0 \end{cases} \tag{4-73}$$

where $c \geq 0$, $c > \sigma_c$, with σ_c an abscissa of convergence of $X(s)$.

To evaluate (4-73) we may make use of the *residue theorem* of the theory of complex variables.† The residue theorem pertains to functions of a complex variable $F(s)$ that contain a finite number of isolated singular points in the complex s-plane. Near such a singular point, say $s = s_0$, the function can be represented as a series of the form (referred to as a *Laurent series*)

$$F(s) = \sum_{n=-\infty}^{\infty} K_n(s - s_0)^n$$

The K_n's are expansion coefficients which can be found by various means. The coefficient K_{-1} is referred to as the *residue* of $F(s)$ at the singular point $s = s_0$.

With this introduction, we may now state the residue theorem.

Residue Theorem

Let C be a closed path in the complex s-plane within and on which a function $F(s)$ is analytic‡ except for the isolated singular points s_1, s_2, . . . s_n with

---

†See Appendix C for a short summary of some pertinent definitions and theorems of complex variable theory.

‡Analyticity refers to the differentiability of a function of a complex variable, but is much stronger than requiring only that the function be differentiable at a point (see Appendix C).

residues $K_1, K_2, \ldots, K_n$, respectively. Then

$$\int_C F(s)\, ds = 2\pi j(K_1 + K_2 + \cdots + K_n) \tag{4-74}$$

where $\int_C (\cdot)\, ds$ is understood to mean the line integral in a counterclockwise sense along C.

How do we make use of the residue theorem to evaluate the inverse Laplace transform of a function of a complex variable, say $X(s)$? Equation (4-73) states that we are to evaluate the integral of $X(s)e^{st}$ along a line parallel to the $j\omega$-axis from $-j\infty$ to $j\infty$. We can apply the residue theorem by choosing a semicircular path in the complex plane as shown in Figure 4-4. Letting $F(s)$ in the residue theorem (4-74) be $X(s)e^{st}$, we obtain

$$\frac{1}{2\pi j}\int_{C_1 + C_2} X(s)e^{st}\, ds = K_1 + K_2 + \cdots + K_n \tag{4-75}$$

where $K_1, K_2, \ldots, K_n$ are the residues of $X(s)e^{st}$, and $C_1 + C_2$ means the closed path consisting of the subpaths C_1 and C_2. Further, noting that the integral over the closed path $C_1 + C_2$ can be written as the sum of the integrals over the separate paths C_1 and C_2, and that the integral over C_1 is an integral between the limits $c - jR$ to $c + jR$, we obtain

$$\frac{1}{2\pi j}\int_{c - jR}^{c + jR} X(s)e^{st}\, ds + \frac{1}{2\pi j}\int_{C_2} X(s)e^{st}\, ds = \sum_{m=1}^{n} K_n \tag{4-76}$$

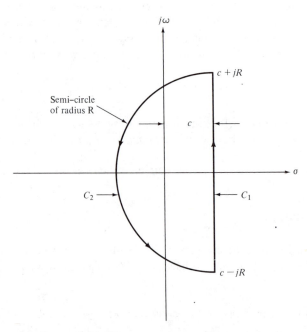

FIGURE 4-4. Closed contour in the complex plane that is appropriate for evaluating the inverse Laplace transform integral.

It can be shown that for $t > 0$ the second integral on the left-hand side contributes nothing to the result provided that $X(s) \to 0$ uniformly as $R \to \infty$. (By uniformly, it is meant that the limit is approached at the same rate for all angles of s within the range defined by C_2.) For example, if $X(s)$ is the ratio of two polynomials, it is sufficient for the degree of the denominator to exceed that of the numerator by one or more. Given this restriction on $X(s)$, it follows that the left-hand side of (4-76) yields the inverse Laplace transform, as defined by (4-73), in the limit as $R \to \infty$. That is, for $t > 0$,

$$x(t) = \sum \text{residues of } X(s)e^{st} \text{ at the finite singularities of } X(s) \quad (4\text{-}77)$$

EXAMPLE 4-14

As an example, consider the function of s given by (4-25) which was inverse transformed in Example 4-4. The function $X(s)$ has two singularities which are first-order poles, given by

$$s_{1,2} = -3 \pm j2$$

Using the Heaviside expansion technique, we write

$$X(s) \triangleq \frac{s + 8}{(s + 3)^2 + 4} = \frac{A_1}{s + 3 + j2} + \frac{A_2}{s + 3 - j2}$$

where

$$A_1 = (s + 3 + j2)X(s) \big|_{s = -3 - j2}$$

$$= \frac{5 - j2}{-j4} = \frac{1}{2} + j\frac{5}{4}$$

and

$$A_2 = (s + 3 - j2)X(s) \big|_{s = -3 + j2}$$

$$= \frac{1}{2} - j\frac{5}{4}$$

We require the residues of

$$X(s)e^{st} = \frac{s + 8}{(s + 3)^2 + 4} e^{st} = \frac{A_1 e^{st}}{s + 3 + j2} + \frac{A_2 e^{st}}{s + 3 - j2}$$

at the poles of $X(s)$. That is, we want the coefficient K_{-1} of the expansion

$$\frac{A_1 e^{st}}{s + 3 + j2} = \sum_{n = -\infty}^{\infty} K_n(s + 3 + j2)^n$$

and the coefficient K'_{-1} of the expansion

$$\frac{A_2 e^{st}}{s + 3 - j2} = \sum_{n = -\infty}^{\infty} K'_n(s + 3 - j2)^n$$

Now since $e^{st} = \sum_{n=0}^{\infty} s^n t^n / n!$, it is apparent that e^{st} has no singularities in the finite s-plane because no negative-power s-terms are present in its Laurent series expansion. It follows, therefore, that in this case we can find the coefficients K_{-1} and K'_{-1} simply by applying the Heaviside technique. That is,

$$K_{-1} = (s + 3 + j2)X(s)e^{st}\big|_{s = -3-j2}$$

$$= \frac{5 - j2}{-j4} e^{(-3-j2)t}$$

$$= \left(\frac{1}{2} + j\frac{5}{4}\right)e^{-(3+j2)t}$$

and

$$K'_{-1} = (s + 3 - j2)X(s)e^{st}\big|_{s = -3+j2}$$

$$= \left(\frac{1}{2} - j\frac{5}{4}\right)e^{-(3-j2)t}$$

From (4-77) we obtain, for $t > 0$,

$$x(t) = \left(\frac{1}{2} + j\frac{5}{4}\right)e^{-(3+j2)t} + \left(\frac{1}{2} - j\frac{5}{4}\right)e^{-(3-j2)t}$$

$$= e^{-3t}\left(\frac{e^{j2t} + e^{-j2t}}{2} + \frac{5}{2}\frac{e^{j2t} - e^{-j2t}}{2j}\right)$$

$$= e^{-3t}(\cos 2t + \tfrac{5}{2}\sin(2t)), \qquad t > 0$$

which is the same result as obtained in Example 4-4. ∎

Generally, the inversion integral would not be used for a problem such as this. It is easier to use partial-fraction expansion and a table of Laplace transform pairs.

*4-6

THE DOUBLE-SIDED LAPLACE TRANSFORM

In the discussion leading to (4-1) it was assumed that the signal $x(t)$ was zero for $t < 0$. This resulted in the single-sided Laplace transform which was perfectly adequate for our purposes. Removing this restriction on $x(t)$ and following the same reasoning that resulted in (4-1), we obtain the *double-sided Laplace transform* of a signal $x(t)$, which is

$$X(s) = \int_{-\infty}^{\infty} x(t)e^{-st}\, dt \qquad (4\text{-}78)$$

The inversion integral remains the same as before with an added restriction on choosing the path of integration which will be discussed shortly. The integral

(4-78) converges absolutely if

$$\int_{-\infty}^{\infty} \left| x(t)e^{-st} \right| dt < \infty$$

But

$$\left| x(t)e^{-st} \right| = \left| x(t) \right| e^{-\sigma t}$$

and the double-sided Laplace transform integral therefore converges absolutely if

$$\int_{-\infty}^{\infty} \left| x(t) \right| e^{-\sigma t} dt < \infty \qquad (4\text{-}79)$$

A sufficient condition on $x(t)$ for (4-79) to hold is that there exist a real positive number A so that, for some real b and c, $\left| x(t) \right|$ is bounded by

$$\left| x(t) \right| \leq \begin{cases} Ae^{ct}, & t > 0 \\ Ae^{bt}, & t < 0 \end{cases} \qquad (4\text{-}80)$$

Then (4-78) is absolutely convergent for

$$c < \sigma < b \qquad (4\text{-}81)$$

When using contour integration to evaluate the inverse double-sided Laplace transform, the path of integration is chosen within this convergence strip and closed by a semicircular arc to the left for $t > 0$ and to the right for $t < 0$. To show (4-81), we break the integral (4-78) up into two parts as

$$X(s) = \int_{-\infty}^{0} x(t)e^{-st} dt + \int_{0}^{\infty} x(t)e^{-st} dt \qquad (4\text{-}82)$$

and use (4-80) to bound $\left| X(s) \right|$ by

$$\left| X(s) \right| \leq \int_{-\infty}^{0} Ae^{(b-s)t} dt + \int_{0}^{\infty} Ae^{(c-s)t} dt$$

$$\leq A \left[\frac{e^{(b-s)t}}{b-s} \Big|_{-\infty}^{0} + \frac{e^{(c-s)t}}{c-s} \Big|_{0}^{\infty} \right] \qquad (4\text{-}83)$$

The first integral will converge for Re $(s) < b$ (the lower limit yields zero when substituted), and the second integral will converge for Re $(s) > c$ (the upper limit yields zero when substituted). It follows that the singularities of $X(s)$ due to the *positive-time* half of $x(t)$ lie to the *left* of the line $s = c$, while the singularities of $X(s)$ due to the *negative-time* half of $x(t)$ lie to the *right* of the line $s = b$.

In contrast to the single-sided Laplace transform, the region of convergence must be specified together with a two-sided Laplace transform in order to identify the corresponding inverse transform uniquely. To illustrate this, consider the two-sided Laplace transforms of the signals

$$x_1(t) = e^{\alpha t} u(t)$$

and

$$x_2(t) = -e^{\alpha t}u(-t)$$

Both have the two-sided Laplace transform $1/(s - \alpha)$. However, their regions of convergence are different. In particular,

$$X_1(s) = \frac{1}{s - \alpha} \qquad \text{for } \sigma > \alpha$$

and

$$X_2(s) = \frac{1}{s - \alpha} \qquad \text{for } \sigma < \alpha$$

where $X_1(s)$ and $X_2(s)$ are the Laplace transforms of $x_1(t)$ and $x_2(t)$, respectively.

We may obtain the two-sided Laplace transform of a signal through use of the single-sided Laplace transform by separating the signal into its positive- and negative-time components. In particular, assume that $x(t)$ is defined for $-\infty < t < \infty$. We may represent it as

$$x(t) = x_1(t) + x_2(t) \tag{4-84}$$

where

$$x_1(t) = x(t)\, u(t) \tag{4-85a}$$

and

$$x_2(t) = x(t)\, u(-t) \tag{4-85b}$$

The two-sided Laplace transform of $x(t)$ is

$$\mathcal{L}_d[x(t)] = \int_{-\infty}^{\infty} x(t)e^{-st}\, dt$$

$$= \int_{-\infty}^{0} x_2(t)e^{-st}\, dt + \int_{0}^{\infty} x_1(t)e^{-st}\, dt \tag{4-86}$$

where $\mathcal{L}_d[\cdot]$ denotes the double-sided Laplace transform. Changing variables in the first integral to $t' = -t$, we obtain

$$\mathcal{L}_d[x(t)] = \int_{0}^{\infty} x_2(-t')e^{st'}\, dt' + \int_{0}^{\infty} x_1(t)e^{-st}\, dt$$

$$= X_2(-s) + X_1(s) \tag{4-87}$$

where

$$X_1(s) = \mathcal{L}_s[x_1(t)] \tag{4-88a}$$

and

$$X_2(s) = \mathcal{L}_s[x_2(-t)] \tag{4-88b}$$

The notation $\mathcal{L}_s[\cdot]$ denotes the single-sided Laplace transform. The signal $x_2(-t)$ is, of course, the reflection (mirror image) of the negative-time portion of $x(t)$ about the $t = 0$ axis. Equation (4-87) tells us to take the single-sided Laplace transform of this signal, replace s by $-s$ in the result, and add this new function of s to the single-sided Laplace transform of $x_1(t)$, which is the positive-time portion of $x(t)$.

EXAMPLE 4-15

Find the two-sided Laplace transform of

$$x(t) = e^{-3t}u(t) + e^{2t}u(-t)$$

Solution: The signals $x_1(t)$ and $x_2(-t)$ are

$$x_1(t) = e^{-3t}u(t)$$

and

$$x_2(-t) = e^{-2t}u(t)$$

Their single-sided Laplace transforms are

$$X_1(s) = \frac{1}{s + 3}, \qquad \sigma > -3$$

and

$$X_2(s) = \frac{1}{s + 2}, \qquad \sigma > 2$$

Replacing s by $-s$ in $X_2(s)$, we obtain

$$X_2(-s) = \frac{1}{-s + 2}, \qquad \text{Re}\,(-s) = -\sigma > 2 \quad \text{or} \quad \sigma < -2$$

Therefore, from (4-87), we obtain

$$X(s) = \frac{1}{s + 3} + \frac{1}{-s + 2} = \frac{-5}{s^2 + s - 6}, \qquad -3 < \sigma < -2 \qquad \blacksquare$$

EXAMPLE 4-16

Find the inverse double-sided Laplace transform of

$$X(s) = \frac{2s}{(s + 3)(s + 1)}, \qquad -3 < \sigma < -1$$

Solution: The partial-fraction expansion of $X(s)$ is

$$X(s) = \frac{3}{s + 3} - \frac{1}{s + 1}, \qquad -3 < \sigma < -1$$

Since the pole at $s = -3$ lies to the *left* of the convergence region, the term $3/(s + 3)$ corresponds to the Laplace transform of the *positive-time* portion of $\mathcal{L}^{-1}\,[X(s)]$. Similarly, since the pole at $s = -1$ lies to the *right* of the convergence region, the term $-1/(s + 1)$ corresponds to the Laplace transform of the *negative-time* portion of $\mathcal{L}^{-1}\,[X(s)]$. Using (4-87), (4-88a), and (4-88b), we obtain

$$x_1(t) = \mathcal{L}^{-1}_s \left[\frac{3}{s + 3} \right] = 3e^{-3t}u(t)$$

and

$$x_2(-t) = \mathcal{L}^{-1}_s \left[-\frac{1}{-s+1} \right]$$

$$= -e^t u(t)$$

Putting these results together, we obtain

$$x(t) = 3e^{-3t}u(t) - e^{-t}u(-t) \qquad \blacksquare$$

4·7

SUMMARY

In this chapter the basic tool for the analysis of lumped, fixed, linear systems has been introduced, namely, the Laplace transform. It is advantageous in that its application provides both the transient and forced responses of the system. To employ it for systems analysis, one proceeds by transforming the governing differential equations of the system into the complex frequency, or s domain. The Laplace-transformed output or response of the system can then be solved for algebraically. Inversion to the time domain for a large share of the problems encountered is accomplished by using partial-fraction expansion of the Laplace transform of the system response and using a table of Laplace transform pairs.

An alternative procedure for obtaining the Laplace transformed response is to use a Laplace-transformed equivalent circuit. Analysis then proceeds in an analogous fashion to using phasors and impedances in sinusoidal steady-state analysis. We consider this approach to systems analysis in Chapter 5.

FURTHER READING

R. V. CHURCHILL, J. W. BROWN, and R. F. VERHEY, *Complex Variables and Applications,* 3rd ed. New York: McGraw-Hill, 1976.
   A revision of the second edition published in 1960. An excellent treatment of complex variables and applications aimed at the senior–first-year graduate level.

E. KREYZIG, *Advanced Engineering Mathematics.* New York: Wiley, 1962.
   The Laplace transform is introduced in Chapter 4 and complex variables are treated in Chapter 10 of this reference.

B. P. LATHI, *Signals, Systems, and Communications.* New York: Wiley, 1965.
   Chapter 5 contains an introduction to the Laplace transform which begins with double-sided and specializes to the single-sided Laplace transform.

C. D. McGILLEM and G. R. COOPER, *Continuous and Discrete Signal and System Analysis.* New York: Holt, Rinehart and Winston, 1974.
   The treatment of Laplace transforms given in Chapter 6 of this book is similar in scope and level to the coverage of this text.

R. J. SCHWARZ and B. FRIEDLAND, *Linear Systems.* New York: McGraw-Hill, 1965.
   The Laplace transform is covered in Chapter 6 of this reference. Both the inversion integral and inversion by partial-fraction expansion are considered.

S. SESHU and N. BALABANIAN *Linear Network Analysis.* New York: Wiley, 1959.
   The appendix of this book gives a very concise treatment of complex-variable theory and inversion of the Laplace transform by contour integration. The applications to

network analysis given in the text are at a level somewhat above those considered here.

M. E. Van Valkenburg, *Network Analysis*, 3rd ed. Englewood Cliffs, N.J. Prentice-Hall, 1974.

A very readable and short introduction to Laplace transforms is given in Chapter 7. Example applications to network analysis are given.

H. H. Skilling, *Electrical Engineering Circuits*, 2nd ed. New York: Wiley, 1965.

Another very readable, elementary treatment of Laplace transforms is given in Chapter 16 at about the same level as Van Valkenburg.

C. A. Desoer and E. S. Kuh, *Basic Circuit Theory*. New York: McGraw-Hill, 1969.

This excellent text on circuit theory is mentioned last because of its completeness. The Laplace transform is treated in Chapter 13 with a $t = 0^-$ lower limit employed. It is at a slightly more difficult level than the preceding two references.

PROBLEMS

SECTIONS 4-1 and 4-2

4-1. Obtain the Laplace transforms of the following signals:

(a) $(1 - e^{-2t})u(t)$

(b) $(e^{-2t} - e^{-10t})u(t)$

(c) $u(t) - u(t - 10)$

(d) $\delta(t) - \delta(t - 10)$

SECTION 4-3

4-2. Obtain the Laplace transforms of the following signals:

(a) $\cos 200\pi t$

(b) $\sin 200\pi t$

(c) $\sqrt{2} \cos (200\pi t - \pi/4)$. [*Hint:* Relate this signal to those given in parts (a) and (b).]

4-3. Obtain the Laplace transform of the triangular signal $x(t) = \Lambda(t - 1)$ by using the differentiation theorem, the time delay theorem, and expressing dx/dt in terms of unit steps.

4-4. Solve the following differential equations by means of the Laplace transform:

(a) $\dfrac{d^2x(t)}{dt^2} + 6\dfrac{dx(t)}{dt} + 5x(t) = e^{-7t}u(t)$ with $x(0) = 0$ and $\left[\dfrac{dx(t)}{dt}\right]_{t=0} = 0$

(b) $\dfrac{dx(t)}{dt} + 3x(t) + 2y(t) = u(t)$ and $\dfrac{dy(t)}{dt} - x(t) = 0$ with $x(0) = y(0) = 0$. Solve for $x(t)$ and $y(t)$. [*Hint:* Note that it is not necessary to obtain equations in terms of $x(t)$ and $y(t)$ alone. You may solve for their Laplace transforms directly and inverse transform them.]

4-5. Solve for the current in the circuit shown if the switch changes as indicated by the arrow at $t = 0$. Assume that the switch has been in position 1 since $t = -\infty$.

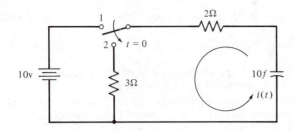

4-6. Obtain the voltage across the resistor as a function of time for $t > 0$. Assume that $i(0) = v_c(0) = 0$.

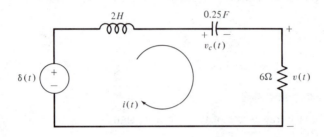

4-7. Find the inverse Laplace transforms of the functions of s given below;

(a) $\dfrac{s + 10}{s^2 + 8s + 20}$ (b) $\dfrac{s + 3}{s^2 + 4s + 5}$

(c) $\dfrac{s}{s^2 + 6s + 18}$ (d) $\dfrac{10}{s^2 + 10s + 34}$

(*Hint:* Use the *s*-shift theorem.)

4-8. Prove the *s*-shift theorem as stated by (4-21) and (4-22).

4-9. Prove the final value theorem by patterning your proof after the proof of the initial value theorem.

4-10. Find the initial and final values, if they exist, of the signals with Laplace transforms given below.

(a) $\dfrac{s + 10}{s^2 + 3s + 2}$ (b) $\dfrac{5}{s^3 + s^2 + 9s + 9}$

(c) $\dfrac{s^2 + 5s + 7}{s^2 + 3s + 2}$ (d) $\dfrac{s + 3}{s^2 + 2s}$

SECTION 4-4

4-11. Obtain the inverse Laplace transforms of the functions of s given in Problem 4.10. Do the initial and final values found in Problem 4.10 agree with the results obtained in this problem?

4-12. With $x(t) = \delta(t)$, $\omega_n = 2$, and $v_c(0^-) = 0$, obtain the inverse Laplace transform of $Y(s)$ as given by (4-34) for:

(a) $\zeta = \frac{5}{4}$

(b) $\zeta = 1$

(c) $\zeta = \dfrac{1}{\sqrt{2}}$

4-13. Obtain the inverse Laplace transform of

(a) $X(s) = \dfrac{7s^3 + 20s^2 + 33s + 82}{(s^2 + 4)(s + 2)(s + 3)}$

(b) $X(s) = \dfrac{2s^3 + 9s^2 + 22s + 23}{[(s + 1)^2 + 4](s + 1)(s + 3)}$

4-14. Obtain the inverse Laplace transform of

$$X(s) = \dfrac{7s^2 + 15s + 10}{(s + 1)^2(s + 3)}$$

4-15. Obtain the inverse Laplace transform of

$$X(s) = \dfrac{s^2(s + 9)}{(s + 3)^3 (s + 1)}$$

4-16. Obtain the inverse Laplace transform of

(a) $X(s) = \dfrac{s^4 + 8s^2 + s + 17}{(s^2 + 4)^2(s + 1)}$ **(b)** $X(s) = \dfrac{s^2 + 3}{[(s + 1)^2 + 4]^2 (s + 1)}$

4-17. Derive pairs 1 and 3 of Table 4-3.

4-18. **(a)** Derive pair 10 of Table 4-3. Use the derivation of (4-71) as a guide.

     (b) Derive pairs 11 and 12 of Table 4-3.

4-19. One way to define the *effective delay* t_0 and *duration* τ of a signal $x(t)$, where $x(t) = 0$, $t < 0$, is

$$t_0 = \dfrac{\displaystyle\int_0^\infty tx(t)\, dt}{\displaystyle\int_0^\infty x(t)\, dt}$$

and

$$\tau^2 = \dfrac{\displaystyle\int_0^\infty (t - t_0)^2 x(t)\, dt}{\displaystyle\int_0^\infty x(t)\, dt}$$

(a) If $X(s)$ is the Laplace transform of $x(t)$, show that

$$t_0 = -\dfrac{X'(0)}{X(0)}$$

and

$$\tau^2 = \frac{X''(0)}{X(0)} - \left[\frac{X'(0)}{X(0)}\right]^2$$

where

$$X'(s) = \frac{dX(s)}{ds}$$

and

$$X''(s) = \frac{d^2X(s)}{ds^2}$$

(b) Suppose that $x(t) = x_1(t) * x_2(t)$ and $X(s) = X_1(s)X_2(s)$. Let t_0, t_{01}, t_{02}, and τ, τ_1, τ_2 be the delay and duration times of $x(t)$, $x_1(t)$, and $x_2(t)$, respectively. Show that $t_0 = t_{01} + t_{02}$ and $\tau^2 = \tau_1^2 + \tau_2^2$

4-20. Derive the following theorems concerning the Laplace transform:

(a) $\mathcal{L}[x(t/a)] = aX(as), a > 0$

(b) $\mathcal{L}[tx(t)] = -\dfrac{dX(s)}{ds}$

(c) Generalize part (b) to $\mathcal{L}[t^n x(t)]$.

(d) Use part (c) to derive pair 3, Table 4-3.

4-21. (a) Obtain the Laplace transform of the triangle shown by applying the differentiation theorem, the time delay theorem, and using the Laplace transform of the unit step.

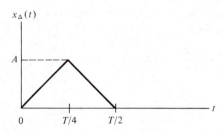

(b) Obtain the one-sided Laplace transform of the triangular waveform shown.

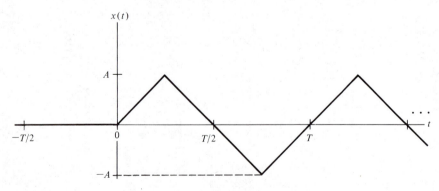

4-22. In the circuits shown, the capacitor is initially charged to V volts and the switch is closed at $t = 0$. Obtain the current, $i(t)$, for $t \geq 0$ using the Laplace transform.

(a)

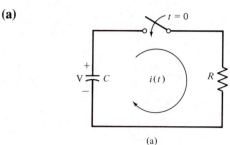

(a)

(b)

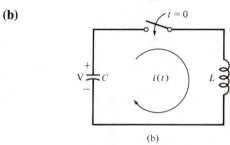

(b)

Answers: (a) $(V/R) \exp(-t/RC)u(t)$; (b) $\sqrt{\dfrac{C}{L}} \, V \sin(t/\sqrt{LC})u(t)$.

4-23. In the circuits shown, the switch is moved in accordance with the arrow at $t = 0$, having been in the top position for a long time. Solve for the current, $i(t)$, for $t \geq 0$ using the Laplace transform.

(a)

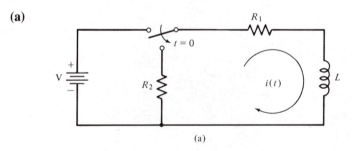

(a)

(b)

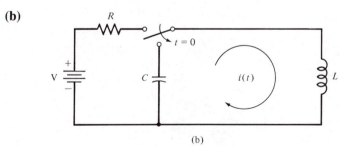

(b)

Answers: (a) $(V/R_1) \exp[-(R_1 + R_2)t/L]u(t)$; (b) $(V/R) \cos(t/\sqrt{LC})u(t)$.

4-24. Which of the following signals have Laplace transforms only, and which have both Fourier and Laplace transforms? Why?

(a) $e^{-10t}u(t)$

(b) $e^{10t}u(t)$

(c) $e^{-10|t|}$

(d) $r(t)$, where $r(t)$ is the unit ramp

(e) $te^{-10t}u(t)$

Applications of the Laplace Transform

INTRODUCTION

In Chapter 4 we developed the Laplace transform and several of its properties and showed how to obtain inverse Laplace transforms using partial-fraction expansion. In this chapter we make use of these tools to analyze electrical networks. We introduce the idea of the Laplace-transformed network, which provides a convenient framework for solution of problems involving both initial conditions and sources.

The latter portion of the chapter contains several topics of interest in the analysis of more general systems. The concept of a transfer function is discussed in some detail, and the Routh test for determining stability is presented. Bode plots are introduced as a means of conveniently displaying the frequency response of a system. Finally, the block diagram is presented as a convenient means of representing a system as a combination of mathematical and schematic models particularly useful for more complicated systems.

NETWORK ANALYSIS USING THE LAPLACE TRANSFORM

In analyzing the circuit of Figure 4-3 (repeated in Figure 5-1) we first wrote down the differential equation expressing Kirchhoff's voltage law and performed the Laplace transform of this equation for various forcing functions, $x(t)$, to find the Laplace transform of the voltage across the resistance R. This approach provided the rational functions for inverse Laplace transform used as Examples 4-8 through 4-11. There is another approach that could have been taken. We now use the circuit of Figure 5-1 to explain this approach.

Each circuit element in Figure 5-1 is first replaced by its Laplace-transformed equivalent, including initial conditions. To see how this is done, consider the capacitor. Its voltage–current relationship is

$$v(t) = \frac{1}{C} \int_{-\infty}^{t} i(\lambda) \, d\lambda \qquad (5\text{-}1)$$

where $v(t)$ is the voltage across it with reference direction in the same sense as the reference direction for the current $i(t)$ through it (across and through variables). We may rewrite (5-1) as

$$v(t) = \frac{1}{C} \int_{0^-}^{t} i(\lambda) \, d\lambda + v(0^-) \qquad (5\text{-}2)$$

where

$$v(0^-) = \frac{1}{C} \int_{-\infty}^{0^-} i(\lambda) \, d\lambda = \frac{q(0^-)}{C} \qquad (5\text{-}3)$$

is the initial voltage across the capacitor. The Laplace transform of (5-2) is

$$V(s) = \frac{1}{sC} I(s) + \frac{v(0^-)}{s} \qquad (5\text{-}4)$$

which suggests the Laplace-transformed KVL equivalent circuit shown in Figure 5-2a. The capacitor is represented by an *impedance* of value $1/sC$, and the initial condition is represented by a voltage generator of value $v_c(0^-)/s$.

Alternatively, (5-1) can be differentiated to produce the relationship

$$i(t) = C \frac{dv}{dt} \qquad (5\text{-}5)$$

which can be Laplace transformed with the aid of the differentiation theorem of Table 4-2. This results in

$$I(s) = sCV(s) - Cv(0^-) \qquad (5\text{-}6)$$

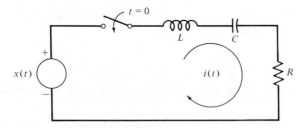

FIGURE 5-1.   Series *RLC* circuit.

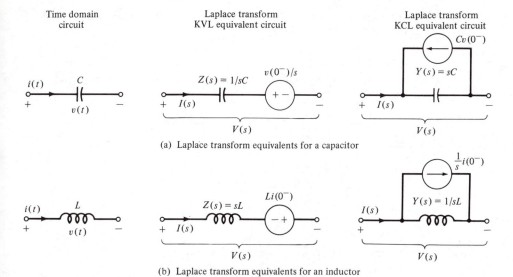

Time domain circuit   **Laplace transform KVL equivalent circuit**   **Laplace transform KCL equivalent circuit**

(a) Laplace transform equivalents for a capacitor

(b) Laplace transform equivalents for an inductor

FIGURE 5-2. Laplace transformed equivalent-circuit elements, which include initial conditions.

which can be interpreted as the Laplace-transformed KCL equivalent circuit shown in Figure 5-2a. Equation (5-6) shows that $I(s)$ can be viewed as being composed of the difference of two currents—that through the capacitor, which is the product of the *admittance sC* and $V(s)$, minus that due to the initial condition, which is $Cv(0^-)$.

Consider next an inductor for which the voltage–current relationship may be written as

$$v(t) = L\frac{di}{dt} \tag{5-7}$$

with the Laplace transformed equivalent

$$V(s) = sLI(s) - Li(0^-) \tag{5-8}$$

which can be rearranged to yield

$$I(s) = \frac{1}{sL}V(s) + \frac{1}{s}i(0^-) \tag{5-9}$$

These equations suggest the equivalent circuits shown in Figure 5-2b.

Since the voltage-current characteristic of a resistor is

$$v(t) = Ri(t) \tag{5-10}$$

its Laplace-transformed equivalent is

$$V(s) = RI(s) \tag{5-11}$$

Thus the Laplace-transformed equivalent circuit for a resistor is simply a resistor with current $I(s)$ through it and voltage $V(s)$ across it.

We now consider an example that illustrates the application of the models just developed to circuit analysis.

EXAMPLE 5-1

We again use the by now familiar circuit of Figure 5-1 as an example. Suppose that

$$x(t) = (-0.5 \cos t + 2.5 \sin t)u(t)$$

as in Example 4-9. This forcing function has the Laplace transform

$$X(s) = -\frac{0.5s}{s^2 + 1} + \frac{2.5}{s^2 + 1}$$

$$= \frac{-0.5s + 2.5}{s^2 + 1}$$

Other parameter values, which were used in Example 4.9, are

$$i_L(0^-) = 0 \text{ A}$$

$$v_c(0^-) = -2 \text{ V}$$

$$\omega_n^2 \triangleq \frac{1}{LC} = 16 \text{ s}^{-2}$$

and

$$2\zeta\omega_n = \frac{R}{L} = 10 \text{ s}^{-1}$$

For simplicity, we now assume that $R = 1 \ \Omega$. This means that the loop current and the voltage across the resistor, which was the output variable considered previously, are the same. Since $R/L = 10$, we have $L = 1/10$ H with $R = 1 \ \Omega$. Thus $C = 1/(16L) = \frac{5}{8}$ F.

Using these circuit parameter values and the assumed $X(s)$ as well as the initial conditions $v_c(0^-) = -2$V and $i_L(0^-) = 0$ A, we may now draw the Laplace-transformed equivalent circuit as shown in Figure 5-3. Writing Kirchhoff's voltage law around the loop in terms of Laplace-transformed voltages and currents yields

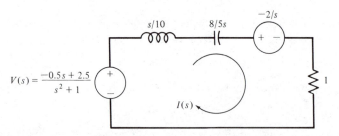

FIGURE 5-3. Laplace-transformed equivalent circuit for Example 5.1.

$$\frac{-0.5s + 2.5}{s^2 + 1} = \left(\frac{s}{10} + \frac{8}{5s} + 1\right)I(s) - 2/s$$

Solving for $I(s)$, we find that the Laplace transform of the current is

$$I(s) = \frac{\dfrac{(-0.5s + 2.5)}{s^2 + 1} + 2/s}{s/10 + 8/5s + 1}$$

Clearing fractions in the numerator and denominator, we obtain

$$I(s) = \frac{15s^2 + 25s + 20}{(s^2 + 1)(s^2 + 10s + 16)}$$

which is the same as (4-41), the result for Example 4-9. ∎

MUTUAL INDUCTANCE

Circuits involving mutual inductance may be handled by using an equivalent circuit for the coupled coils called a T-equivalent. Consider the transformer model shown diagrammatically in Figure 5-4. With the reference directions for the terminal voltages and currents as shown in Figure 5-4, it is defined by the voltage–current relationships

$$v_1 = L_1 \frac{di_1}{dt} + M \frac{di_2}{dt} \tag{5-12a}$$

$$v_2 = M \frac{di_1}{dt} + L_2 \frac{di_2}{dt} \tag{5-12b}$$

The parameters L_1 and L_2 are referred to as the *self-inductances* and the parameter M is called the *mutual inductance*. Whereas self-inductances are always positive quantities, mutual inductance may be positive or negative depending on the polarity marks, usually indicated by dots, used on the transformer circuit diagram. The value of M will be positive if each current reference is directed toward or away from the polarity marks on the schematic diagram, whereas M is negative if one current reference is directed away and one current reference is directed toward the corresponding dot-marked terminal. This is determined by how the transformer is wound. Thus Figure 5-4a shows a case where M is positive, and Figure 5-4b illustrates a case for M negative.

An equivalent representation of the transformer defined by (5-12a) and (5.12b) may be obtained by rewriting them as

$$v_1 = (L_1 - M) \frac{di_1}{dt} + M \frac{d}{dt} (i_1 + i_2) \tag{5-13a}$$

and

$$v_2 = M \frac{d}{dt} (i_1 + i_2) + (L_2 - M) \frac{di_2}{dt} \tag{5-13b}$$

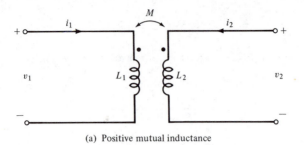

(a) Positive mutual inductance

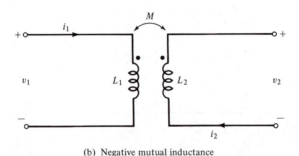

(b) Negative mutual inductance

FIGURE 5-4. **Schematic representation of a transformer.**

These equations suggest the T-equivalent circuit shown in Figure 5-5. This equivalent circuit may be used to replace a transformer in a circuit by equivalent inductances and the analysis of the circuit then proceeds as in the case of an ordinary circuit containing only R, L, and C elements. The next example illustrates the use of the T-equivalent model shown in Figure 5-5.

EXAMPLE 5-2

The coils of a large horseshoe type of electromagnet used in a magnetohydrodynamic generator are represented by the circuit model shown in Figure 5-6. (a) Find the current provided by the power supply after the switch is closed; (b) a lug drops off one of the coils accidentally causing a break in the circuit as indicated by the dots. Find the voltage induced in L_1 by the current discontinuity in L_2.

Solution: (a) We replace the circuit of Figure 5-6 with the Laplace-transformed equivalent circuit of Figure 5-6, which employs the equivalent T-representation of the transformer. Since $i_1(0^-) = i_2(0^-) = 0$, there are no initial condition generators. Writing loop equations, we obtain

$$\frac{100}{s} = (5 + 10s)I_1(s) - (5 + 5s)I_2(s)$$

$$0 = -(5 + 5s)I_1(s) + (10 + 10s)I_2(s)$$

Multiplying the first equation by 2 and adding, we obtain

$$\frac{200}{s} = (5 + 15s)I_1(s)$$

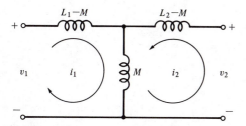

FIGURE 5-5. T-equivalent circuit for a transformer.

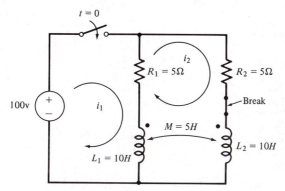

(a) Time–domain equivalent circuit of a large electromagnet

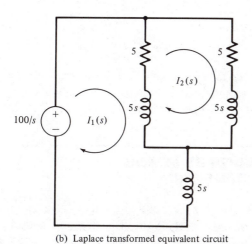

(b) Laplace transformed equivalent circuit

FIGURE 5-6. Circuits for Example 5-2.

or

$$I_1(s) = \frac{40}{3s(s + \frac{1}{3})}$$

$$= \frac{A}{s} + \frac{B}{s + \frac{1}{3}}$$

where $A = 40$ and $B = -40$. Thus $i_1(t) = 40(1 - e^{-t/3})u(t)$.

(b) For the second part of the problem, we let the time at which the lug drops off be $t = 0$. The initial conditions for $i_1(t)$ and $i_2(t)$ are the steady-state values that occur when the switch has been closed for a long time. They are $i_1(0^-) = 40$ A, $i_2(0^-) = 20$ A. The $t = 0^-$ inductor currents for the equivalent circuit of Figure 5-6 are then easily found to be 20 A for each of the inductors $L_1 - M$ and $L_2 - M$, and 40 A for M. With initial-condition generators included for these inductor currents, the equivalent circuit after the lug drops off is as shown in Figure 5-7. The loop equation for $I_1(s)$ is now

$$\frac{100}{s} + 300 = (5 + 10s)I_1(s)$$

Solving for $I_1(s)$, we obtain

$$I_1(s) = \frac{30(s + \frac{1}{3})}{s(s + \frac{1}{2})}$$

The Laplace transform of the voltage induced in L_1 is

$$V_{L_1}(s) = \frac{30(s + \frac{1}{3})}{s(s + \frac{1}{2})}(10s) - 300$$

$$= \frac{300(s + \frac{1}{3})}{(s + \frac{1}{2})} - 300$$

$$= -\frac{50}{s + \frac{1}{2}}$$

Thus the induced voltage is

$$V_{L_1}(t) = -50e^{-t/2}u(t)$$

The student should show that there is an impulse voltage across the R_2L_2 combination due to the lug dropping off. This impulse (actually a large voltage pulse) will probably cause failure of the insulation in the magnet coil. ◼

5-4

ANALYSIS OF CIRCUITS BY MEANS OF THE LAPLACE TRANSFORM

Recall that the governing equations for a planar, fixed, linear network consisting of inductances, capacitances, resistances, ideal voltage sources, and ideal current sources can be written in two forms. These two forms are obtained by writing Kirchhoff's voltage law around properly chosen loops (loop analysis), and by

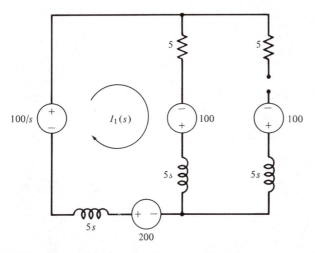

FIGURE 5-7. Equivalent circuit for Example 5-2b.

using Kirchhoff's current law at a properly chosen set of nodes (nodal analysis). These two techniques will now be reviewed and the Laplace transform used to simplify their application.

Loop Analysis

For a network consisting of N_b branches and N_v nodes we found that $N_b - (N_v - 1)$ independent equations could be written by employing KVL around $N_b - (N_v - 1)$ properly chosen loops. For planar networks, one such proper choice for these loops is the meshes or "windows" of the network. In this case, the equations have the form (dependent sources excluded for the moment):

$$Z_{11}(D)i_1(t) + Z_{12}(D)i_2(t) + \cdots + Z_{1n}(D)i_n(t) = e_1(t)$$

$$Z_{21}(D)i_1(t) + Z_{22}(D)i_2(t) + \cdots + Z_{2n}(D)i_n(t) = e_2(t) \qquad (5\text{-}14)$$

$$\cdot$$
$$\cdot$$
$$\cdot$$

$$Z_{n1}(D)i_1(t) + Z_{n2}(D)i_2(t) + \cdots + Z_{nn}(D)i_n(t) = e_n(t)$$

where $n = N_b - (N_v - 1)$. For a clockwise reference direction choice of the mesh currents, the Z_{ii}'s have the form

$$Z_{ii}(D) = L_{ii}D + R_{ii} + S_{ii}D^{-1} \qquad (5\text{-}15)$$

where L_{ii} = total series inductance around ith mesh

R_{ii} = total series resistance around ith mesh

S_{ii} = sum of all reciprocal capacitances around ith mesh ($S = C^{-1}$)

Also,

$$Z_{ij}(D) = -L_{ij}D - R_{ij} - S_{ij}D^{-1} \tag{5-16}$$

where L_{ij} = sum of all inductances common to meshes i and j

$\quad\quad R_{ij}$ = sum of all resistances common to meshes i and j

$\quad\quad S_{ij}$ = sum of all reciprocal capacitances common to
meshes i and j

$$D = \frac{d}{dt}$$

$$D^{-1} = \int_{-\infty}^{t} (\cdot) \, d\lambda$$

$\quad\quad i_j(t)$ = jth mesh current

$\quad\quad e_j(t)$ = algebraic sum of the voltage sources around mesh j

Taking the Laplace transform of (5-14), we obtain

$$\mathbf{Z}(s)\mathbf{I}(s) = \mathbf{E}(s) + \mathbf{V}_i(s) \tag{5-17}$$

where $\mathbf{I}(s)$ is a column matrix of Laplace-transformed loop currents, $\mathbf{E}(s)$ is the Laplace-transformed source voltage matrix, $\mathbf{V}_i(s)$ is a Laplace-transformed initial condition generator matrix, and the matrix

$$\mathbf{Z}(s) = [Z_{ij}(s)] \tag{5-18}$$

has elements given by

$$Z_{ii}(s) = sL_{ii} + R_{ii} + \frac{S_{ii}}{s} \tag{5-19}$$

$$Z_{ij}(s) = sL_{ij} + R_{ij} + \frac{S_{ij}}{s}$$

$\mathbf{Z}(s)$ is called the *loop impedance matrix*. If no dependent sources are present, it is symmetric. The *loop current matrix,* $\mathbf{I}(s)$, in (5-17) is defined as

$$\mathbf{I}(s) = \begin{bmatrix} \mathcal{L}[i_1(t)] \\ \mathcal{L}[i_2(t)] \\ \cdot \\ \cdot \\ \cdot \\ \mathcal{L}[i_n(t)] \end{bmatrix} \triangleq \mathcal{L} \begin{bmatrix} i_1(t) \\ i_2(t) \\ \cdot \\ \cdot \\ \cdot \\ i_n(t) \end{bmatrix} \tag{5-20}$$

and the *source voltage matrix* $\mathbf{E}(s)$ is defined as

$$\mathbf{E}(s) = \mathcal{L} \begin{bmatrix} e_1(t) \\ e_2(t) \\ \cdot \\ \cdot \\ \cdot \\ e_n(t) \end{bmatrix} \tag{5-21}$$

The second term on the right-hand side of (5-17) is the *initial-condition matrix*. By using the equivalent circuits in the second column of Figure 5-2, the initial conditions can be easily included with the source voltage matrix. Just how this is accomplished will be illustrated by an example.

It is emphasized that the above equations only hold if there are no *dependent sources* present in the network. With dependent sources present one must go through writing the KVL equations for each mesh from the beginning, the end result being that some elements of $\mathbf{Z}(s)$ will have additional terms present due to the dependent sources. Thus $\mathbf{Z}(s)$ is not symmetric if dependent sources are present.

To illustrate the use of (5-17) through (5-21), we now consider an example.

EXAMPLE 5-3.

Consider the circuit of Figure 5-8 with an initial current through the inductor of $i_c(0^-) = i_0$ and an initial voltage across the capacitor of $v_c(0^-) = v_0$. (a) First, assume that $v_c(t)$ is an independent source. We may write down the governing equations on a loop basis in the time domain immediately and Laplace transform them, or we may work with the Laplace-transformed equivalent circuit. We choose the latter course and can immediately write down the matrix equations as

$$
\begin{bmatrix}
R_1 + R_2 + sL & -R_2 & 0 \\[2mm]
-R_2 & R_2 + \dfrac{1}{sC} & -\dfrac{1}{sC} \\[3mm]
0 & -\dfrac{1}{sC} & R_3 + R_4 + \dfrac{1}{sC}
\end{bmatrix}
\begin{bmatrix}
I_1(s) \\[2mm] I_2(s) \\[3mm] I_3(s)
\end{bmatrix}
$$

$$
= \begin{bmatrix}
Li_0 + V_s(s) \\[2mm]
\dfrac{-v_0}{s} \\[3mm]
\dfrac{v_0}{s} + V_c(s)
\end{bmatrix}
\tag{5-22}
$$

Note that the loop impedance matrix is symmetric. The diagonal elements are the sums of the impedances around each loop. The off-diagonal elements have magnitudes equal to the impedance common to the loops involved and are preceded by a minus sign; for example, R_2 is common to loops 1 and 2 so it appears in the matrix as the element Z_{12} with a minus sign because $i_1(t)$ and $i_2(t)$ were chosen to have opposite reference directions through R_1. (b) Now assume that $v_c(t)$ is a dependent source; in particular, let

$$v_c(t) = Ki_1(t)$$

where K is a constant of proportionality. By writing down the loop equations and rearranging, the student should show that the element $Z_{13}(s)$ of the loop impedance matrix in (5-22) becomes $-K$; that is, now

5-4 ANALYSIS OF CIRCUITS BY MEANS OF THE LAPLACE TRANSFORM

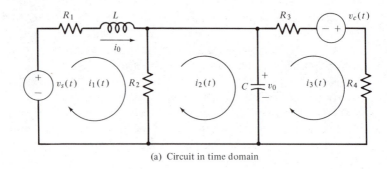

(a) Circuit in time domain

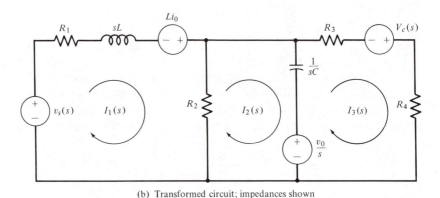

(b) Transformed circuit; impedances shown

FIGURE 5-8. Circuits for the illustration of loop analysis.

$$\mathbf{Z}(s) = \begin{bmatrix} R_1 + R_2 + sL & -R_2 & 0 \\ -R_2 & R_2 + \dfrac{1}{sC} & -\dfrac{1}{sC} \\ -K & -\dfrac{1}{sC} & R_3 + R_4 + \dfrac{1}{sC} \end{bmatrix} \qquad (5\text{-}23)$$

Hence the presence of the dependent source means that $\mathbf{Z}(s)$ is no longer symmetric. ∎

Nodal Analysis

For a network consisting of N_b branches and N_v nodes, we know that $N_v - 1$ independent equations can be written by writing KCL at $N_v - 1$ properly chosen nodes. If the node for which we do not write a KCL equation is the same one that we choose as a reference node (i.e., "ground"), the equations have the following symmetric form:

$$Y_{11}(D)v_1(t) + Y_{12}(D)v_2(t) + \cdots + Y_{1n}(D)v_n(t) = i_1(t)$$

$$Y_{21}(D)v_1(t) + Y_{22}(D)v_2(t) + \cdots + Y_{2n}(D)v_n(t) = i_2(t) \qquad (5\text{-}24)$$

$$\vdots$$

$$Y_{n1}(D)v_1(t) + Y_{n2}(D)v_2(t) + \cdots + Y_{nn}(D)v_n(t) = i_n(t)$$

where $Y_{ii}(D) = C_{ii}D + G_{ii} + \Gamma_{ii}D^{-1}$ $\qquad\qquad$ (5-25)

$\quad C_{ii} = $ sum of all capacitances between node i and all other nodes

$\quad G_{ii} = $ sum of all conductances between node i and all other nodes

$\quad \Gamma_{ii} = $ sum of all reciprocal inductances between node i and all other nodes ($\Gamma = L^{-1}$)

$$D = \frac{d}{dt}, \qquad D^{-1} = \int_{-\infty}^{t} (\cdot)\, d\lambda$$

Also,

$$Y_{ij}(D) = C_{ij}D + G_{ij} + \Gamma_{ij}D^{-1} \qquad (5\text{-}26)$$

where $G_{ij} = $ sum of all conductances between nodes i and j

$\quad C_{ij} = $ sum of all capacitances between nodes i and j

$\quad \Gamma_{ij} = $ sum of all reciprocal inductances between nodes i and j

$\quad v_j(t) = $ voltage drop between node j and the reference node

$\quad i_j(t) = $ total source current flowing into node j

Taking the Laplace transform of (5-24), we obtain

$$\mathbf{Y}(s)\mathbf{V}(s) = \mathbf{I}(s) + \mathbf{J}(s) \qquad (5\text{-}27)$$

where $\mathbf{Y}(s)$ is called the *node admittance matrix* and has elements given by

$$Y_{ii}(s) = sC_{ii} + G_{ii} + \frac{\Gamma_{ii}}{s} \qquad (5\text{-}28)$$

$$Y_{ij}(s) = sC_{ij} + G_{ij} + \frac{\Gamma_{ij}}{s} \qquad (5\text{-}29)$$

If no *dependent sources* are present, $\mathbf{Y}(s)$ is a symmetric matrix. The other matrices in (5-27) are defined as

$$\mathbf{V}(s) = \mathcal{L} \begin{bmatrix} v_1(t) \\ v_2(t) \\ \vdots \\ v_n(t) \end{bmatrix} \qquad (n = N_v - 1) \qquad (5\text{-}30)$$

$$\mathbf{I}(s) = \mathcal{L} \begin{bmatrix} i_1(t) \\ i_2(t) \\ \vdots \\ i_n(t) \end{bmatrix} \qquad (n = N_v - 1) \qquad (5\text{-}31)$$

and $\mathbf{J}(s)$ is the *initial condition matrix* which results when the derivatives and integrals of (5-24) are Laplace transformed. It can be included directly with the *source current matrix*, $\mathbf{I}(s)$, by using the Laplace-transformed equivalent circuits of Figure 5-2 for capacitive and inductive elements with nonzero initial conditions.

EXAMPLE 5-4

We consider the same circuit as Figure 5-8, except that the two voltage sources and the series resistors are replaced by their Norton equivalents.† The resulting circuit is shown in Figure 5-9. Working with the Laplace-transformed equivalent circuit of Figure 5-9, where initial conditions have been replaced by equivalent-current generators, we may immediately write down the Laplace-transformed matrix equations on a nodal basis. They are

$$\begin{bmatrix} \dfrac{1}{R_1} + \dfrac{1}{sL} & -\dfrac{1}{sL} & 0 \\[2mm] -\dfrac{1}{sL} & \dfrac{1}{R_2} + \dfrac{1}{R_3} + sC + \dfrac{1}{sL} & -\dfrac{1}{R_3} \\[2mm] 0 & -\dfrac{1}{R_3} & \dfrac{1}{R_3} + \dfrac{1}{R_4} \end{bmatrix} \begin{bmatrix} V_1(s) \\[2mm] V_2(s) \\[2mm] V_3(s) \end{bmatrix}$$

$$= \begin{bmatrix} \dfrac{V_s(s)}{R_1} - \dfrac{i_0}{s} \\[3mm] \dfrac{i_0}{s} + Cv_0 - \dfrac{V_c(s)}{R_3} \\[3mm] \dfrac{V_c(s)}{R_3} \end{bmatrix} \qquad (5\text{-}32)$$

We note that the admittance matrix, $\mathbf{Y}(s)$ is symmetric and can be written down directly from the Laplace-transformed circuit diagram by inspection. Also,

†This step need not be carried out, but in that case the equilibrium equations cannot be written down by inspection in matrix form.

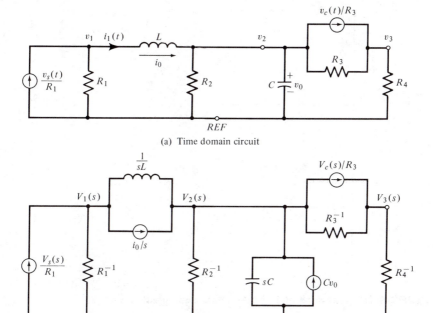

(a) Time domain circuit

(b) Laplace transformed circuit including initial condition generators; admittances shown

FIGURE 5-9.   Circuit for Example 5-4.

by using the initial-condition generators of Figure 5-2, the initial conditions are included automatically with the source current matrix. Note also that a convenient way to handle voltage sources in nodal analysis is to convert them to Norton equivalents.

The student should write down the KCL equations for this circuit assuming that $v_c(t)$ is a dependent source; for example, $V_c(s) = \alpha R_3 I_1(s)$. The admittance matrix is no longer symmetric in this case. ∎

5-5

TRANSFER FUNCTIONS AND FREQUENCY RESPONSE

The concept of a *transfer function* is of great importance in system studies. It provides a method of completely specifying the behavior of a system subjected to inputs. We will define this important quantity shortly. First, however, consider the following simple example.

EXAMPLE 5-5

Determine $v_o(t)$ for the network shown in Figure 5-10a assuming zero initial voltage on the capacitor.

Following the ideas of the preceding section, we may use the Laplace-trans-

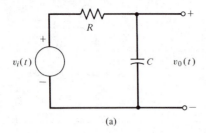

(a)

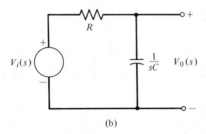

(b)

FIGURE 5-10 Network and transformed equivalent network.

formed equivalent network of Figure 5-10b. From this network we easily obtain

$$V_o(s) = \frac{V_i(s)}{sRC + 1}$$

and

$$v_o(t) = \mathcal{L}^{-1}\left[\frac{V_i(s)}{sRC + 1}\right]$$

We cannot proceed further until $v_i(t)$ is specified, but note that these relations are valid regardless of the specific $v_i(t)$ chosen. ∎

Taking a cue from our example, we note that we can write

$$\frac{V_o(s)}{V_i(s)} = \frac{1}{sRC + 1}$$

This expression gives a relationship between the Laplace transform of the input and the Laplace transform of the output which is valid for any input. This is the transfer function for the circuit of the example (sometimes referred to as the system function). There are some things that we should note about it:

1. The transfer function is independent of the particular input and is a property of the circuit only.
2. The transfer function is obtained for the case of zero initial conditions.
3. The transfer function for this lumped, linear, fixed circuit is a rational function of s.

We may generalize now and define the *transfer function* of a fixed, linear system as the ratio of the Laplace transform of the system output to the Laplace transform of the system input when all initial conditions are zero. If we denote $y(t)$ as the output and $x(t)$ as the input, we may express the transfer function, $H(s)$, as

$$H(s) = \frac{Y(s)}{X(s)} \bigg|_{\text{all initial conditions zero}}$$

This relation must hold true for any input. Suppose that $x(t) = \delta(t)$. Then it is obvious that $H(s)$ is the Laplace transform of the unit impulse response of the system since $X(s) = 1$.

If the system is lumped, in addition to being linear and fixed, the transfer function may be written as

$$H(s) = \frac{b_m s^m + b_{m-1} s^{m-1} + \cdots + b_0}{a_n s^n + a_{n-1} s^{n-1} + \cdots + a_0} \triangleq \frac{N(s)}{D(s)} \qquad (5\text{-}33)$$

where the a_n's and b_m's are the coefficients of the differential equation that describes the system:

$$a_n \frac{d^n y(t)}{dt^n} + a_{n-1} \frac{d^{n-1} y(t)}{dt^{n-1}} + \cdots + a_0 y(t)$$

$$= b_m \frac{d^m x(t)}{dt^m} + b_{m-1} \frac{d^{m-1} x(t)}{dt^{m-1}} + \cdots + b_0 x(t) \qquad (5\text{-}34)$$

The a_n's and b_m's are real because they result from real system components such as resistors, capacitors, inductors, springs, masses, and so on. Thus the numerator and denominator polynomials, $N(s)$ and $D(s)$, have real coefficients. This implies that their roots are either real or occur in complex-conjugate pairs. The roots of $D(s)$ are referred to as the *poles* of the transfer function $H(s)$ and the roots of $N(s)$ are referred to as the *zeros* of $H(s)$.

Since $H(s)$ is the Laplace transform of the impulse response, it follows that the roots of $N(s)$ and $D(s)$ must satisfy certain conditions if $h(t)$ represents a response from a stable system.

First, requiring the system to be bounded-input, bounded-output (BIBO) stable, we note that the degree of $N(s)$ must be less than or equal to the degree of $D(s)$, for if this were not the case, long division of $D(s)$ into $N(s)$ results in the following form for $H(s)$:

$$H(s) = C_k s^k + C_{k-1} s^{k-1} + \cdots C_1 s + C_0 + \frac{N'(s)}{D(s)}$$

Multiplication of $H(s)$ by the Laplace transform of a bounded input signal (e.g., a unit step) and inverse Laplace transformation will then result in a response that is not bounded. For example, if the term $C_1 s$ alone is present, which would result if $N(s)$ were of degree one more than $D(s)$, the output in response to a unit step would contain the term $\mathcal{L}^{-1}[(C_1 s) \cdot (1/s)] = C_1 \delta(t)$. Also, all roots of $D(s)$—the poles of $H(s)$—must have real parts which are negative; that is, the poles must lie in the left-half of the s-plane. The reason for this is apparent if one considers the inverse Laplace transform of a partial-fraction expansion of $Y(s) = X(s)H(s)$. Assume that $H(s)$ contains the simple complex-conjugate pair of poles $s_1 = -\alpha + j\omega_0$ and $s_2 = -\alpha - j\omega_0$, giving the denominator quadratic factor $(s + \alpha)^2 + \omega_0^2$. Incomplete partial-fraction expansion of $Y(s)$ then gives

$$Y(s) = \frac{As + B}{(s + \alpha)^2 + \omega_0^2} + Y_1(s)$$

The first term of $Y(s)$, according to Table 4-3, corresponds to terms in $y(t)$ of the form

$$\exp(-\alpha t) \cos \omega_0 t \, u(t)$$

and

$$\exp(-\alpha t) \sin \omega_0 t \, u(t)$$

Since $s_{1,2} = -\alpha \pm j\omega_0$ corresponds to the complex-conjugate pole locations, we see that α must be positive (poles in the left-half plane) if $\exp(-\alpha t)$ is to decrease with increasing t. Otherwise, the output will be unbounded regardless of whether the input is bounded or not. Similar considerations of pole factors of the form $[(s + \alpha)^2 + \omega_0^2]^n$ in $H(s)$ result in time-domain terms of the form

$$At^{n-1} \exp(-\alpha t) \cos \omega_0 t \, u(t)$$

or

$$At^{n-1} \exp(-\alpha t) \sin \omega_0 t \, u(t)$$

Clearly, if $\alpha > 0$, a bounded response results. That is, poles in the left-hand s-plane need not be simple.

If $\alpha = 0$, a bounded output will not result, even if $n = 1$, when the system is forced with a sinusoid of frequency ω_0. This can be seen since $Y(s)$ will then have a denominator factor of $(s^2 + \omega_0^2)^2$. This will give a term in the partial-fraction expansion of the form $A/(s^2 + \omega_0^2)^2$, which Table 4-3 indicates will give an unbounded term in $y(t)$. Therefore, BIBO stability requires strictly left-half-plane poles.

With these considerations of the permissible pole locations for a stable system, we are naturally curious about restrictions, if any, on zero locations. With no condition other than BIBO stability placed on a transfer function, there are no restrictions on zero locations. If we impose an additional condition on the transfer function, defined as *minimum phase*, zeros must be located in the left-half plane also. A minimum phase transfer function has the property that the absolute value of its phase is the smallest possible of all transfer functions with the same poles.

Because the locations of the poles of a system are important in regard to its time-domain response, it is necessary that we have at our disposal means of determining whether they lie in the right- or left-half planes. The next section describes such a procedure. However, we wish to first examine the frequency response of a system in relation to its pole–zero diagram. We recall from Chapter 3 that the frequency response of the system was provided by the magnitude and argument, referred to as the amplitude and phase response, respectively, of the

Fourier transform, $\tilde{H}(j\omega)$, of the impulse response. Since for a realizable system, $h(t) = 0$ for $t < 0$, we see that

$$\tilde{H}(j\omega) = \int_0^\infty h(t)e^{-j\omega t}\, dt = H(s)\big|_{s=j\omega}$$

That is, the amplitude response is provided by the magnitude of the transfer function $H(s)$ along the $j\omega$-axis, and the phase response is provided by the argument of $H(s)$ along the $j\omega$-axis. We can obtain a quick plot of $|\tilde{H}(j\omega)|$ and $\underline{/\tilde{H}(j\omega)}$ by considering a pole-zero plot of $H(s)$ and graphically evaluating these two functions. The procedure will be illustrated by an example.

EXAMPLE 5-6

Consider the transfer function

$$H(s) = \frac{s+2}{(s+1)^2 + 1} = \frac{s+2}{(s+1+j)(s+1-j)}$$

which corresponds to the pole–zero diagram shown in Figure 5-11. We wish to compute the magnitude and argument of $H(s)$ with $s = j\omega$. Letting $s = j\omega$, we write $\tilde{H}(j\omega)$ in terms of magnitude and phase as

$$\tilde{H}(j\omega) = \frac{|j\omega + 2|\,\underline{/j\omega + 2}}{|j\omega + (1+j)|\,|j\omega + (1-j)|\,\underline{/j\omega + (1+j)}\,\underline{/j\omega + (1-j)}}$$

$$\triangleq \frac{A_1}{A_2 A_3}\,\underline{/\theta_1 - \theta_2 - \theta_3} = |H(j\omega)|\,\underline{/\theta(j\omega)}$$

$$\text{where } A_1 = |j\omega + 2|$$

$$A_2 = |j\omega + 1 + j|$$

$$A_3 = |j\omega + 1 - j|$$

$$\theta_1 = \underline{/j\omega + 2}$$

$$\theta_2 = \underline{/j\omega + (1+j)}$$

$$\theta_3 = \underline{/j\omega + (1-j)}$$

Picking an arbitrary point on the $j\omega$-axis, we see that

$$|\tilde{H}(j\omega)| = \frac{A_1}{A_2 A_3}$$

can be viewed as the ratio of the magnitudes of the phasor labeled 1 to those of the phasors labeled 2 and 3. Furthermore,

$$\theta(j\omega) = \underline{/\tilde{H}(j\omega)} = \theta_1 - \theta_2 - \theta_3$$

can be obtained from Figure 5-11 simply by measuring the angles θ_1, θ_2, and θ_3 and computing $\theta_1 - \theta_2 - \theta_3$. Both $|\tilde{H}(j\omega)|$ and $\theta(j\omega)$ can be computed

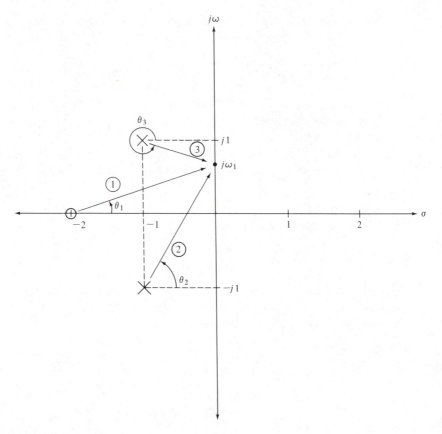

FIGURE 5-11. Pole–zero diagram for Example 5-6.

graphically or with a calculator. Results for various values of ω are given in Table 5-1. ∎

5-6

STABILITY AND THE ROUTH ARRAY

The concept of stability is extremely important in system studies. This is true for the simple reason that stability is, in almost all cases, the minimum condition that a system must meet in order to behave acceptably. On the other hand, it does not follow that a system will exhibit acceptable behavior just because it is stable.

There are two concepts of stability in common use. The first notion of stability involves the behavior of the system when it is subjected to general inputs. The concept is quite simple. If the system output is bounded for all time for every bounded input, the system is said to be bounded-input, bounded-output (BIBO) stable, as discussed in Section 5-5.

The second concept of stability involves the behavior of the system after it is subjected to small disturbances of short duration. Let us assume that the

TABLE 5-1. Frequency-Response Values for Example 5-6

ω (rad/s)	$|\tilde{H}(j\omega)|$	$\theta(j\omega)$ (deg)
0.0	1.00	0.00
0.2	1.00	-5.82
0.4	1.02	-12.19
0.6	1.028	-19.49
0.8	1.026	-27.83
1.0	1.00	-36.87
1.2	0.95	-45.90
1.4	0.87	-54.10
1.6	0.79	-62.27
1.8	0.71	-67.02
2.0	0.63	-71.57
2.5	0.49	-79.02
3.0	0.39	-83.09
4.0	0.28	-86.82
5.0	0.22	-88.30
10.0	0.10	-89.78
100.0	0.01	-90.00

system is in some equilibrium condition and that some disturbance causes it to be moved from its operating point to some other operating point. This second operating point is assumed to be not much different from the original one. Now suppose that the disturbance is removed and we observe the system behavior after this time. If the system output always remains bounded and eventually tends toward the original operating point, the system is said to be *stable*. If the system output grows without bound, the system is said to be *unstable*. If the output remains bounded but does not tend toward the original operating point, the system is said to be *marginally stable*. For example, a marginally stable system may exhibit a sustained oscillation or a constant output. These concepts are illustrated in Figure 5-12, where the horizontal position of the ball represents the output. As is readily seen, this concept of stability is quite different from the previous one. However, for lumped, linear, time-invariant systems it can be shown that the two concepts of stability are equivalent; that is, one implies the other.

It is well known that a necessary and sufficient condition for the stability of a linear, time-invariant system is that the system poles all lie in the left-half s-plane. The system poles are simply the roots of the characteristic polynomial, which is the denominator of the overall system transfer function. The validity of the stability condition is readily established. The system response will contain terms of the form $e^{p_i t}$, where the p_i are the system poles. If any of the poles have positive real parts, the response will clearly increase without bound. If the real part is zero, the response will, at best, oscillate with constant amplitude or maintain a nonzero dc value. Thus the response for a small initial disturbance does not decay to zero. On the other hand, if all poles have negative real parts (i.e., are in the left-half s-plane) all terms in the response ultimately decay to zero.

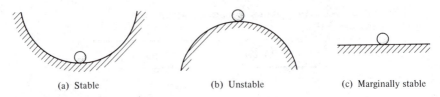

(a) Stable   (b) Unstable   (c) Marginally stable

FIGURE 5-12. Illustration of stability concepts.

We have established that the determination of system stability (for linear, time-invariant systems) is equivalent to determining whether or not all poles lie in the left-half s-plane (l.h.p.). How can we do this? One approach would be to factor the characteristic polynomial. This would tell us the exact root locations. However, this involves factoring an nth-degree polynomial in s, which is no easy task. Furthermore, the exact root locations really give more information than is required—we only need to know if all the roots are in the l.h.p. Thus we turn to methods other than factoring the characteristic polynomial.

A necessary, but not sufficient, condition for a polynomial to have strictly l.h.p. roots is that all its coefficients be strictly positive (i.e., greater than zero). The reason for this is fairly easy to see. Let us write the polynomial in factored form and then multiply it out so as to examine the form of the coefficients. We have

$$p(s) = (s + p_1)(s + p_2)(s + p_3) \cdots (s + p_n)$$

$$= s^n + (p_1 + p_2 + \cdots + p_n)s^{n-1} + \left(\begin{array}{c} \text{sum of products} \\ \text{of } p\text{'s taken two} \\ \text{at a time} \end{array} \; s^{n-1} \right)$$

$$+ \cdots + \left(\begin{array}{c} \text{sum of products} \\ \text{of } p\text{'s taken } n - 1 \\ \text{at at time} \end{array} \right) s + p_1 p_2 \cdots p_n$$

From the way in which the coefficients are constructed it is clear that if a negative or zero coefficient is present, it must arise due to one or more of the p_i having negative or zero real part. Since the p_i are the negatives of the roots, this means that at least one root lies in the r.h.p. or on the imaginary axis. Thus the system is not stable. Unfortunately, having all the coefficients positive does not guarantee stability. However, this condition is sometimes useful in that it allows us to immediately recognize as unstable a system whose characteristic polynomial contains nonpositive coefficients.

A stability test that is both necessary and sufficient is clearly desirable. Such a test is provided by the *Routh criterion*. It is sometimes called the Routh–Hurwitz criterion because Hurwitz developed an equivalent test formulated in terms of determinants. The Routh criterion makes use of an array of coefficients generated from the coefficients of the characteristic polynomial. When the array is complete, the number of sign changes in the first column is equal to the number of r.h.p. roots. For a stable system there must be no such roots (i.e., no first-column sign changes).

The Routh array is constructed in a straightforward manner from the coefficients of the characteristic polynomial. Consider a polynomial

$$a_n s^n + a_{n-1} s^{n-1} + a_{n-2} s^{n-2} + \cdots + a_1 s + a_0 \qquad (5\text{-}35)$$

The Routh array is formed as shown in Figure 5-13. The first two rows are obtained directly from the polynomial coefficients. The coefficients of the third row are then generated from those of the first two rows as follows:

$$b_1 = \frac{a_{n-1} a_{n-2} - a_n a_{n-3}}{a_{n-1}}, \qquad b_2 = \frac{a_{n-1} a_{n-4} - a_n a_{n-5}}{a_{n-1}} \qquad (5\text{-}36)$$

and so on until there are no more a terms from which to compute coefficients. The multiplication is indicated by arrows in Figure 5-13. Once the s^{n-2} row is obtained, it is used with the coefficients of the s^{n-1} row to compute the coefficients of the next (s^{n-3}) row. This is done in exactly the same manner; that is,

$$c_1 = \frac{b_1 a_{n-3} - b_2 a_{n-1}}{b_1}, \qquad c_2 = \frac{b_1 a_{n-5} - b_3 a_{n-1}}{b_1} \qquad (5\text{-}37)$$

and so on, until this row stops. Succeeding rows are computed in the same way until the array terminates with only one element in the s^0 row. It is important to label the power of s corresponding to each row so that we know when the array terminates. Finally, when the array is complete, the elements of the first column are examined for sign changes. The number of sign changes is equal to

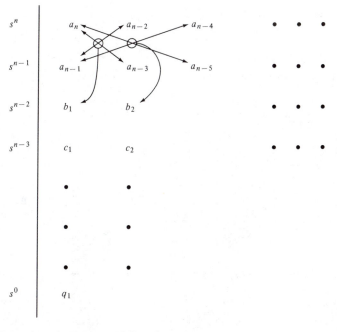

FIGURE 5-13.   Formation of Routh array.

the number of r.h.p. roots of the polynomial from whose coefficients the array was generated. The process sounds a good deal more complicated than it actually is. The following example will illustrate the procedure.

EXAMPLE 5-7

Consider the following polynomial:

$$p(s) = s^3 + 14s^2 + 41s - 56$$

Find the number of r.h.p. roots.

We can readily see that there is at least one root in the r.h.p. because of the negative coefficient, but let us apply the Routh test and see how many r.h.p. roots there are. The Routh array is found to be

s^3	1	41
s^2	14	-56
s	$\dfrac{41(14) + 56}{14} = 45$	
s_0	-56	

We see that there is one sign change in the first column, so there is one r.h.p. root. The result can be checked for this simple case by factoring the polynomial as

$$p(s) = (s - 1)(s + 7)(s + 8)$$

Clearly, the r.h.p. root is at $s = 1$. ∎

EXAMPLE 5-8

Find the number of r.h.p. roots of

$$p(s) = s^4 + 5s^3 + s^2 + 10s + 1$$

The Routh array is

s^4	1	1	1
s^3	5	10	
s^2	-1	1	
s	15		
s^0	1		

Here we have two sign changes ($+5$ to -1 and -1 to $+15$), so there are two r.h.p. roots. ∎

There are two cases in which development of the Routh array requires slight modification of the procedure discussed. Both pertain to zero entries in the array.

The first case is one in which the first element of a row is zero, but there is at least one nonzero element in the row. The zero element causes difficulty because we must divide by it in order to compute the elements of the next row. To avoid the difficulty, we replace the zero element by a small positive number ϵ and continue. When the array is completed we let ϵ tend toward zero and examine the sign changes in the first column. This technique is illustrated in Example 5-9.

EXAMPLE 5-9

Determine the stability of a system whose characteristic polynomial is

$$p(s) = s^4 + s^3 + s^2 + s + 3$$

The Routh array is begun as usual, but we find a zero first element in the s^2 row with another element in that row nonzero. So we replace the zero by ϵ and continue as shown below.

s^4	1	1	3
s^3	1	1	
s^2	$\cancel{0}\,\epsilon$	3	
s	$\dfrac{\epsilon - 3}{\epsilon}$		
s^0	3		

We then let ϵ tend toward zero. All terms in the first column remain positive except the term in the s row, which becomes very large and negative and leads to two sign changes. Hence the system is unstable since it has two r.h.p. poles. ∎

The second case requiring special treatment is one in which an entire row of the array is zero. This situation arises whenever some or all of the roots are arranged symmetrically about the origin. The zero row is always one corresponding to an odd power of s. The preceding row contains coefficients which may be used to form a polynomial that is even in s. This polynomial is known as the *auxiliary polynomial,* and it contains the roots that cause the zero row. In addition, the degree of the auxiliary polynomial is equal to the number of roots that occur in symmetric pairs. To continue with the array, we differentiate the auxiliary polynomial and use the resulting coefficients in place of the row of zeros. The remaining portion of the array is computed as usual. the procedure is best illustrated by an example.

EXAMPLE 5-10

Find the number of r.h.p. roots of

$$p(s) = s^7 + 3s^6 + 3s^5 + s^4 + s^3 + 3s^2 + 3s + 1$$

The Routh array is formed as shown below.

s^7	1	3	1	3
s^6	3	1	3	1
s^5	$\dfrac{8}{3}$	0	$\dfrac{8}{3}$	
s^4	1	0	1	← auxiliary polynomial $s^4 + 1$
s^3	$\cancel{0}4$	0		$\leftarrow \dfrac{d}{ds}(s^4 + 1) = 4s^3$
s^2	$\cancel{0}\epsilon$	1		
s	$-\dfrac{4}{\epsilon}$			
s^0	1			

We see that two r.h.p. roots are predicted. The roots that cause the zero row are contained in the auxiliary polynomial and are found to be $\pm\sqrt{2}/2 \pm j\,\sqrt{2}/2$. We may write $p(s)$ as

$$p(s) = (s + 1)^3 (s^4 + 1)$$

so that there are in fact two r.h.p. roots—the two contained in $(s^4 + 1)$. ■

A common application of the Routh criterion is that of determining the range of some system parameter for which the system is stable, that is, for which the characteristic polynomial has no r.h.p. roots. Such a situation arises frequently in the study of control systems, where the variable parameter is most frequently an amplifier gain. The procedure is illustrated in Example 5-11.

EXAMPLE 5-11
Find the range of K for which

$$s^3 + 3s^2 + 3s + 1 + K = 0$$

has no r.h.p. roots. The Routh array is

s^3	1	3
s^2	3	$1 + K$
s	$\dfrac{8 - K}{3}$	
s^0	$1 + K$	

For the polynomial to have no r.h.p. roots, there must be no sign changes in the first column, that is,

$$\frac{8 - K}{3} > 0, \qquad 1 + K > 0$$

Then the condition on K is

$$-1 < K < 8$$

in order to have no r.h.p. roots. ∎

BODE PLOTS

Frequently, it is desirable to display graphically the frequency response of a system. Such graphical displays are useful in control system analysis and design and in filter design. The frequency response is a sinusoidal steady-state concept and is obtained from the system transfer function by letting $s = j\omega$. Usually, we are interested in plotting both the magnitude and the angle of the transfer function versus the angular frequency ω. It is conventional to plot the magnitude in decibels (dB) and use a log scale for ω. [The magnitude of a function H in decibels, denoted $|H|_{dB}$, is given by $|H|_{dB} = 20 \log_{10}|H|$.] There are two reasons for using this choice of scales. First, since rather large ranges of frequency and magnitude may be involved, the log scales may help to compress the data so that they can be displayed on a conventionally sized page. Second, the nature of the transfer function is such that the choice of a decibel scale leads to computational simplicity.

Recall that for a linear, lumped, time-invariant system, the transfer function is a rational function of s, that is,

$$H(s) = \frac{P(s)}{Q(s)}$$

where P and Q are polynomials in s. Therefore, $H(s)$ can be written as a product of factors of the following forms:

1. Constant factor, K.
2. Poles or zeros at the origin, $s^{\pm N}$.
3. Real poles or zeros, $(Ts + 1)^{\pm N}$.
4. Complex-conjugate poles or zeros, $(T^2 s^2 + 2\zeta Ts + 1)^{\pm N}$, where ζ is the damping ratio and $0 < \zeta < 1$.

Since

$$|H|_{dB} = 20 \log_{10}|H|$$

plotting the magnitude of H in decibels allows us to plot the factors individually and then sum the results to obtain the complete magnitude plot. A frequency-response plot obtained in this way is known as a *Bode plot,* after H. W. Bode, who invented and used such plots to analyze feedback amplifiers in connection with his work at Bell Laboratories.

The phase plot is also obtained by summing the plots of the phase of the individual factors. This follows directly from the fact that the angle of the product of two complex numbers is the sum of the angles of the two numbers. Since we can obtain both the magnitude and phase plots for the complete transfer function by summing the plots for the individual factors, we now turn to developing the Bode plot for each of the factors of interest.

1. *Constant Factor, K.* For the constant factor K, we have

$$|K|_{dB} = 20 \log_{10} |K| \tag{5-38}$$

which is just a constant. Therefore, the magnitude plot is just a horizontal line. The angle is zero or $\pm 180°$, depending on whether K is positive or negative, respectively. Therefore, the phase curve is a horizontal line.

2. *Poles or Zeros at the Origin, $s^{\pm N}$.* In this case

$$H(j\omega) = (j\omega)^{\pm N}$$

so

$$|H|_{dB} = 20 \log_{10} \omega^{\pm N} = \pm 20N \log_{10} \omega \tag{5-39}$$

Since a log scale is used for ω, (5-39) describes a family of straight lines that intersect the ω-axis (0 dB) at $\omega = 1$ and which have slope $\pm 20N$ dB/decade. (A decade is a factor of 10 in ω, for which $\log_{10} \omega$ changes by 1 unit.) These lines are shown in Figure 5-14.

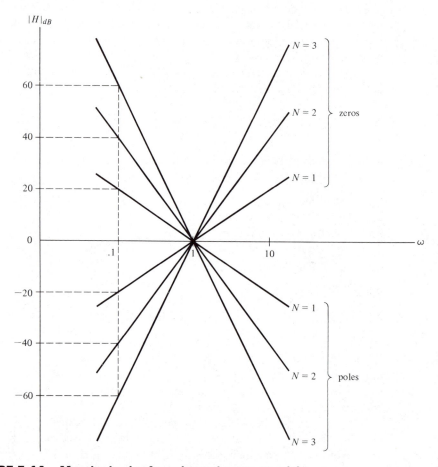

FIGURE 5-14. **Magnitude plot for poles and zeros at origin.**

The angle of H is given by

$$\underline{/H} = \pm N90°$$ (5-40)

so poles or zeros at the origin simply introduce a constant phase curve as a function of ω.

3. *Real Poles or Zeros*, $(Ts + 1)^{\pm N}$. Here we have amplitude and phase curves which are not straight lines; however, they are asymptotic to straight lines. Hence our strategy will be to develop these asymptotes and make corrections as necessary to obtain the actual curves. For the magnitude we obtain

$$|H|_{dB} = 20 \log_{10} (\sqrt{(\omega T)^2 + 1})^{\pm N}$$

$$= \pm 20N \log_{10} \sqrt{(\omega T)^2 + 1}$$ (5-41)

$$= \pm 10N \log_{10} (\omega^2 T^2 + 1)$$

When ω is very small (i.e., $\omega \ll 1/T$), (5-41) becommes

$$|H|_{db} \simeq \pm 10N \log_{10} 1 = 0 \text{ dB}$$

Therefore, the low-frequency asymptote is 0 dB. When $\omega \gg 1/T$ we obtain

$$|H|_{dB} = \pm 20N \log_{10} \omega T = \pm 20N \left(\log_{10} \omega - \log_{10} \frac{1}{T} \right)$$ (5-42)

which is the equation of the high-frequency asymptote. It describes a straight line of slope $\pm 20N$ dB/decade which intersects 0 dB when $\omega = 1/T$. The actual curve lies almost on the asymptotes except within about one-half decade either side of $\omega = 1/T$, which is called the *corner frequency*. In this region, the magnitude varies smoothly. At the corner frequency

$$|H|_{dB} = \pm 10N \log_{10} 2 \simeq \pm 3N \text{ dB}$$ (5-43)

This gives us enough information to sketch the magnitude curve quite closely. A sketch is shown for a real pole factor of order N in Figure 5-15. A zero factor would have a magnitude plot which is the mirror image through the ω-axis of Figure 5-15.

The phase curve is obtained from

$$\underline{/H} = \pm N \tan^{-1} \omega T$$ (5-44)

At low frequencies the angle of H is about $0°$, whereas at high frequencies it approaches $\pm N90°$. At the corner frequency the angle is $\pm N45°$. The phase curve for a pole of order N is shown in Figure 5-16 together with an approximation consisting of straight lines. Note that the sloped-line asymptote starts one decade before the corner frequency and ends a decade after it. The slope is $-N45°$/decade. For a zero factor, the curve would be the mirror image through the ω-axis of Figure 5-16.

4. *Complex-Conjugate Poles or Zeros* $(T^2 s^2 + 2\zeta Ts + 1)^{\pm N}$. For simplicity we consider only the case of a single pair of complex-conjugate poles,

$$H(s) = \frac{1}{T^2 s^2 + 2\zeta Ts + 1}$$ (5-45)

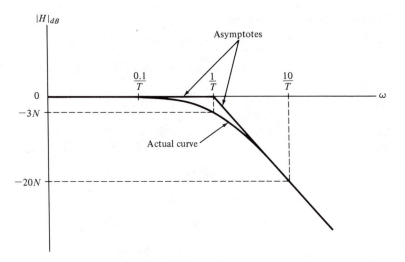

FIGURE 5-15. Asymptotic and actual magnitude curves for a real pole factor of order **N**.

If the poles are repeated N times, all coordinates on the curves we generate are simply multiplied by N. Their basic shape remains unchanged. Of course, if we have zeros instead of poles, curves are mirror images through the ω-axis of the curves corresponding to poles.

From (5-45) we may write the magnitude of H in decibels as

$$|H|_{dB} = -20 \log_{10} [(1 - T^2\omega^2)^2 + 4\zeta^2T^2\omega^2]^{1/2}$$

$$= -10 \log_{10} [(1 - T^2\omega^2)^2 + 4\zeta^2T^2\omega^2]$$

(5-46)

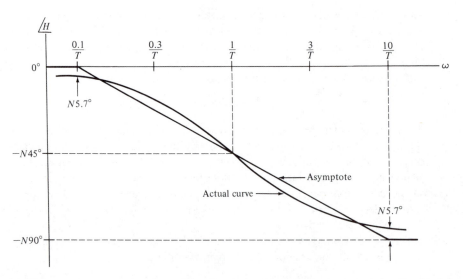

FIGURE 5-16. Asymptotic and actual phase curves for a real pole factor of order **N**.

Again we consider the asymptotic cases. For very small ω ($\omega \ll 1/T$) we obtain

$$|H|_{dB} \simeq -10 \log_{10} 1 = 0 \text{ dB} \qquad (5\text{-}47)$$

Therefore, the curve is asymptotic to the ω-axis at low frequencies. For $\omega \gg 1/T$ we may write (5-46) as

$$
\begin{aligned}
|H|_{dB} &\simeq -10 \log_{10} (\omega^4 T^4 + 4\zeta^2 T^2 \omega^2) \\
&\simeq -10 \log_{10} \omega^4 T^4 = -40 \log_{10} \omega T \qquad (5\text{-}48) \\
&= -40 \left(\log_{10} \omega - \log_{10} \frac{1}{T} \right)
\end{aligned}
$$

Note that we have neglected the ω^2 term compared to the ω^4 term because ω is large. Equation (5-48) describes a straight line of slope -40 dB/decade. The line intersects the ω-axis at the corner frequency $\omega = 1/T$. The actual magnitude curve is asymptotic to this line at high frequencies. In the general vicinity of the corner frequency the actual curve does not lie on the asymptotes but varies smoothly. However, the manner in which it varies depends on the damping ratio. For $\zeta < \sqrt{2}/2 = 0.707$, the curve exhibits a maximum. The magnitude of this resonant peak (in decibels) is given by $20 \log_{10} (1/2\zeta \sqrt{1 - \zeta^2})$. The frequency at which this maximum occurs, called the *resonant frequency*, is $\omega_p = (1/T) \sqrt{1 - 2\zeta^2}$. These values give enough information to sketch the magnitude curve. Figure 5-17 shows the asymptotic curve and actual magnitude curves for several values of the damping ratio.

The phase curve for the complex pole factor may be obtained from

$$\underline{/H} = -\tan^{-1} \frac{2\zeta T \omega}{1 - \omega^2 T^2} \qquad (5\text{-}49)$$

However, if we wish to use the principal value of the inverse tangent, (5-49) is valid only for $0 < \omega < 1/T$. For $\omega > 1/T$ the curve lies below $-90°$ and approaches $-180°$ for large ω since two poles are present. An expression valid for $\omega > 1/T$ is

$$\underline{/H} = -180° + \tan^{-1} \frac{2\zeta T \omega}{\omega^2 T^2 - 1} \qquad (5\text{-}50)$$

Equations (5-49) and (5-50) may be used to generate phase curves. Note that these curves depend on the damping ratio. We may still use a straight-line approximation beginning at $\omega = 0.1/T$ and ending at $\omega = 10/T$ with slope $-90°$/decade, but the error between the approximation and the actual curve increases as ζ is made small.† Figure 5-18 shows the asymptotic curves and actual phase curves for several values of ζ.

Now that we know how to obtain Bode plots for the individual factors which may appear in a transfer function, we are in a position to construct a Bode plot for almost any rational transfer function. The usual strategy is to plot the asymp-

---

†The asymptotic approximation given here is common in the study of control systems. Other straight-line approximations would be more accurate and are sometimes encountered.

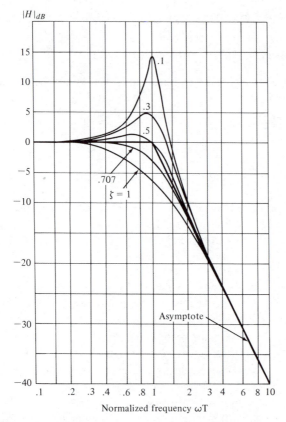

FIGURE 5-17. Asymptotic and actual magnitude curves for a complex pole factor for various ζ.

totes for the magnitude and phase curves of the individual factors and then add to obtain the asymptotic curves for the complete transfer function. Finally, the corrected curves are sketched. We close this section with a detailed example which illustrates the complete procedure.

EXAMPLE 5-12

Construct a Bode plot for the transfer function

$$H(s) = \frac{100 \, (s + 2)}{s(s^2 + 5s + 25)}$$

First we must convert this to the standard form in which the factors appear as $(sT + 1)$ and $(s^2T^2 + 2\zeta Ts + 1)$. Factoring 2 from the numerator and 25 from the denominator yields

$$H(s) = \frac{8(0.5 \, s + 1)}{s[(0.2)^2 s^2 + 2(0.5)(0.2) \, s + 1]}$$

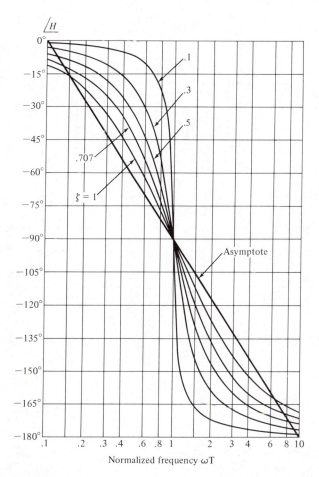

FIGURE 5-18. Asymptotic and actual phase curves for a complex pole factor for various ζ.

We can recognize the following factors:

1. Constant factor, $K = 8$ or 18.06 dB.
2. Simple zero factor $(0.5s + 1)$, corner frequency $\omega = 2$.
3. Simple pole factor at origin, $1/s$.
4. Complex pole factor $[(0.2)^2 s^2 + 2(0.5)(0.2)s + 1]$, corner frequency $\omega = 5$, damping ratio $\zeta = 0.5$.

The asymptotes for these factors are obtained as previously outlined and are shown in Figure 5-19 and 5-20. Note that the constant factor introduces no contribution to the phase curve, while the pole at the origin yields $-90°$ regardless of frequency. The complete asymptotic magnitude and phase curves are shown as dashed lines. They are obtained by adding together the individual curves. Finally, the corrected curves ar the solid smooth curves shown. These are constructed by using the results shown in Figure 5-14 through 5-18 and adding the various corrections to the final asymptotic curve. ∎

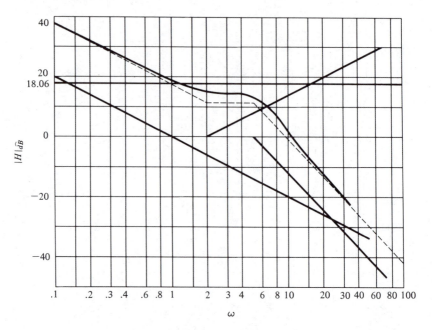

FIGURE 5-19. Asymptotic and actual magnitude curves for Example 5-12.

BLOCK DIAGRAMS

We close this chapter with a discussion of a special form of system model which is very useful in the analysis of large systems—the block diagram. The block diagram is a combination of mathematical and schematic models of the system. As such, it gives not only the equations of each of the parts of the system, but also shows how these parts are interconnected. Therefore, the block diagram gives considerable insight as to the system structure and provides all necessary information to determine the system equations or transfer function.

The basic element of block diagrams is the single block shown in Figure 5-21. Signal transmission is assumed possible only in the direction of the arrows. In Figure 5-21 $X(s)$ is the Laplace transform of the input, $Y(s)$ is the Laplace transform of the output, and $G(s)$ is the transfer function discussed in Section 5-5. Accordingly, the output is given by

$$Y(s) = G(s)X(s) \qquad (5\text{-}51)$$

and this relation is assumed to hold *regardless of what is connected to the output of the block*. Such an assumption is equivalent to neglecting all loading effects. Obviously, there will be cases in which this assumption is not valid. In such cases, we must develop the transfer function under the conditions of system operation (i.e., we must take into account the effects of loading). Once we do this, we may then proceed as though loading were not present. From this point on, then, we will assume that the effects of loading have already been taken

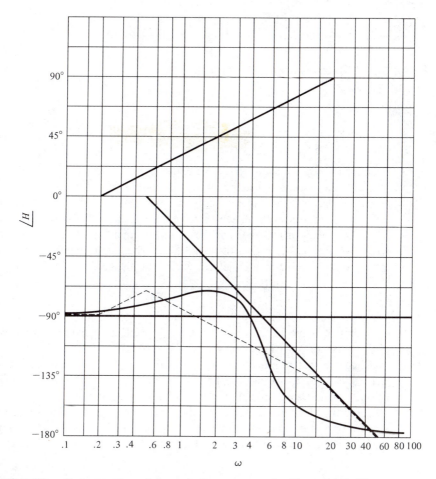

FIGURE 5-20. Asymptotic and actual phase curves for Example 5-12.

into account in the development of the transfer function and need not be considered further.

The simplest interconnection of blocks is the cascade connection shown in Figure 5-22. The transfer function can be found easily. Note that

$$Y(s) = G_2(s)W(s), \qquad W(s) = G_1(s)X(s)$$

Hence

$$Y(s) = G_2(s)G_1(s)X(s)$$

FIGURE 5-21. Basic element of block diagram.

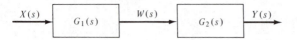

FIGURE 5-22. Cascade connection of blocks.

or

$$G(s) = \frac{Y(s)}{X(s)} = G_1(s)G_2(s) \tag{5-52}$$

The result is readily extended to several blocks in cascade: The transfer function of blocks in cascade is the product of the transfer functions of the individual blocks. Rules such as this become very important in simplifying complicated block diagrams, as we shall see shortly. Before doing this, however, we need to define one further element of the block diagram—the summer. A summer with several inputs is shown in Figure 5-23. The output is just the sum of the inputs, each with the sign indicated; that is, for the case shown

$$Y(s) = X_1(s) - X_2(s) - X_3(s) + X_4(s) \tag{5-53}$$

Summers are usually used to model the comparison function so essential to the operation of a feedback control system, as well as to indicate points where disturbances or other inputs may enter the system.

When the system block diagram is very complicated it becomes tedious to write equations, as we did for the cascade connection, and solve them for the overall system transfer function. It is helpful to have some way of systematically reducing the block diagram to some form we can recognize. The most common simple form is the single-loop feedback system of Figure 5-24. We can obtain the system transfer function easily. First note that

$$C(s) = G(s)E(s) \tag{5-54}$$

Furthermore,

$$E(s) = R(s) - H(s)C(s) \tag{5-55}$$

Substitution of (5-55) into (5-54) yields

$$C(s) = G(s)R(s) - G(s)H(s)C(s)$$

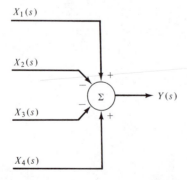

FIGURE 5-23. Symbol for summer.

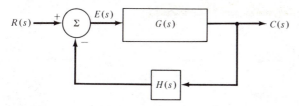

FIGURE 5-24. Single-loop feedback system.

or

$$C(s)[1 + G(s)H(s)] = G(s)R(s) \qquad (5\text{-}56)$$

Rearranging, we obtain the transfer function as

$$T(s) = \frac{C(s)}{R(s)} = \frac{G(s)}{1 + G(s)H(s)} \qquad (5\text{-}57)$$

Thus the single-loop system may be replaced by a single block whose transfer function is given by (5-57). Several relations useful for simplification of block diagrams are given in Table 5-2. It is easy to verify these relations by writing down the algebraic relations for each case and showing that the same transfer functions result.

When we reduce a block diagram we make use of the relations shown in Table 5-2 to combine blocks in cascade or parallel, move summing points and pick off points and simplify loops until only one loop is left. At that point we can simply write down the answer using the last entry of the table. Although it sounds complicated, the procedure is really very straightforward. It is best discussed with a particular example in mind.

EXAMPLE 5-13

Find $T(s) = Y(s)/X(s)$ for the block diagram of Figure 5-25.

First, we replace the cascade connection of G_2 and G_4 by its equivalent block with transfer function $G_2 G_4$. This new block is then in parallel with G_3, so we perform a further simplification, yielding a block with transfer function $G_3 + G_2 G_4$. At this stage the system appears as shown in Figure 5-26. Next we replace the inner loop consisting of G_1 and H_1 by its equivalent transfer function $G_1/(1 + G_1 H_1)$ and further simplify by combining this block with $G_3 + G_2 G_4$. Now the system is reduced to the single loop shown in Figure 5-27. We can immediately write the transfer function of this single-loop system as

$$T(s) = \frac{Y(s)}{X(s)} = \frac{G_1(G_3 + G_2 G_4)/(1 + G_1 H_1)}{1 + H_2 G_1(G_3 + G_2 G_4)/(1 + G_1 H_1)}$$

The result may be simplified somewhat to yield

$$T(s) = \frac{G_1(G_3 + G_2 G_4)}{1 + G_1 H_1 + H_2 G_1 (G_3 + G_2 G_4)}$$

■

TABLE 5-2. Relations for Block Diagram Reduction

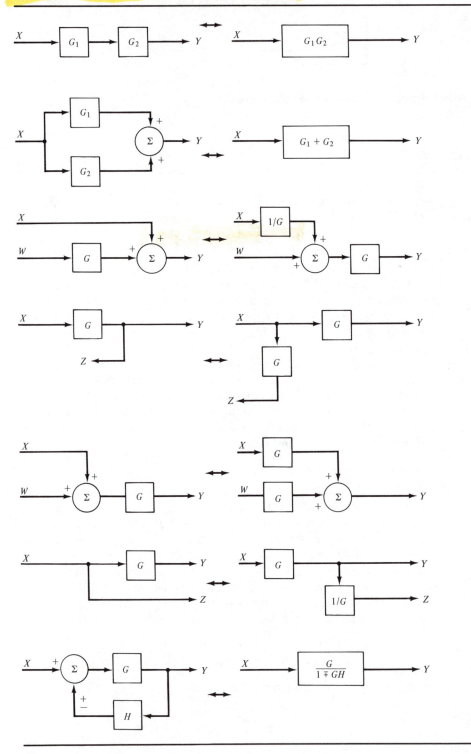

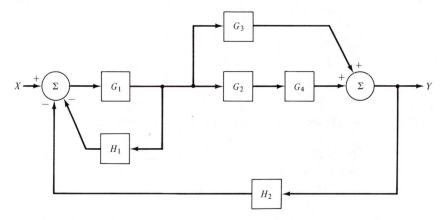

FIGURE 5-25. Block diagram to be reduced.

Although this example has not used all the equivalencies of Table 5-2, it does indicate the general procedure to be followed in reducing block diagrams.

We close this section with a perhaps simpler, but considerably more realistic, example.

EXAMPLE 5-14

The armature-controlled dc servo motor is a common element in position control systems. We wish to find the transfer function between the input voltage (armature voltage E_a) and the output position (angular shaft position θ).

First, we must define some parameters of the motor and obtain equations governing its behavior. Once we have done this we can construct a block diagram and then determine the transfer function. This is an approach commonly employed in the development of models for physical systems. We assume that the motor has inertia J and friction B. Then the torque T produced by the motor and the shaft position θ are related by

$$J\ddot{\theta} + B\dot{\theta} = T$$

Laplace transforming with zero initial conditions and rearranging, we obtain

$$\frac{\theta(s)}{T(s)} = \frac{1}{s(sJ + B)} \tag{5-58}$$

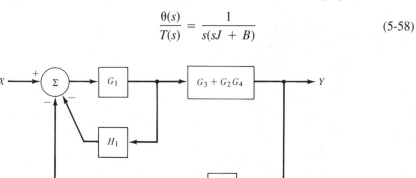

FIGURE 5-26. Partially simplified block diagram.

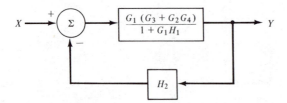

FIGURE 5-27. System after reduction to single-loop form.

as the transfer function between torque and shaft position. In an armature-controlled dc motor the torque produced is proportional to the armature current,

$$T = K_m I_a \tag{5-59}$$

where K_m is called the *motor constant* and depends on the particular motor being used. To obtain an expression for the armature current, two factors must be considered. First, the armature contains resistance and inductance R_a and L_a, respectively. Thus its impedance is

$$Z_a(s) = sL_a + R_a$$

Second, a *back emf* (electromotive force) appears in the armature circuit as a result of the speed of the motor:

$$v_b(t) = K_B \dot{\theta}(t)$$

or

$$V_b(s) = K_B s\theta(s) \tag{5-60}$$

where K_B is a constant of proportionality. This back emf is in opposition to the applied voltge E_a, so the voltage appearing across the armature impedance is

$$V_a(s) = E_a(s) - V_b(s) \tag{5-61}$$

The armature current is then given by

$$I_a(s) = \frac{V_a(s)}{sL_a + R_a} \tag{5-62}$$

Equations (5-58) through (5-62) define a linear model of the armature-controlled dc motor. A block diagram representing these equations is shown in Figure 5-28. This is a single-loop negative-feedback system with $G(s)$ given by the

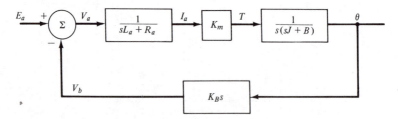

FIGURE 5-28. Block diagram of armature-controlled dc motor.

product of the forward-loop transfer functions; that is,

$$G(s) = \frac{K_m}{s(sL_a + R_a)(sJ + B)}$$

and

$$H(s) = K_B s$$

Therefore,

$$\frac{\theta(s)}{E_a(s)} = \frac{G}{1 + GH} = \frac{K_m/[s(sL_a + R_a)(sJ + B)]}{1 + K_m K_B/[(sL_a + R_a)(sJ + B)]}$$

or

$$\frac{\theta(s)}{E_a(s)} = \frac{K_m}{s[(sL_a + R_a)(sJ + B) + K_m K_B]}$$

It is perhaps worth pointing out that in most small motors L_a may be neglected, so that

$$\frac{\theta(s)}{E_a(s)} \simeq \frac{K_m}{s[sJR_a + (R_a B + K_m K_B)]}$$

By grouping the various parameters in an obvious way we could also obtain the following representation as an approximation that is valid for many small motors:

$$\frac{\theta(s)}{E_a(s)} \simeq \frac{K'_m}{s(s\tau_m + 1)} \qquad (5\text{-}63) \quad \blacksquare$$

In Example 5-11 we showed how to use the Routh criterion to determine the range of a system parameter in order to ensure stability. We are now in a position to see how such a problem might come about.

EXAMPLE 5-15
The system shown in Figure 5-29 is a block diagram of a position-control system which, if it is stable, will follow a ramp angular position input in steady state. The motor is modeled by (5-63) with $K'_m = 5$ and $\tau_m = 1$ s. The controller includes a variable gain K. We wish to know the range of K, if any, which yields a stable system.

Straightforward block diagram reduction yields the system transfer function as (show this)

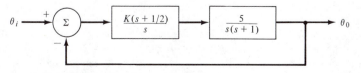

FIGURE 5-29. Position-control system.

$$\frac{\theta_0(s)}{\theta_i(s)} = \frac{5K(s + \frac{1}{2})}{s^3 + s^2 + 5Ks + 2.5K}$$

The characteristic polynomial is just

$$s^3 + s^2 + 5Ks + 2.5K$$

and must have no r.h.p. roots if the system is to be stable. The Routh array is

s^3	1	$5K$
s^2	1	$2.5K$
s	$2.5K$	
s^0	1	

For stability, there must be no sign changes in the first column; therefore, any $K > 0$ will yield a stable system. In actual design of such a system it would be necessary to choose a value of K that led to a suitable transient response. Such topics are discussed in detail in courses concerning control systems. ∎

5-9

SUMMARY

In this chapter we have treated a number of important applications of the Laplace transform. We have seen how the Laplace transform may be used to provide a general analysis technique for linear, fixed electrical networks which may contain nonzero initial conditions. The advantage of the Laplace transform is that it allows simultaneous determination of the forced and the transient responses of the system.

The concept of a transfer function was introduced and various means of examining its properties were discussed. Bode plots were developed as a convenient means of displaying frequency-response properties and the Routh test was advanced for the study of stability. Finally, the block diagram was used to provide a combination mathematical and pictorial representation of a system, and rules were given for simplifying block diagrams in order to determine system transfer functions.

FURTHER READING

For further reading on applications of the Laplace transform in systems analysis, the reader is referred to the references given in Chapter 4.

K. OGATA, *Modern Control Engineering*. Englewood Cliffs, N.J.: Prentice-Hall, 1970.
This book contains good treatments of the Routh test, block diagrams, and Bode plots, as do most control systems texts. The level is about the same as the present text, but many more examples (all oriented toward control systems) are given. The solved problems at the end of the chapters are most interesting.

PROBLEMS

SECTIONS 5-2 AND 5-3

5-1. Sketch the Laplace-transformed equivalents, including initial condition generators, of the circuits shown. Choose initial condition generators appropriate for writing loop equations.

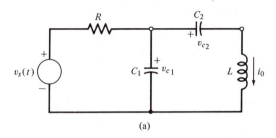

(a)

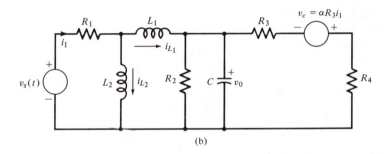

(b)

5-2. Write the Laplace-transformed loop equations for the circuits of Problem 5-1 by inspection. Change to matrix notation. Include initial conditions.

5-3. Sketch Laplace-transformed equivalent circuits for Problem 5-1 using initial condition generators appropriate for writing node equations.

5-4. Write by inspection the Laplace-transformed node equations for the circuits of Problem 5-1. Change to matrix notation. Include initial conditions.

5-5. Find $v_0(t)$ for the circuit shown. All initial conditions are zero.

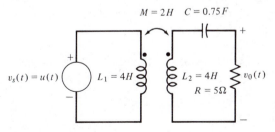

5-6. Obtain the current in the circuit shown, as a function of ω. Identify the steady-state and transient components. Plot the magnitude of its steady-state component versus ω.

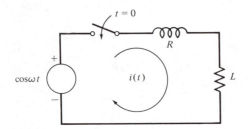

5-7. If more than two coupled coils are present in the same circuit, it is necessary to show polarity markings with more than one set of symbols (e.g., dots and triangles). An example of such a circuit is shown.

(a) Assuming mutual inductances of M_{12}, M_{13}, and M_{23} between the coils with self-inductances L_1, L_2 and L_3, respectively, write the governing differential equations on a loop basis for this circuit.

(b) Laplace transform the differential equations found in part (a) assuming zero initial conditions.

(c) Discuss the feasibility of replacing the coupled coils with equivalent T-circuits.

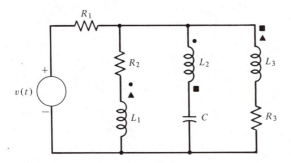

5-8. Repeat Example 5-4 assuming that $V_c(s) = \alpha R_3 I_1(s)$.

SECTION 5-5

5-9. Show that (5-33) results by Laplace transforming (5-34), assuming zero initial conditions, and taking the ratio of $Y(s)$ to $X(s)$, where $Y(s)$ is the Laplace transform of the output and $X(s)$ is the Laplace transform of the input.

5-10. Assume that the rational functions of s in Problem 4-7 are transfer functions. Plot the magnitude of the frequency-response function for each by using the technique discussed in Section 5-4.

5-11. Assume that the rational functions of s in Problem 4-10 are transfer functions. Plot the magnitude of the frequency response for each by using the technique discussed in Section 5-4.

SECTION 5-6

5-12. Determine the number of r.h.p. roots of the following polynomials.
(a) $s^3 + 6s^2 + 3s + 2$
(b) $s^4 + 3s^3 + 12s^2 + 12s + 36$

(c) $s^4 + s^3 - s - 1$

(d) $4s^5 + s^4 + 4s^3 + s^2 + 15s + 10$

5-13. Find the range, if any, of the parameter K for stability.

(a) $s^3 + 7s^2 + 11s + 5 + K$

(b) $s^4 + 2s^3 + 2s^2 + s + K$

5-14. Find the condition on the constants a, b, and c for the polynomial shown to have no r.h.p. roots.

$$s^3 + as^2 + bs + c$$

5-15. *Singular cases with initial conditions.* If the following situations arise in circuits, we have the so-called "singular" cases, resulting in discontinuous capacitance voltages or inductor currents:

(1) Loops consisting entirely of capacitors or capacitors and voltage generators.

(2) Junctions of inductors or inductors and current generators.

Obtain $i_1(t)$ and $i_2(t)$ in the circuit shown, using two methods:

(a) Employ $t = 0^-$ initial conditions in the Laplace transform.

(b) Invoke KVL and conservation of charge through $t = 0$ around each loop. Show that both procedures give the same result.

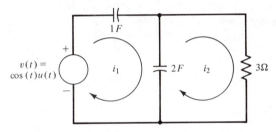

Both capacitors discharged at $t = 0$.

Answer:

$$i_2(t) = \tfrac{1}{82} \left(\tfrac{1}{9}e^{-t/9} + 9 \cos t - \sin t\right)u(t)$$

$$i_1(t) = \tfrac{2}{3} \delta(t) + \tfrac{1}{82} \left(\tfrac{1}{27} e^{-t/9} + 3 \cos t - 55 \sin t\right)u(t)$$

Note that $v_{c_1}(0^+) = \tfrac{2}{3} v$; $v_{c_2}(0^+) = \tfrac{1}{3} v$.

5-16. *Singular example for inductor currents.* Referring to the discussion in Problem 5-15, obtain $v_1(t)$ and $v_2(t)$ in the circuit shown by using two methods:

(a) Employ $t = 0^-$ initial conditions in the Laplace transform.

(b) Invoke KCL and require conservation of flux linkages $[Li(t)]$ through $t = 0$ at each node.

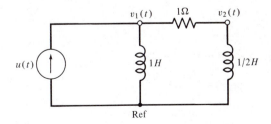

Both inductors have zero current flowing at $t = 0^-$.

Answer:

$$v_1(t) = \tfrac{1}{3}\,\delta(t) + \tfrac{4}{9}\,e^{-2t/3}u(t)$$

$$v_2(t) = \tfrac{1}{3}\,\delta(t) - \tfrac{2}{9}\,e^{-2t/3}u(t)$$

Note that the voltage impulses are required to cause the currents to change from 0 to $i_{L_1}(0^+) = \tfrac{1}{3}$ A and $i_{L_2}(0^+) = \tfrac{2}{3}$ A.

5-17. *Duality.* For planar networks, the following rules allow us to find a dual of a network; that is, the loop equations for one network are the same as the node equations for the other network except for the symbols:

1. Assign a node for the dual network to be constructed to each mesh of the given network plus an additional reference node.

2. For each *inductor* of L henries in the original network, assign a *capacitor* of L farads in the dual. For each *capacitor* of C farads in the original network, assign an *inductor* of C henries in the dual. For each *resistance* of R ohms in the original network, assign a *conductance* of R mhos in the dual. For each *current* (*voltage*) generator in the original network, assign a *voltage* (*current*) generator in the dual of the same numerical value.

3. The mesh (node) equations for the original network will now be identical to the node (mesh) equations of the dual network with the following replacements:

$$\text{node equations} \leftrightarrow \text{mesh equations}$$

$$v \leftrightarrow i$$

$$L \leftrightarrow C$$

$$R \leftrightarrow G$$

$$\text{open circuit} \leftrightarrow \text{short circuit}$$

Obtain the duals of the networks given in Problems 5-15 and 5-16. Show that they have equivalent equilibrium equations on a mesh and node basis.

SECTION 5-7

5-18. Assume that the rational functions of s in Problem 4-10 are transfer functions. Construct amplitude and phase Bode plots (both asymptotic annd corrected).

5-19. Construct a Bode plot for the transfer function

$$G(s) = \frac{250(s + 2)}{s(s^2 + 12s + 100)}$$

5-20. Construct a Bode plot for the transfer function

$$G(s) = \frac{1000(s^2 + s + 1)}{(s + 6)(s + 10)^2}$$

5-21. Show that the complex pole factor given by (5-45) will have no peak in the magnitude curve if $\zeta > \sqrt{2}/2 = 0.707$.

SECTION 5-8

5-22. Obtain $Y(s)$ as a function of $X_1(s)$ and $X_2(s)$ for the system shown.

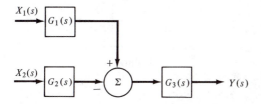

5-23. Find the transfer function $Y(s)/R(s)$.

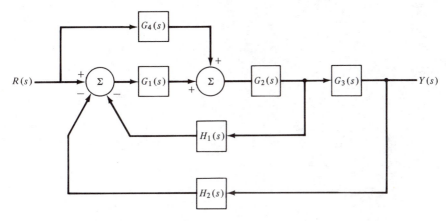

5-24. In the following system, let $G(s) = 6/(s + 1)(s + 3)$. $R(s) = a/s$ and $D(s) = b/s$, with a and b constants. Show that if K is made large, then $\lim_{t\to\infty} y(t) \simeq a$; that is, the disturbance has little effect on the system output in steady state.

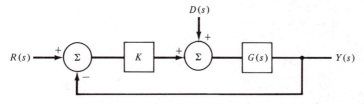

5-25. **(a)** Obtain the transfer function of the two-stage RC-coupled amplifier shown.

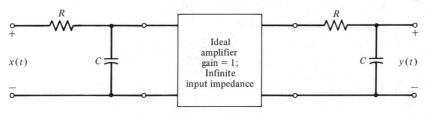

(b) Generalize the result of part (a) to n stages and show that

$$H_n(s) = (1 + RCs)^{-n}$$

where $H_n(s)$ is the transfer function of n RC stages coupled as shown.

(c) Show that the impulse response of the n-stage RC-coupled amplifier is

$$h_n(t) = \underbrace{\left[\frac{1}{RC} e^{-t/RC} u(t) \right] * \cdots * \left[\frac{1}{RC} e^{-t/RC} u(t) \right]}_{n \text{ convolutions}}$$

(d) Using the results of Problem 4-19, show that the effective delay and duration for the impulse response of the n-stage RC-coupled amplifier are given by $t_0 = nRC$ and $\tau = \sqrt{n}\, RC$, respectively.

CHAPTER 6

State-Variable Techniques

INTRODUCTION

So far the techniques of system analysis we have discussed have been applied primarily to systems with only a single input and a single output. Although such systems are often good models for individual parts of a system, they are rarely sufficient for modeling the overall system. Most engineering systems of interest have several inputs and several outputs. Although such a system may be modeled as an interconnection of single-input, single-output systems, it is not always advantageous to do so. As we shall see, the state-variable method provides a convenient formulation procedure. It allows us to handle systems with many inputs and outputs within precisely the same notational framework that we will use for single-input, single-output systems.

The state-variable formulation has other advantages as well. In our previous discussion we have been concerned with transfer functions between the system input and output. No information concerning the internal behavior of the system is readily available using this formulation. On the other hand, the state-variable formulation allows us to determine the internal behavior of the system easily while still giving the input–output information we desire. Furthermore, the state-variable formulation is often the most efficient form from the standpoint of computer simulation of the system. This characteristic alone leads us to prefer the state-variable formulation for highly complex systems.

In this chapter we develop the concepts and formulation of the state-variable approach to system analysis. We solve the state equations using both time-domain and Laplace transform techniques and examine some properties of the solution. Finally, we show how the state-variable method may be applied to circuit analysis.

6-2

STATE-VARIABLE CONCEPTS

There are several concepts which, if understood, will make our further discussion both easier and more meaningful. The most fundamental concept is that of the state of a system. Rather than give a detailed mathematical definition of the state of a system, we prefer to follow a less formal approach which we believe yields more insight. We may say that the state of a system at time t_0 includes the minimum information necessary to specify completely the condition of the system at time t_0 and allow determination of all system outputs at time $t > t_0$ when inputs up to time t are specified. In short, the state of a system at time t_0 tells us everything that is important about the system at time t_0. In a linear, time-invariant electric network, for example, knowledge of all capacitor voltages and inductor currents uniquely specifies the network condition at any particular time. Furthermore, this information and knowledge of the input allow us to find the condition of the network at any future time. Thus for this example the values of the capacitor voltages and inductor currents at some time t_0 constitute the state of the system at time t_0.

The state of a system at time t_0 is simply the set of values, at time t_0, of an appropriately chosen set of variables. These variables are called the *state variables*. For the electric network example above, the state variables are thus the capacitor voltages and inductor currents. The number of state variables is equal to the order of the system and is usually just the number of energy storage elements. The choice of state variables is not unique. In fact, there are infinitely many choices for any given system, but not all are equally convenient. Usually, state variables are chosen so that they correspond to physically measurable quantities or lead to particularly simplified calculations.

It is often advantageous to think of an n-dimensional space in which each coordinate is defined by one of the state variables $x_1, x_2, \ldots, x_n$, where n is the order of the system. This space is called the *state space*. The *state vector* is an n-vector $\mathbf{x}$ whose elements are the state variables. Then at any time t the state vector defines a point in the state space (i.e., the state of the system at that time). As time progresses and the system state changes, a set of points will be defined. This set of points, the locus of the tip of the state vector as time progresses, is called a *trajectory* of the system.

The concepts and terminology of this section are used throughout the remainder of this chapter. The student may find it helpful to refer to this section from time to time.

FORM OF THE STATE EQUATIONS

An nth order *linear differential system* with m inputs and p outputs may always be represented by n first-order differential equations and p output equations in the following manner†:

$$\begin{aligned}
\dot{x}_1 &= a_{11}x_1 + a_{12}x_2 + \cdots + a_{1n}x_n \\
&\quad + b_{11}u_1 + b_{12}u_2 + \cdots + b_{1m}u_m \\
\dot{x}_2 &= a_{21}x_1 + a_{22}x_2 + \cdots + a_{2n}x_n \\
&\quad + b_{21}u_1 + b_{22}u_2 + \cdots + b_{2m}u_m
\end{aligned} \tag{6-1}$$

$$\begin{aligned}
\vdots
\end{aligned}$$

$$\begin{aligned}
\dot{x}_n &= a_{n1}x_1 + a_{n2}x_2 + \cdots + a_{nn}x_n \\
&\quad + b_{n1}u_1 + b_{n2}u_2 + \cdots + b_{nm}u_m
\end{aligned} \tag{6-2}$$

and

$$\begin{aligned}
y_1 &= c_{11}x_1 + c_{12}x_2 + \cdots + c_{1n}x_n \\
&\quad + d_{11}u_1 + d_{12}u_2 + \cdots + d_{1m}u_m
\end{aligned}$$

$$\begin{aligned}
\vdots
\end{aligned}$$

$$\begin{aligned}
y_p &= c_{p1}x_1 + c_{p2}x_2 + \cdots + c_{pn}x_n \\
&\quad + d_{p1}u_1 + d_{p2}u_2 + \cdots + d_{pm}u_m
\end{aligned}$$

where the dot indicates time differentiation ($\dot{x} = dx/dt$); the u_i, $i = 1, 2, \ldots, m$, are the system inputs, and the y_i, $i = 1, 2, \ldots, p$, are the system outputs. The x_i, $i = 1, 2 \ldots, n$, are called the state variables. Equations (6-1) are often called the *state equations,* while equations (6-2) comprise the *output equations.* Together, the $n + p$ equations constitute a state-equation model for the system. Many books call these the normal-form equations. In the most general case the a's, b's, c's, and d's may be functions of time. However, the solution of such a set of time-varying state equations is exceedingly difficult and must almost always be carried out with the aid of a computer. For this reason we restrict our attention to time-invariant systems, ones for which all the coefficients are constant. In this case we can obtain solutions without too much difficulty, as we shall see in the following section.

At this point it is reasonable to ask two questions. First, what are the advantages of this method, if any? Second, how do we find the state equations for a system? We consider the second question in a later section. For now, we concentrate on providing answers to the first question. The most obvious advantage of the state-variable representation is that it allows us to treat the case of multiple inputs and multiple outputs. This is something that can be done only with

†For convenience we have omitted the explicit time dependence. We will follow this approach unless understanding would be enhanced by including (t).

considerable difficulty using the other analysis techniques we have discussed. Of course, the single-input, single-output system is included as a special case of (6-1) and (6-2). For this situation $m = p = 1$.

There are other advantages of the state model as well. The model is given in the time domain, and it is straightforward to obtain a simulation diagram for the equations. This is extremely useful if we wish to use computer simulation methods to study the system. Furthermore, an extremely compact matrix notation can be used for the state model. Using the laws of matrix algebra, it becomes much less cumbersome to manipulate the equations.

Let us turn our attention to the development of simulation diagrams for state equations. For brevity we consider a two-input, one-output second-order system given by

$$\dot{x}_1 = a_{11}x_1 + a_{12}x_2 + b_{11}u_1 + b_{12}u_2$$

$$\dot{x}_2 = a_{21}x_1 + a_{22}x_2 + b_{21}u_1 + b_{22}u_2 \qquad (6\text{-}3)$$

$$y = c_1x_1 + c_2x_2 + d_1u_1 + d_2u_2$$

It is evident that if we knew the states, we could find the output, for the inputs and all coefficients are assumed known. Therefore, consider only the first two equations. If we knew $\dot{x}_1$ and $\dot{x}_2$ we could obtain x_1 and x_2 by simple integration. Hence $\dot{x}_1$ and $\dot{x}_2$ should be the inputs of two integrators. The corresponding integrator outputs are x_1 and x_2. This leaves only the problem of obtaining $\dot{x}_1$ and $\dot{x}_2$ for use as inputs to the integrators. In fact, this is already specified by the state equations: $\dot{x}_1$ and $\dot{x}_2$ are simply the appropriate linear combinations of x_1, x_2, u_1 and u_2 specified by the first two of equations (6-3). Once we do this we simply use the inputs u_1 and u_2 and the integrator outputs x_1 and x_2 to form the system output y. The completed diagram is shown in Figure 6-1. Those familiar with analog-computer simulation techniques will recognize this as essentially an analog computer program for (6-3). Although the development has been for a relatively simple system, the extension to more complicated systems is obvious. One need only employ more integrators and provide the block diagram elements to reflect the mathematical statements of the state model. Of course, the diagram becomes cluttered and more complicated, but this is to be expected—so are the equations.

An examination of the simulation diagram reveals the significance of the various constants in the state model. The a_{ij} essentially specify the internal system connection and behavior, even in the absence of inputs, but with nonzero initial values of the states. The b_{ij} specify the manner in which inputs affect the internal behavior of the system. The c_{ij} specify the manner in which the internal workings of the system affect the output, while the d_{ij} represent direct transmission paths from input to output. If the state variables are chosen to be physically measurable variables of interest, considerable insight into system behavior can be obtained from the simulation diagram.

It is perhaps worth noting that the state equations can be obtained directly from a simulation diagram. All that is required is to write an expression for the

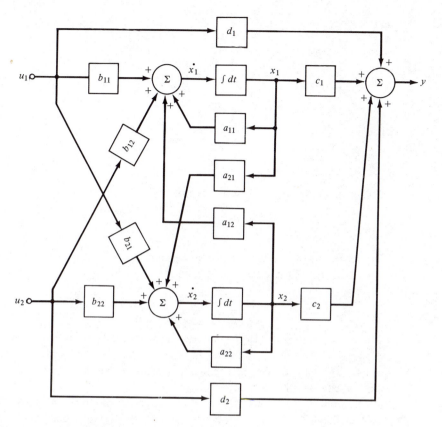

FIGURE 6-1. Simulation diagram of the system of (6-3).

output of each summer in terms of its inputs. In Figure 6-1 the inputs to the summer whose output is $\dot{x}_1$ are $b_{11}u_1$, $b_{12}u_2$, $a_{11}x_1$ and $a_{12}x_2$. Summing these yields the first of equations (6-3). Similar equations at the other summers yield the remaining two equations.

It should be abundantly clear that manipulations of (6-1) and (6-2) will present unnecessary difficulty unless some compact notation can be developed. Fortunately, such a notation is already available through the use of vectors and matrices. Let us define vectors

$$\mathbf{x} = \begin{bmatrix} x_1 \\ x_2 \\ \cdot \\ \cdot \\ \cdot \\ x_n \end{bmatrix}, \quad \mathbf{u} = \begin{bmatrix} u_1 \\ u_2 \\ \cdot \\ \cdot \\ \cdot \\ u_m \end{bmatrix}, \quad \mathbf{y} = \begin{bmatrix} y_1 \\ y_2 \\ \cdot \\ \cdot \\ \cdot \\ y_p \end{bmatrix} \qquad (6\text{-}4)$$

and matrices

$$
A = \begin{bmatrix} a_{11} & a_{12} & \cdots & a_{1n} \\ a_{21} & a_{22} & \cdots & a_{2n} \\ \cdot & \cdot & & \cdot \\ \cdot & \cdot & & \cdot \\ \cdot & \cdot & & \cdot \\ a_{n1} & a_{n2} & \cdots & a_{nn} \end{bmatrix}, \quad
B = \begin{bmatrix} b_{11} & b_{12} & \cdots & b_{1m} \\ b_{21} & b_{22} & \cdots & b_{2m} \\ \cdot & \cdot & & \cdot \\ \cdot & \cdot & & \cdot \\ \cdot & \cdot & & \cdot \\ b_{n1} & b_{n2} & \cdots & b_{nm} \end{bmatrix},
$$

$$
C = \begin{bmatrix} c_{11} & c_{12} & \cdots & c_{1n} \\ c_{21} & c_{22} & \cdots & c_{2n} \\ \cdot & \cdot & & \cdot \\ \cdot & \cdot & & \cdot \\ \cdot & \cdot & & \cdot \\ c_{p1} & c_{p2} & \cdots & c_{pn} \end{bmatrix}, \quad
D = \begin{bmatrix} d_{11} & d_{12} & \cdots & d_{1m} \\ d_{21} & d_{22} & \cdots & d_{pm} \\ \cdot & \cdot & & \cdot \\ \cdot & \cdot & & \cdot \\ \cdot & \cdot & & \cdot \\ d_{p1} & d_{p2} & \cdots & d_{pm} \end{bmatrix}
$$

(6-5)

Then using the usual rules of matrix algebra† it is obvious that the state model of (6-1) and (6-2) may be written compactly as

$$\dot{x} = Ax + Bu \tag{6-6}$$

$$y = Cx + Du$$

These equations may also be indicated schematically by a block diagram as shown in Figure 6-2. The double lines are used to indicate a multiple-variable signal flow path of the appropriate dimension. The blocks represent matrix multiplication of the appropriate vectors and matrices. The block containing the integrator in fact contains n integrators with appropriate connections specified by the **A** and **B** matrices. Of course, to obtain a useful simulation model we must know all the elements of all the matrices, but at least this block diagram indicates the essential structure of the system and the signal flow paths.

A similar statement can be made regarding the state equations themselves. Before we can obtain numerical solutions in any particular case we must know all elements of the four matrices **A**, **B**, **C** and **D**. However, as in most problems, it is generally most convenient to do the majority of the work in terms of arbitrary matrices. Particular numerical data may then be substituted as a last step. Furthermore, the compact matrix notation provides a convenient theoretical tool on those occasions when one is interested only in general properties of the system, without regard to particular numerical data. We exploit this notation in the following section to obtain solutions to the state equations.

6-4

TIME-DOMAIN SOLUTION OF THE STATE EQUATIONS

For the purpose of developing the solution of the state equations, let us consider (6-6) repeated here for convenience:

$$\dot{x} = Ax + Bu \tag{6-7}$$

$$y = Cx + Du$$

†For those unfamiliar with matrix algebra, Appendix A provides a summary of the necessary results.

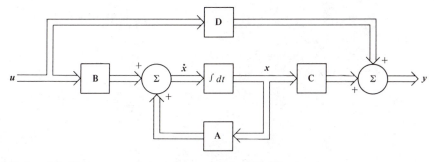

FIGURE 6-2. Block diagram of the state model of (6-6).

together with the known initial condition

$$\mathbf{x}(t_0) = \mathbf{x}_0 \tag{6-8}$$

We assume that the inputs are specified for all $t \geq t_0$. It is evident that if the state vector can be found, the output vector $\mathbf{y}$ can be determined easily from the second of equations (6-7). The problem then becomes that of solving the first of equations (6-7) subject to the initial condition of (6-8).

An intuitively pleasing approach to the solution of the state equations is to simply build up the solution as time progresses. Suppose that we divide the time interval $t \geq t_0$ into an infinite number of subintervals, each of duration Δt. If Δt is small, then

$$\dot{\mathbf{x}}(t_1) \simeq \frac{\mathbf{x}(t_1 + \Delta t) - \mathbf{x}(t_1)}{\Delta t} \triangleq \frac{\Delta \mathbf{x}}{\Delta t} \tag{6-9}$$

where $\Delta \mathbf{x}$ is the change in $\mathbf{x}$ in the time interval from t_1 to $t_1 + \Delta t$ for any $t_1 = t_0 + k \Delta t$, $k = 0, 1, 2, \dots$. Equation (6-9) suggests that, in the time interval t_0 to $t_0 + \Delta t$, $\mathbf{x}$ changes by the amount

$$\Delta \mathbf{x} = \dot{\mathbf{x}}(t_0) \, \Delta t$$

so that at $t_0 + \Delta t$,

$$\mathbf{x}(t_0 + \Delta t) \simeq \mathbf{x}(t_0) + \dot{\mathbf{x}}(t_0) \, \Delta t \tag{6-10}$$

Now $\dot{\mathbf{x}}(t_0)$ can be found from (6-7) as

$$\dot{\mathbf{x}}(t_0) = \mathbf{A}\mathbf{x}(t_0) + \mathbf{B}\mathbf{u}(t_0) = \mathbf{A}\mathbf{x}_0 + \mathbf{B}\mathbf{u}(t_0) \tag{6-11}$$

since all quantities on the right-hand side of (6-11) are known. We thus find $\mathbf{x}(t_0 + \Delta t)$. Repeating, we can find $\dot{\mathbf{x}}(t_0 + \Delta t)$ from this knowledge and use it to predict $\mathbf{x}(t_0 + 2\Delta t)$. Continuing the process, we can build up the solution one step at a time for as long as we wish. Of course, the result is only approximate. Its accuracy depends on how small Δt is chosen. As Δt tends toward zero, the approximate solution approaches the actual one. This method of solution, or modifications of it, is useful for digital computation.

Actually, the solution process can be expressed concisely as a recursion relation. Let us write (6-9) as

$$\dot{\mathbf{x}}(t_0 + k\Delta t) = \frac{\mathbf{x}(t_0 + (k + 1)\Delta t) - \mathbf{x}(t_0 + k\Delta t)}{\Delta t} \qquad (6\text{-}12)$$

where k is any nonnegative integer. Upon rewriting the state equations at time $t = t_0 + k\,\Delta t$, we obtain

$$\frac{\mathbf{x}(t_0 + (k + 1)\Delta t) - \mathbf{x}(t_0 + k\Delta t)}{\Delta t} = \mathbf{Ax}(t_0 + k\Delta t)$$

$$+ \mathbf{Bu}(t_0 + k\Delta t)$$

or

$$\mathbf{x}(t_0 + (k + 1)\Delta t) = \mathbf{x}(t_0 + k\Delta t)$$

$$+ \Delta t[\mathbf{Ax}(t_0 + k\Delta t)$$

$$+ \mathbf{Bu}(t_0 + k\Delta t)] \qquad (6\text{-}13)$$

Equation (6-13) provides the desired recursion relation. If $\mathbf{x}(t_0 + k\Delta t)$ is considered the present value of $\mathbf{x}$, then the next value may be found. Letting $k = 0, 1, 2, \ldots$ allows the solution to be constructed numerically. This equation is an example of a difference equation. We shall study such equations in some detail in the following chapter.

Although the solution technique described above has a great deal of intuitive appeal and is computationally convenient, it suffers the disadvantage that it does not provide a closed-form solution. Frequently, a closed-form solution is desired, so we must determine how to solve the equations to yield this result. Of course, we consider only

$$\dot{\mathbf{x}} = \mathbf{Ax} + \mathbf{Bu}, \qquad \mathbf{x}(t_0) = \mathbf{x}_0 \qquad (6\text{-}14)$$

realizing that once $\mathbf{x}$ is known, $\mathbf{y}$ can be easily found. Using the approach familiar to us from differential equations, we assume that the solution has two parts,

$$\mathbf{x} = \mathbf{x}_h + \mathbf{x}_p$$

where $\mathbf{x}_h$ is the solution to the homogeneous equation and $\mathbf{x}_p$ is a particular solution. The homogeneous solution must satisfy the initial condition.

The homogeneous equation is given by

$$\dot{\mathbf{x}}_h = \mathbf{Ax}_h, \qquad \mathbf{x}_h(t_0) = \mathbf{x}_0 \qquad (6\text{-}15)$$

By analogy to the scalar equation

$$\dot{x} = ax, \qquad x(t_0) = x_0$$

which has solution

$$x(t) = e^{a(t - t_0)}x_0$$

we might guess the solution of (6-15) to be

$$\mathbf{x}_h(t) = e^{\mathbf{A}(t - t_0)}\mathbf{x}_0 \qquad (6\text{-}16)$$

Immediately a question arises as to the meaning of the matrix exponential. We shall have more to say about it later, but for now let us assume that it can be differentiated in the same fashion as the scalar exponential, that it is an $n \times n$ matrix, and that it satisfies

$$e^{A0} = I \tag{6-17}$$

where I is an $n \times n$ identity matrix. Then

$$\dot{x}_h(t) = Ae^{A(t-t_0)}x_0 = Ax_h$$

and the assumed solution satisfies the homogeneous equation. Furthermore, in view of (6-17),

$$x_h(t_0) = e^{A \cdot 0}x_0 = x_0$$

Hence (6-16) gives the homogeneous solution.

Before determining the particular solution we digress to note some properties of the matrix exponential e^{At}. For our purposes we may define e^{At} in terms of its infinite series expansion as follows:

$$e^{At} = I + At + \frac{A^2t^2}{2!} + \frac{A^3t^3}{3!} + \cdots \tag{6-18}$$

Clearly, e^{At} is an $n \times n$ matrix. The assumptions we have made regarding e^{At} may be easily verified directly from (6-18). The inverse of e^{At} may also be found as

$$(e^{At})^{-1} = e^{-At} \tag{6-19}$$

Now let us complete the solution of the state equations by determining the particular solution. We adopt the method of variation of parameters and assume a particular solution of the form

$$x_p(t) = e^{A(t-t_0)}z(t) \tag{6-20}$$

where $z(t)$ is a vector function to be found. Differentiate (6-20) and require that x_p satisfy the original equation:

$$\dot{x}_p(t) = Ae^{A(t-t_0)}z(t) + e^{A(t-t_0)}\dot{z}(t) = Ax_p(t) + Bu(t)$$

$$= Ae^{A(t-t_0)}z(t) + Bu(t)$$

Then

$$e^{A(t-t_0)}\dot{z}(t) = Bu(t)$$

or

$$\dot{z}(t) = e^{A(t_0-t)}Bu(t) \tag{6-21}$$

Equation (6-21) can be solved by direct integration to yield

$$z(t) - z(t_0) = \int_{t_0}^{t} e^{A(t_0-\lambda)}Bu(\lambda)\, d\lambda \tag{6-22}$$

Now the particular solution must be such that

$$\mathbf{x}_p(t_0) = \mathbf{0}$$

since the homogeneous solution already satisfies the initial condition. It can be shown that $e^{A(t-t_0)}$ is nonsingular; therefore, from (6-20) we must have $\mathbf{z}(t_0) = \mathbf{0}$. Then

$$\mathbf{z}(t) = \int_{t_0}^{t} e^{A(t_0 - \lambda)} \mathbf{B}\mathbf{u}(\lambda) \, d\lambda$$

Using this result in (6-20) then yields the particular solution as

$$\mathbf{x}_p(t) = e^{A(t - t_0)} \int_{t_0}^{t} e^{A(t_0 - \lambda)} \mathbf{B}\mathbf{u}(\lambda) \, d\lambda$$

or

$$\mathbf{x}_p(t) = \int_{t_0}^{t} e^{A(t - \lambda)} \mathbf{B}\mathbf{u}(\lambda) \, d\lambda \tag{6-23}$$

The complete solution is obtained by adding the homogeneous solution and the particular solution to yield

$$\mathbf{x}(t) = e^{A(t - t_0)}\mathbf{x}_0 + \int_{t_0}^{t} e^{A(t - \lambda)} \mathbf{B}\mathbf{u}(\lambda) \, d\lambda \tag{6-24}$$

The matrix exponential arises so often that it is usually given a special name and symbol. It is called the *state transition matrix* and denoted by $\mathbf{\Phi}(t)$. Using this notation we may write the solution as

$$\mathbf{x}(t) = \mathbf{\Phi}(t - t_0)\mathbf{x}_0 + \int_{t_0}^{t} \mathbf{\Phi}(t - \lambda)\mathbf{B}\mathbf{u}(\lambda) \, d\lambda \tag{6-25}$$

The reason for calling $\mathbf{\Phi}(t)$ the state transition matrix is clear. Once the input is specified, the state transition matrix uniquely describes the manner in which the system state changes from its value at $t = t_0$ to its value at any time $t > t_0$. After we examine an alternative method of solving the state equations in the following section, we shall describe some techniques of finding the state transition matrix. Using (6-25), we may write the complete system response as

$$\mathbf{y}(t) = \mathbf{C}\mathbf{\Phi}(t - t_0)\mathbf{x}_0 + \int_{t_0}^{t} \mathbf{C}\mathbf{\Phi}(t - \lambda)\mathbf{B}\mathbf{u}(\lambda) \, d\lambda + \mathbf{D}\mathbf{u}(t) \tag{6-26}$$

6-5

FREQUENCY-DOMAIN SOLUTION OF THE STATE EQUATIONS

An alternative method of solving the state equation makes use of the Laplace transform technique discussed in Chapter 5. This method is quite attractive from an analytical point of view. Again we are concerned primarily with

$$\dot{x} = Ax + Bu, \qquad x(t_0) = x_0 \qquad (6\text{-}27)$$

For convenience, we shall assume that $t_0 = 0$. Laplace transforming (6-27) yields

$$sX(s) - x_0 = AX(s) + BU(s) \qquad (6\text{-}28)$$

Rearranging yields

$$(sI - A)X(s) = x_0 + BU(s)$$

or

$$X(s) = (sI - A)^{-1}x_0 + (sI - A)^{-1}BU(s) \qquad (6\text{-}29)$$

When we take the inverse transform we must obtain precisely the $x(t)$ given by our previous analysis, with $t_0 = 0$. From the first term of (6-29) it is evident that

$$\mathcal{L}^{-1}[(sI - A)^{-1}] = \Phi(t) \qquad (6\text{-}30)$$

or

$$\Phi(s) = (sI - A)^{-1} \qquad (6\text{-}31)$$

where $\Phi(s)$ is the Laplace transform of the state transition matrix.

The second term of (6-29) must yield the integral term of (6-25), which is consistent with the convolution property of a product of Laplace transforms,

$$\mathcal{L}^{-1}[(sI - A)^{-1}BU(s)] = \mathcal{L}^{-1}[\Phi(s)Bu(s)]$$

$$= \Phi(t) * Bu(t)$$

$$= \int_0^t \Phi(t - \lambda)Bu(\lambda)\, d\lambda \qquad (6\text{-}32)$$

Therefore, the time-domain solution is exactly as we have written in the preceding section. In the frequency domain, the response of the system is given by

$$Y(s) = C(sI - A)^{-1}x_0 + C(sI - A)^{-1}BU(s) + DU(s)$$

$$= C(sI - A)^{-1}x_0 + [C(sI - A)^{-1}B + D]U(s) \qquad (6\text{-}33)$$

Equation (6-33) is simply the Laplace transform of (6-26).

For single-input, single-output systems we defined the transfer function as the ratio of the Laplace transform of the output to the Laplace transform of the input, under the requirement of zero initial conditions. We can extend this concept to the multiple-variable case. For zero initial conditions (6-33) becomes

$$Y(s) = [C(sI - A)^{-1}B + D]U(s) \qquad (6\text{-}34)$$

It is evident that the quantity in brackets plays the role of a transfer function. Accordingly, we define a transfer function matrix as

$$H(s) \triangleq C(sI - A)^{-1}B + D = C\Phi(s)B + D \qquad (6\text{-}35)$$

Hence we may write

$$\mathbf{Y}(s) = \mathbf{H}(s)\mathbf{U}(s) \qquad (6\text{-}36)$$

Since there are m inputs and p outputs, $\mathbf{H}$ is a $p \times m$ matrix having elements $H_{ij}(s)$; $i = 1, 2, \ldots, p$; $j = 1, 2, \ldots, m$. The element $H_{ij}(s)$ is the transfer function between the jth input u_j and the ith output y_i.

We may also define an impulse response matrix $\mathbf{H}(t)$ as

$$\mathbf{H}(t) \triangleq \mathcal{L}^{-1}[\mathbf{H}(s)] \qquad (6\text{-}37)$$

Hence $\mathbf{H}(t)$ may be written as

$$\mathbf{H}(t) = \mathbf{C}\boldsymbol{\Phi}(t)\mathbf{B} + \mathbf{D}\boldsymbol{\delta}(t)$$

$$= \mathbf{C}e^{\mathbf{A}t}\mathbf{B} + \mathbf{D}\boldsymbol{\delta}(t) \qquad (6\text{-}38)$$

where $\boldsymbol{\delta}(t)$ is an m-vector of unit impulses. $\mathbf{H}(t)$ has the same dimension as $\mathbf{H}(s)$ and its elements $h_{ij}(t)$ satisfy

$$h_{ij}(t) = \mathcal{L}^{-1}[H_{ij}(s)] \qquad (6\text{-}39)$$

where $h_{ij}(t)$ is equal to $y_i(t)$ when a unit impulse is applied at the jth input.

The formal solution of the state equations is now complete. We have an expression that gives the system output for any input and any set of initial conditions. The matrix multiplications and the integral of (6-26), although perhaps tedious, are straightforward. The integral may be evaluated either directly or with the aid of the Laplace transform form of (6-32), and employing partial-fraction expansions. If we use a time-domain approach, it is evident that we must know the state transition matrix $\boldsymbol{\Phi}(t) = e^{\mathbf{A}t}$. We now turn to this problem.

6-6

FINDING THE STATE TRANSITION MATRIX

Two techniques for determining the state transition matrix of a given system have already been mentioned, although not specifically identified as such. These make use of the series definition of $e^{\mathbf{A}t}$ and of the Laplace transform method mentioned in Section 6.5. Recall that we defined the state transition matrix in terms of its infinite series as

$$\boldsymbol{\Phi}(t) = e^{\mathbf{A}t} = \mathbf{I} + \mathbf{A}t + \frac{\mathbf{A}^2 t^2}{2!} + \frac{\mathbf{A}^3 t^3}{3!} + \cdots \qquad (6\text{-}40)$$

For a given $\mathbf{A}$, (6-40) may be evaluated to any desired degree of accuracy at any specified time t by performing the indicated matrix multiplications and summation. A simple example will illustrate the procedure.

EXAMPLE 6-1

Find the state transition matrix if

$$\mathbf{A} = \begin{bmatrix} -1 & 0 \\ 0 & -2 \end{bmatrix}$$

It is immediately found that

$$\mathbf{A}^2 = \begin{bmatrix} 1 & 0 \\ 0 & 4 \end{bmatrix}, \quad \mathbf{A}^3 = \begin{bmatrix} -1 & 0 \\ 0 & -8 \end{bmatrix}, \quad \ldots, \quad \mathbf{A}^n = \begin{bmatrix} (-1)^n & 0 \\ 0 & (-2)^n \end{bmatrix}$$

Therefore,

$$\mathbf{e}^{\mathbf{A}t} = \begin{bmatrix} 1 & 0 \\ 0 & 1 \end{bmatrix} + \begin{bmatrix} -1 & 0 \\ 0 & -2 \end{bmatrix} t + \begin{bmatrix} 1 & 0 \\ 0 & 4 \end{bmatrix} \frac{t^2}{2} + \begin{bmatrix} -1 & 0 \\ 0 & -8 \end{bmatrix} \frac{t^3}{6} + \cdots$$

$$= \begin{bmatrix} 1 - t + \dfrac{t^2}{2} + \dfrac{t^3}{6} + \cdots & 0 \\ 0 & 1 - 2t + \dfrac{4t^2}{2} - \dfrac{8t^3}{6} + \cdots \end{bmatrix}$$

We immediately recognize the series representations of e^{-t} and e^{-2t}, so

$$\mathbf{e}^{\mathbf{A}t} = \begin{bmatrix} e^{-t} & 0 \\ 0 & e^{-2t} \end{bmatrix} \qquad \blacksquare$$

This approach may be satisfactory for numerical computations in which the computer can carry out the calculations to whatever degree of accuracy is required. However, except in extremely simple cases, it is unsatisfactory for analytical work. The problem revolves around recognizing the closed-form expression of the infinite series. The following example should prove convincing.

EXAMPLE 6-2

Find the state transition matrix for

$$\mathbf{A} = \begin{bmatrix} 0 & 1 \\ -6 & -5 \end{bmatrix}$$

We have

$$\mathbf{A}^2 = \begin{bmatrix} -6 & -5 \\ 30 & 19 \end{bmatrix}, \quad \mathbf{A}^3 = \begin{bmatrix} 30 & 19 \\ -114 & -65 \end{bmatrix}, \quad \mathbf{A}^4 = \begin{bmatrix} -114 & -65 \\ 390 & 211 \end{bmatrix}$$

and so on. Using this result in (6-40) yields

$$
\mathbf{e}^{\mathbf{A}t} =
\begin{bmatrix}
1 - \dfrac{6t^2}{2!} + \dfrac{30t^3}{3!} - \dfrac{114t^4}{4!} + \cdots & t - \dfrac{5t^2}{2!} + \dfrac{19t^3}{3!} - \dfrac{65t^4}{4!} + \cdots \\[2ex]
-6t + \dfrac{30t^2}{2!} - \dfrac{114t^3}{3!} + \dfrac{390t^4}{4!} + \cdots & 1 - 5t + \dfrac{19t^2}{2!} - \dfrac{65t^3}{3!} + \dfrac{211t^4}{4!} + \cdots
\end{bmatrix}
$$

This expression for $\mathbf{e}^{\mathbf{A}t}$ can be shown to be the first five terms of

$$
\mathbf{e}^{\mathbf{A}t} =
\begin{bmatrix}
3e^{-2t} - 2e^{-3t} & e^{-2t} - e^{-3t} \\
-6e^{-2t} + 6e^{-3t} & -2e^{-2t} + 3e^{-3t}
\end{bmatrix}
$$

but it is highly unlikely that we would realize it unless we already knew the answer. ∎

The second method of determining the state transition matrix makes use of the fact that

$$
\mathbf{\Phi}(s) = (s\mathbf{I} - \mathbf{A})^{-1} \tag{6-41}
$$

We simply form $(s\mathbf{I} - \mathbf{A})$, take the matrix inverse, and then take the inverse Laplace transform to find $\mathbf{\Phi}(t) = \mathbf{e}^{\mathbf{A}t}$. Partial-fraction-expansion methods are usually helpful in taking the inverse transform. An example will illustrate this method.

EXAMPLE 6-3

Use the Laplace transform method to find the state transition matrix if

$$
\mathbf{A} =
\begin{bmatrix}
0 & 1 \\
-6 & -5
\end{bmatrix}
$$

We have

$$
(s\mathbf{I} - \mathbf{A}) =
\begin{bmatrix}
s & -1 \\
6 & s + 5
\end{bmatrix}
$$

and

$$(s\mathbf{I} - \mathbf{A})^{-1} = \begin{bmatrix} \dfrac{s+5}{(s+2)(s+3)} & \dfrac{1}{(s+2)(s+3)} \\[3mm] \dfrac{-6}{(s+2)(s+3)} & \dfrac{s}{(s+2)(s+3)} \end{bmatrix}$$

Performing a partial-fraction expansion of each term and inverse transforming yields

$$\mathbf{e}^{\mathbf{A}t} = \mathcal{L}^{-1}[(s\mathbf{I} - \mathbf{A})^{-1}] = \begin{bmatrix} 3e^{-2t} - 2e^{-3t} & e^{-2t} - e^{-3t} \\[2mm] -6e^{-2t} + 6e^{-3t} & -2e^{-2t} + 3e^{-3t} \end{bmatrix} u(t)$$

as promised in Example 6.2. ∎

The Laplace transform method is quite convenient for analytical work since it yields answers in closed form. However, it is not very useful for machine computation. There exist other techniques for finding $\mathbf{e}^{\mathbf{A}t}$ which are often useful. One such technique involves finding a suitable coordinate transformation which takes $\mathbf{A}$ to diagonal or Jordan canonical form. Once this is done the state transition matrix for the new matrix can be written by inspection. Then $\mathbf{e}^{\mathbf{A}t}$ is found by applying the inverse coordinate transformation. A fourth technique makes use of the fact that $\mathbf{e}^{\mathbf{A}t}$ may be expressed *exactly* by a *finite* power series in $\mathbf{A}$. The coefficients of the series expansion are found by solving a set of simultaneous algebraic equations. These results are based on the Cayley–Hamilton theorem (see Appendix A). The coefficients of the finite power series representation become functions of time. Therefore, it is possible to write

$$\mathbf{\Phi}(t) = \mathbf{e}^{\mathbf{A}t} = \alpha_0(t)\mathbf{I} + \alpha_1(t)\mathbf{A} + \alpha_2(t)\mathbf{A}^2 + \cdots + \alpha_{n-1}(t)\mathbf{A}^{n-1} \quad (6\text{-}42)$$

When the eigenvalues of $\mathbf{A}$ are distinct it is straightforward to determine the coefficients $\alpha_i(t)$. Suppose that the eigenvalues $\lambda_i, i = 1, 2, \ldots n$, are distinct. Then we use (6-42) with $\mathbf{A}$ replaced by each λ_i in turn to obtain a set of n equations:

$$e^{\lambda_1 t} = \alpha_0(t) + \alpha_1(t)\lambda_1 + \alpha_2(t)\lambda_1^2 + \cdots + \alpha_{n-1}(t)\lambda_1^{n-1}$$

$$e^{\lambda_2 t} = \alpha_0(t) + \alpha_1(t)\lambda_2 + \alpha_2(t)\lambda_2^2 + \cdots + \alpha_{n-1}(t)\lambda_2^{n-1} \quad (6\text{-}43)$$

$$\cdot$$
$$\cdot$$
$$\cdot$$

$$e^{\lambda_n t} = \alpha_0(t) + \alpha_1(t)\lambda_n + \alpha_2(t)\lambda_n^2 + \cdots + \alpha_{n-1}(t)\lambda_n^{n-1}$$

Equations (6-43) may be solved for the coefficients $\alpha_i(t)$, $i = 0, 1, 2, \ldots,$ $n-1$. We will use a second-order example to illustrate the procedure.

EXAMPLE 6-4

Use the Cayley–Hamilton approach to find the state transition matrix for

$$\mathbf{A} = \begin{bmatrix} 0 & 1 \\ -6 & -5 \end{bmatrix}$$

The characteristic equation is

$$|\lambda\mathbf{I} - \mathbf{A}| = \begin{vmatrix} \lambda & -1 \\ 6 & \lambda + 5 \end{vmatrix} = \lambda(\lambda + 5) + 6 = \lambda^2 + 5\lambda + 6 = 0$$

Clearly, the eigenvalues are

$$\lambda_1 = -2, \qquad \lambda_2 = -3$$

We have

$$\mathbf{e}^{\mathbf{A}t} = \alpha_0(t)\mathbf{I} + \alpha_1(t)\mathbf{A}$$

and from (6-43) we get

$$e^{-2t} = \alpha_0(t) - 2\alpha_1(t)$$

$$e^{-3t} = \alpha_0(t) - 3\alpha_1(t)$$

Subtracting, we obtain

$$\alpha_1(t) = e^{-2t} - e^{-3t}$$

It follows that

$$\alpha_0(t) = 3e^{-2t} - 2e^{-3t}$$

Then

$$\mathbf{e}^{\mathbf{A}t} = (3e^{-2t} - 2e^{-3t})\begin{bmatrix} 1 & 0 \\ 0 & 1 \end{bmatrix} + (e^{-2t} - e^{-3t})\begin{bmatrix} 0 & 1 \\ -6 & -5 \end{bmatrix}$$

or

$$\mathbf{e}^{\mathbf{A}t} = \begin{bmatrix} 3e^{-2t} - 2e^{-3t} & e^{-2t} - e^{-3t} \\ -6e^{-2t} + 6e^{-3t} & -2e^{-2t} + 3e^{-3t} \end{bmatrix}$$

which agrees with the results of the preceding two examples. ∎

If some or all of the eigenvalues are repeated, the method must be modified somewhat. Since this discussion is not meant to be exhaustive, we refer the interested reader to the references at the end of the chapter.

We now possess all the tools necessary to find the complete response of a system described by a state model. Therefore, we close this section with two examples in which we use the Laplace transform method to solve such a set of equations.

EXAMPLE 6-5

Find the complete response, given

$$
\begin{bmatrix} \dot{x}_1 \\ \dot{x}_2 \end{bmatrix} = \begin{bmatrix} 0 & 1 \\ -2 & -3 \end{bmatrix} \begin{bmatrix} x_1 \\ x_2 \end{bmatrix} + \begin{bmatrix} 0 \\ 1 \end{bmatrix} u
$$

$$
\begin{bmatrix} y_1 \\ y_2 \end{bmatrix} = \begin{bmatrix} 1 & 0 \\ 1 & 1 \end{bmatrix} \begin{bmatrix} x_1 \\ x_2 \end{bmatrix}
$$

where

$$
\mathbf{x}(0) = \begin{bmatrix} 1 \\ 1 \end{bmatrix}
$$

and u is a unit step input. This is a single-input, two-output system. From (6-33) we have the Laplace transform of the response as

$$
\mathbf{Y}(s) = \mathbf{C}(s\mathbf{I} - \mathbf{A})^{-1}\mathbf{x}_0 + \mathbf{C}(s\mathbf{I} - \mathbf{A})^{-1}\mathbf{B}U(s)
$$

First, we obtain $(s\mathbf{I} - \mathbf{A})^{-1}$ as

$$
(s\mathbf{I} - \mathbf{A})^{-1} = \begin{bmatrix} s & -1 \\ 2 & s+3 \end{bmatrix}^{-1} = \frac{1}{s^2 + 3s + 2} \begin{bmatrix} s+3 & 1 \\ -2 & s \end{bmatrix}
$$

Then

$$
\mathbf{Y}(s) = \frac{\begin{bmatrix} 1 & 0 \\ 1 & 1 \end{bmatrix} \begin{bmatrix} s+3 & 1 \\ -2 & s \end{bmatrix} \begin{bmatrix} 1 \\ 1 \end{bmatrix}}{s^2 + 3s + 2} + \frac{\begin{bmatrix} 1 & 0 \\ 1 & 1 \end{bmatrix} \begin{bmatrix} s+3 & 1 \\ -2 & s \end{bmatrix} \begin{bmatrix} 0 \\ 1 \end{bmatrix} \frac{1}{s}}{s^2 + 3s + 2}
$$

$$
= \frac{\begin{bmatrix} 1 & 0 \\ 1 & 1 \end{bmatrix} \begin{bmatrix} s+4 \\ s-2 \end{bmatrix}}{(s+1)(s+2)} + \frac{\begin{bmatrix} 1 & 0 \\ 1 & 1 \end{bmatrix} \begin{bmatrix} 1 \\ s \end{bmatrix}}{s(s+1)(s+2)}
$$

$$
= \begin{bmatrix} \dfrac{s+4}{(s+1)(s+2)} \\[4mm] \dfrac{2(s+1)}{(s+1)(s+2)} \end{bmatrix} + \begin{bmatrix} \dfrac{1}{s(s+1)(s+2)} \\[4mm] \dfrac{s+1}{s(s+1)(s+2)} \end{bmatrix}
$$

Now, canceling the $s + 1$ factors in the bottom two terms and applying the partial fraction expansion yields

$$\mathbf{Y}(s) = \begin{bmatrix} \dfrac{3}{s+1} - \dfrac{2}{s+2} \\[2ex] \dfrac{2}{s+2} \end{bmatrix} + \begin{bmatrix} \dfrac{1}{2}{s} - \dfrac{1}{s+1} + \dfrac{\frac{1}{2}}{s+2} \\[2ex] \dfrac{1}{2}{s} - \dfrac{\frac{1}{2}}{s+2} \end{bmatrix}$$

or

$$\mathbf{y}(t) = \begin{bmatrix} (3e^{-t} - 2e^{-2t})u(t) \\[1ex] 2e^{-2t}u(t) \end{bmatrix} + \begin{bmatrix} (\frac{1}{2} - e^{-t} + \frac{1}{2}e^{-2t})u(t) \\[1ex] (\frac{1}{2} - \frac{1}{2}e^{-2t})u(t) \end{bmatrix}$$

The first term is the initial condition response, while the second is the forced response. Combining the terms yields the complete response in its simplest form as

$$\begin{bmatrix} y_1(t) \\[1ex] y_2(t) \end{bmatrix} = \begin{bmatrix} (\frac{1}{2} + 2e^{-t} - \frac{3}{2}e^{-2t})u(t) \\[1ex] (\frac{1}{2} + \frac{3}{2}e^{-2t})u(t) \end{bmatrix}$$

■

Note that we did not explicitly find $e^{\mathbf{A}t}$, but instead carried through all multiplications in terms of the transforms until the final step. Unless $e^{\mathbf{A}t}$ is explicitly needed, this is usually the easiest procedure, as it leads to somewhat simpler functions which must be inverse transformed.

EXAMPLE 6-6

In this example we consider a system with two inputs and two outputs. Such systems are difficult to handle unless we use state-variable methods. For simplicity, we use a system similar to that in Example 6.5 and let initial conditions be zero. The system equations are assumed to be

$$\begin{bmatrix} \dot{x}_1 \\ \dot{x}_2 \end{bmatrix} = \begin{bmatrix} 0 & 1 \\ -2 & -3 \end{bmatrix} \begin{bmatrix} x_1 \\ x_2 \end{bmatrix} + \begin{bmatrix} 2 & 1 \\ 0 & 1 \end{bmatrix} \begin{bmatrix} u_1 \\ u_2 \end{bmatrix}$$

$$\begin{bmatrix} y_1 \\ y_2 \end{bmatrix} = \begin{bmatrix} 1 & 0 \\ 1 & 1 \end{bmatrix} \begin{bmatrix} x_1 \\ x_2 \end{bmatrix}$$

We assume that $u_1 = u(t)$, a unit step function, and $u_2 = e^{-3t}u(t)$. For $\mathbf{x}(0) = \mathbf{0}$ we obtain from (6-33),

$$\mathbf{Y}(s) = \mathbf{C}(s\mathbf{I} - \mathbf{A})^{-1}\mathbf{B}\mathbf{U}(s)$$

Using $(s\mathbf{I} - \mathbf{A})^{-1}$ from Example 6.5, we obtain

$$\mathbf{Y}(s) = \frac{\begin{bmatrix} 1 & 0 \\ 1 & 1 \end{bmatrix}\begin{bmatrix} s+3 & 1 \\ -2 & s \end{bmatrix}\begin{bmatrix} 2 & 1 \\ 0 & 1 \end{bmatrix}\begin{bmatrix} \dfrac{1}{s} \\[2mm] \dfrac{1}{s+3} \end{bmatrix}}{(s+1)(s+2)}$$

$$= \frac{\begin{bmatrix} s+3 & 1 \\ s+1 & s+1 \end{bmatrix}\begin{bmatrix} \dfrac{2}{s} + \dfrac{1}{s+3} \\[3mm] \dfrac{1}{s+3} \end{bmatrix}}{(s+1)(s+2)}$$

or

$$\mathbf{Y}(s) = \frac{1}{(s+1)(s+2)}\begin{bmatrix} \dfrac{3s^2 + 16s + 18}{s(s+3)} \\[4mm] \dfrac{(s+1)(4s+6)}{s(s+3)} \end{bmatrix}$$

Finally, we obtain

$$\mathbf{Y}(s) = \begin{bmatrix} \dfrac{3s^2 + 16s + 18}{s(s+1)(s+2)(s+3)} \\[4mm] \dfrac{4s+6}{s(s+2)(s+3)} \end{bmatrix}$$

We may take the inverse transform with the aid of the partial-fraction expansion to obtain

$$\begin{bmatrix} y_1(t) \\ y_2(t) \end{bmatrix} = \begin{bmatrix} (3 - \frac{5}{2}e^{-t} - e^{-2t} + \frac{1}{2}e^{-3t})u(t) \\ (1 + e^{-2t} - 2e^{-3t})u(t) \end{bmatrix}$$

■

Two comments are perhaps in order at this point. First, it would be difficult to solve the equations of this example without the use of state-variable methods. The reader should convince himself of this by drawing a block diagram of the equations and considering how it might be used to obtain y_1 and y_2. Second, although we have only treated a second-order system, the same techniques can be used on systems of any order. The algebra, of course, becomes tedious for higher-order systems.

STATE EQUATIONS FOR ELECTRICAL NETWORKS

We now consider the problem of obtaining state equations for systems. We consider only the special case of electrical networks here. However, by the use of suitable analogies, the technique can be extended to other classes of lumped-parameter systems (e.g., mechanical systems).

Earlier in this chapter we hinted that a suitable set of state variables for electrical networks were the capacitor voltages and inductor currents. Physically, this choice is attractive because the capacitor voltages and inductor currents specify the stored energy. These are convenient choices from a mathematical point of view as well. We recall the element relations for capacitors and inductors as

$$i_c = C \frac{dv_c}{dt}, \qquad v_L = L \frac{di_L}{dt}$$

Note that the derivatives of the chosen state variables appear in these expressions. If we can write expressions for the capacitor current and inductor voltage in terms of the state variables and source voltages and/or currents, then we can obtain equations in the form

$$\text{derivative of state variable} = \text{linear combination of state variables and inputs}$$

This is precisely the form of the state equations. The capacitor current and inductor voltage can always be expressed in terms of state variables and source quantities by writing suitable KCL and KVL equations. The output equation can also be obtained in this way. Let us summarize the method in the following algorithm and then consider some examples.

Algorithm:
1. Choose capacitor voltages and inductor currents as state variables.
2. For each capacitor, write a KCL, expressing the capacitor current $C\, dv/dt$ in terms of state variables, source quantities, and other currents as necessary.
3. For each inductor, write a KVL, expressing the inductor voltage $L\, di/dt$ in terms of state variables, source quantities, and other voltages as necessary.
4. Write other KCL, KVL, and element relation equations as necessary to eliminate the "other" currents and voltages in steps 2 and 3 in terms of state variables and source quantities. The resulting equations, after dividing by C and L as appropriate, are the state equations.
5. Write KCL and KVL equations and use element relations as necessary to express the output(s) in terms of state variables and source quantities. The state model is then complete.

We will use this algorithm to obtain state equations for two electrical networks.

EXAMPLE 6-7

Obtain a state model for the network shown in Figure 6-3.

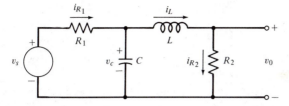

FIGURE 6-3

We choose as state variables the capacitor voltage v_c and the inductor current i_L. Writing a KCL for the capacitor yields

$$C\dot{v}_c = i_{R_1} - i_L$$

A KVL for the inductor gives

$$L\dot{i}_L = v_c - v_{R_2}$$

But following step 4 of the algorithm,

$$i_{R_1} = \frac{v_s - v_c}{R_1}$$

and

$$v_{R_2} = R_2 i_{R_2} = R_2 i_L$$

Hence

$$C\dot{v}_c = \frac{v_s}{R_1} - \frac{v_c}{R_1} - i_L$$

$$L\dot{i}_L = v_c - R_2 i_L$$

or

$$\dot{v}_c = \frac{v_s}{R_1 C} - \frac{v_c}{R_1 C} - \frac{i_L}{C}$$

$$\dot{i}_L = \frac{v_c}{L} - \frac{R_2 i_L}{L}$$

Clearly,

$$v_0 = v_{R_2} = R_2 i_L$$

If we wish we may write the equations in matrix form as

$$\begin{bmatrix} \dot{v}_c \\ \dot{i}_L \end{bmatrix} = \begin{bmatrix} -\dfrac{1}{R_1 C} & -\dfrac{1}{C} \\ \dfrac{1}{L} & -\dfrac{R_2}{L} \end{bmatrix} \begin{bmatrix} v_c \\ i_L \end{bmatrix} + \begin{bmatrix} \dfrac{1}{R_1 C} \\ 0 \end{bmatrix} v_s$$

$$v_0 = [0 \ R_2] \begin{bmatrix} v_c \\ i_L \end{bmatrix}$$

EXAMPLE 6-8

Obtain state equations for the two-input, one-output circuit shown in Figure 6-4. The inputs are v_{s_1} and v_{s_2}, and the output is i_0.

Again we choose v_c and i_L as state variables. A KCL for the capacitor yields

$$C\dot{v}_c = i_0 - i_{R_3}$$

A KVL around the left hand loop gives

$$L\dot{i}_L = v_{s_1} - v_{R_1}$$

Turning our attention to the inductor current equation, we see that we must eliminate v_{R_1}. We have

$$v_{R_1} = R_1 i_{R_1}$$

and writing a loop equation around the loop defined by v_{s_1}, R_1, R_2, and C yields

$$R_1 i_{R_1} + R_2 i_0 + v_c = v_{s_1}$$

Since

$$i_0 = i_{R_1} - i_L$$

we have

$$(R_1 + R_2)i_{R_1} = v_{s_1} - v_c + R_2 i_L$$

and

$$i_{R_1} = \frac{v_{s_1} - v_c + R_2 i_L}{R_1 + R_2}$$

Then

$$v_{R_1} = \frac{v_{s_1} - v_c + R_2 i_L}{R_1 + R_2} R_1$$

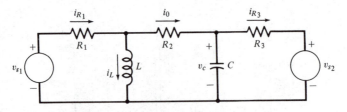

FIGURE 6-4. A two-input two-output system.

and we have

$$L\dot{i}_L = v_{s_1} - \frac{R_1}{R_1 + R_2} v_{s_1} + \frac{R_1}{R_1 + R_2} v_c - \frac{R_1 R_2}{R_1 + R_2} i_L$$

or

$$\dot{i}_L = \frac{R_2}{L(R_1 + R_2)} v_{s_1} + \frac{R_1}{L(R_1 + R_2)} v_c - \frac{R_1 R_2}{L(R_1 + R_2)} i_L$$

as one state equation.

Consider next the capacitor voltage equation

$$C\dot{v}_c = i_0 - i_{R_3}$$

Both i_0 and i_{R_3} must be eliminated. Now

$$i_{R_3} = \frac{v_c - v_{s_2}}{R_3}$$

so we have

$$C\dot{v}_c = i_0 - \frac{v_c}{R_3} + \frac{v_{s_2}}{R_3}$$

We also had

$$i_0 = i_{R_1} - i_L$$

Using the result for i_{R_1} yields

$$i_0 = \frac{v_{s_1}}{R_1 + R_2} - \frac{v_c}{R_1 + R_2} + \frac{R_2}{R_1 + R_2} i_L - i_L$$

or

$$i_0 = \frac{v_{s_1}}{R_1 + R_2} - \frac{v_c}{R_1 + R_2} - \frac{R_1 i_L}{R_1 + R_2}$$

This is the output equation. We may substitute this into the capacitor voltage state equation to yield

$$C\dot{v}_c = \frac{v_{s_1}}{R_1 + R_2} - \frac{v_c}{R_1 + R_2} - \frac{v_c}{R_3} - \frac{R_1 i_L}{R_1 + R_2} + \frac{v_{s_2}}{R_3}$$

Then the corresponding state equation is

$$\dot{v}_c = -\left[\frac{1}{CR_3} + \frac{1}{C(R_1 + R_2)} \right] v_c$$

$$- \frac{R_1}{C(R_1 + R_2)} i_L + \frac{1}{C(R_1 + R_2)} v_{s_1} + \frac{1}{CR_3} v_{s_2}$$

We may put all this in matrix form as

$$\begin{bmatrix} \dot{v}_c \\ \dot{i}_L \end{bmatrix} = \begin{bmatrix} -\dfrac{1}{CR_3} - \dfrac{1}{C(R_1 + R_2)} & -\dfrac{R_1}{C(R_1 + R_2)} \\ \dfrac{R_1}{L(R_1 + R_2)} & -\dfrac{R_1 R_2}{L(R_1 + R_2)} \end{bmatrix} \begin{bmatrix} v_c \\ i_L \end{bmatrix}$$

$$+ \begin{bmatrix} \dfrac{1}{C(R_1 + R_2)} & \dfrac{1}{CR_3} \\ \dfrac{R_2}{L(R_1 + R_2)} & 0 \end{bmatrix} \begin{bmatrix} v_{s_1} \\ v_{s_2} \end{bmatrix}$$

$$i_0 = \begin{bmatrix} -\dfrac{1}{R_1 + R_2} & -\dfrac{R_1}{R_1 + R_2} \end{bmatrix} \begin{bmatrix} v_c \\ i_L \end{bmatrix} + \begin{bmatrix} \dfrac{1}{R_1 + R_2} & 0 \end{bmatrix} \begin{bmatrix} v_{s_1} \\ v_{s_2} \end{bmatrix} \quad \blacksquare$$

While the second of our examples requires considerably more equations to be written in order to eliminate the undesired variables, both of the examples show that the algorithm may be applied to determine the state equations. Of course, more complicated circuits may be handled by the same method. The effort naturally increases.

6-8

STATE EQUATIONS FOR nTH-ORDER DIFFERENTIAL EQUATIONS

We have commented previously that a system which is represented by an nth-order ordinary differential equation can also be represented by n first-order differential equations (i.e., by state equations). We close this chapter with a brief discussion of how this may be accomplished. Only one technique is presented here. The interested reader is referred to Ogata's book for a discussion of other methods.

Let us suppose that a system is represented by the nth-order differential equation

$$x^{(n)} + a_{n-1} x^{(n-1)} + a_{n-2} x^{(n-2)} + \cdots + a_1 \dot{x} + a_0 x$$

$$= b_m u^{(m)} + b_{m-1} u^{(m-1)} + \cdots + b_1 \dot{u} + b_0 u, \quad (6\text{-}44)$$

where u is the scalar input, x is the scalar output and $m < n$. We wish to obtain the state model

$$\dot{\mathbf{x}} = \mathbf{A}\mathbf{x} + \mathbf{b}u$$

$$y = \mathbf{c}\mathbf{x} + du \quad (6\text{-}45)$$

Let us first consider the case where $m = 0$; that is, there are no derivatives of the input. Let us define state variables as

$$x_1 = x, \quad x_2 = \dot{x}, \quad x_3 = \ddot{x}, \quad \ldots, \quad x_n = x^{(n-1)} \tag{6-46}$$

Then

$$\dot{x}_1 = x_2$$
$$\dot{x}_2 = x_3$$

$$\cdot$$
$$\cdot \tag{6-47}$$
$$\cdot$$

$$\dot{x}_{n-1} = x_n$$
$$\dot{x}_n = -a_0 x_1 - a_1 \dot{x} - \cdots - a_{n-1} x^{(n-1)} + b_0 u$$
$$= -a_0 x_1 - a_1 x_2 - \cdots - a_{n-1} x_n + b_0 u$$

where the last equation is obtained from the differential equation (6-44) and the state variable definitions (6-46). We may write (6-47) as

$$\begin{bmatrix} \dot{x}_1 \\ \dot{x}_2 \\ \cdot \\ \cdot \\ \cdot \\ \dot{x}_{n-1} \\ \dot{x}_n \end{bmatrix} = \begin{bmatrix} 0 & 1 & 0 & \cdots & & 0 \\ 0 & 0 & 1 & 0 & \cdots & 0 \\ \cdot & & & & & \cdot \\ \cdot & & & & & \cdot \\ \cdot & & & & & \cdot \\ 0 & 0 & \cdots & & 0 & 1 \\ -a_0 & -a_1 & \cdots & & & -a_{n-1} \end{bmatrix} \begin{bmatrix} x_1 \\ x_2 \\ \cdot \\ \cdot \\ \cdot \\ \\ x_n \end{bmatrix} + \begin{bmatrix} 0 \\ 0 \\ \cdot \\ \cdot \\ \cdot \\ 0 \\ b_0 \end{bmatrix} u \tag{6-48}$$

Of course, since the outut is just x_1, we have

$$y = \begin{bmatrix} 1 & 0 & 0 & \cdots & 0 \end{bmatrix} \begin{bmatrix} x_1 \\ x_2 \\ \cdot \\ \cdot \\ \cdot \\ x_n \end{bmatrix} \tag{6-49}$$

and $d = 0$. Equations (6-48) and (6-49) are the desired state model.

If the original differential equation contains derivatives of the input, the method must be modified somewhat. We may no longer choose state variables simply as the output and its derivatives. It is still possible, however, to choose state variables such that the equations look quite similar to (6-48). Let us consider the next simplest case (i.e., b_0 and b_1 nonzero) and see what happens. Suppose that we choose the state variables as in (6-46). Then all equations are the same as (6-47) except the last one, which is

$$\dot{x}_n = -a_0 x_1 - a_1 x_2 - \cdots - a_{n-1} x_n + b_0 u + b_1 \dot{u}$$

However, the $\dot{u}$ term is not valid for state equations. Hence another choice of state variables must be made. Let us choose

$$x_1 = x, \quad x_2 = \dot{x}, \quad x_3 = \ddot{x}, \quad \ldots,$$

$$x_{n-1} = x^{(n-2)}, \quad x_n = x^{(n-1)} - b_1 u \quad (6\text{-}50)$$

Then

$$\dot{x}_1 = x_2$$

$$\dot{x}_2 = x_3$$

.

.

.

$$\dot{x}_{n-2} = x_{n-1}$$

$$\dot{x}_{n-1} = x^{(n-1)} = x_n + b_1 u$$

$$\dot{x}_n = x^{(n)} - b_1 \dot{u} = -a_0 x_1 - a_1 x_2 - \cdots - a_{n-1} x_n + (b_0 - a_{n-1} b_1) u$$

Thus the state equations are

$$
\begin{bmatrix} \dot{x}_1 \\ \dot{x}_2 \\ . \\ . \\ . \\ \dot{x}_{n-1} \\ \dot{x}_n \end{bmatrix} =
\begin{bmatrix}
0 & 1 & 0 & \cdots & & 0 \\
0 & 0 & 1 & 0 & \cdots & 0 \\
& & & & & . \\
& & & & & . \\
& & & & & . \\
0 & 0 & \cdots & & 0 & 1 \\
-a_0 & -a_1 & \cdots & & & -a_{n-1}
\end{bmatrix}
\begin{bmatrix} x_1 \\ x_2 \\ . \\ . \\ . \\ x_{n-1} \\ x_n \end{bmatrix}
$$

$$
+ \begin{bmatrix} 0 \\ 0 \\ . \\ . \\ . \\ b_1 \\ b_0 - a_{n-1} b_1 \end{bmatrix} u \qquad (6\text{-}51)
$$

Note that only the vector **b** is changed from (6-48). The output equation is identical to (6-49).

For $m > 1$ this approach can be generalized to yield state equations whose **A** matrix remains in the form of (6-51). Other forms are also possible. These details are beyond the scope of this text.

6-9

SUMMARY

In this chapter we have developed state-variable techniques for system analysis. Matrix formulation of the state equations was introduced as a notational aid. We found solutions to state equations using both time-domain and transform techniques, and showed how to obtain state equations for electrical networks and for systems described by scalar differential equations. Although our concern has been with linear, time-invariant systems, the techniques we have discussed are rather powerful and can be extended to the case of time-varying and/or nonlinear systems. For such systems, as well as for complicated linear, time-invariant systems, the state-variable method has much to offer.

FURTHER READING

P. M. DeRusso, R. J. Roy, and C. M. Close, *State Variables for Engineers*. New York: Wiley, 1965.

A rather detailed treatment of state-variable techniques. Good for additional topics and details at a level not much different from the present text.

C. A. Desoer and E. S. Kuh, *Basic Circuit Theory*. New York: McGraw-Hill, 1969.

An excellent introductory circuits text. Gives a thorough treatment of state-space methods applied to circuits, including nonlinear and/or time-varying circuits.

K. Ogata, *Modern Control Engineering*. Englewood Cliffs, N.J.: Prentice-Hall, 1970.

A good introduction to state-variable methods as applied to control systems is given in Chapters 14 through 16. Contains many ideas not introduced here.

D. G. Schultz and J. L. Melsa, *State Functions and Linear Control Systems*. New York: McGraw-Hill, 1967.

A higher-level and more complete treatment of state-variable methods applied to control systems. The first two chapters contain the topics presented here, plus several others.

PROBLEMS

SECTION 6-3

6-1. Construct a simulation diagram for the system described by

$$\dot{x}_1 = -2x_1 + 3x_2 + 2u$$

$$\dot{x}_2 = -x_1 - 6x_2 + u$$

$$y = x_1 + 2x_2 + 4u$$

6-2. Construct a simulation diagram for the system whose state equations are

$$\begin{bmatrix} \dot{x}_1 \\ \dot{x}_2 \\ \dot{x}_3 \end{bmatrix} = \begin{bmatrix} -1 & 0 & 0 \\ 0 & -3 & 0 \\ 0 & 0 & -4 \end{bmatrix} \begin{bmatrix} x_1 \\ x_2 \\ x_3 \end{bmatrix} + \begin{bmatrix} 1 \\ 2 \\ -1 \end{bmatrix} u$$

$$y = \begin{bmatrix} 1 & 3 & -2 \end{bmatrix} \begin{bmatrix} x_1 \\ x_2 \\ x_3 \end{bmatrix}$$

6-3. Write state equations in matrix form which describe the simulation diagram.

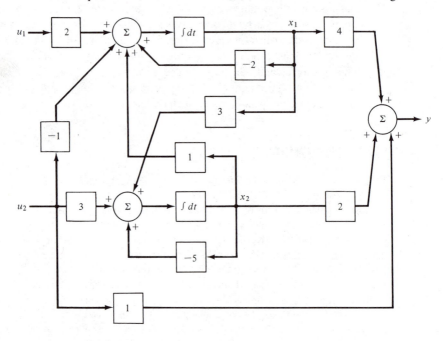

SECTION 6-4 AND 6-6

6-4. Use the series expansion method to compute e^{At} if

$$A = \begin{bmatrix} -1 & 0 \\ 0 & -4 \end{bmatrix}$$

6-5. Use the series expansion method to compute e^{At} if

$$A = \begin{bmatrix} -1 & 1 \\ 0 & -1 \end{bmatrix}$$

6-6. Show that if an $n \times n$ matrix $\mathbf{A}$ is given by

$$
\mathbf{A} = \begin{bmatrix} \lambda_1 & 0 & 0 & \cdots & 0 \\ 0 & \lambda_2 & 0 & \cdots & 0 \\ & \cdot & & & \\ & \cdot & & & \cdot \\ & \cdot & & & \\ 0 & 0 & 0 & \cdots & \lambda_n \end{bmatrix}
$$

then

$$
\mathbf{e}^{\mathbf{A}t} = \begin{bmatrix} e^{\lambda_1 t} & 0 & 0 & \cdots & 0 \\ 0 & e^{\lambda_2 t} & 0 & \cdots & 0 \\ & \cdot & & & \\ & \cdot & & & \cdot \\ & \cdot & & & \\ 0 & 0 & 0 & \cdots & e^{\lambda_n t} \end{bmatrix}
$$

6-7. For the $\mathbf{A}$ of Problem 6-4, find $\mathbf{x}(t)$ if $u(t)$ is a unit step function and

$$
\mathbf{x}(0) = \begin{bmatrix} 2 \\ 1 \end{bmatrix}, \qquad \mathbf{B} = \begin{bmatrix} 1 \\ 3 \end{bmatrix}
$$

6-8. In (6-13) we gave the recursion relation for computing the approximate solution to the state equations. Let the system be described by

$$
\begin{bmatrix} \dot{x}_1 \\ \dot{x}_2 \end{bmatrix} = \begin{bmatrix} -1 & 0 \\ 0 & -0.6 \end{bmatrix} \begin{bmatrix} x_1 \\ x_2 \end{bmatrix}, \qquad \begin{bmatrix} x_1(0) \\ x_2(0) \end{bmatrix} = \begin{bmatrix} 3 \\ 2 \end{bmatrix}
$$

(a) With $\Delta t = \frac{1}{2}$ s, use (6-13) to compute $\mathbf{x}(1 \text{ s})$.
(b) Repeat part (a) with $\Delta t = \frac{1}{10}$ s.
(c) Compute the actual value of $\mathbf{x}(1 \text{ s})$ and compare with parts (a) and (b).

SECTION 6-5 AND 6-6

6-9. Use the Laplace transform method to find $\mathbf{e}^{\mathbf{A}t}$ for the $\mathbf{A}$ given in Problem 6-5.

6-10. Use the Laplace transform method to find $\mathbf{e}^{\mathbf{A}t}$ if

$$
\mathbf{A} = \begin{bmatrix} 0 & 1 & 0 \\ 0 & 0 & 1 \\ -6 & -11 & -6 \end{bmatrix}
$$

6-11. Use the Cayley–Hamilton approach to find e^{At} if

$$A = \begin{bmatrix} 0 & 1 \\ -3 & -4 \end{bmatrix}$$

6-12. Use the Laplace transform approach to find $y(t)$ for the system given by

$$\begin{bmatrix} \dot{x}_1 \\ \dot{x}_2 \end{bmatrix} = \begin{bmatrix} 0 & 1 \\ -4 & -5 \end{bmatrix} \begin{bmatrix} x_1 \\ x_2 \end{bmatrix} + \begin{bmatrix} 0 \\ 2 \end{bmatrix} u$$

$$y = \begin{bmatrix} 1 & 0 \end{bmatrix} \begin{bmatrix} x_1 \\ x_2 \end{bmatrix}$$

where $x_0 = \begin{bmatrix} 1 \\ 1 \end{bmatrix}$ and u is a unit step.

6-13. Find the transfer function matrix for the system

$$\begin{bmatrix} \dot{x}_1 \\ \dot{x}_2 \end{bmatrix} = \begin{bmatrix} 0 & 1 \\ -3 & -4 \end{bmatrix} \begin{bmatrix} x_1 \\ x_2 \end{bmatrix} + \begin{bmatrix} 0 \\ 2 \end{bmatrix} u$$

$$\begin{bmatrix} y_1 \\ y_2 \end{bmatrix} = \begin{bmatrix} 1 & 0 \\ 0 & 1 \end{bmatrix} \begin{bmatrix} x_1 \\ x_2 \end{bmatrix}$$

SECTION 6-7

6-14 Obtain a state model for the network shown.

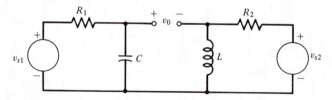

6-15 Obtain a state model for the network shown.

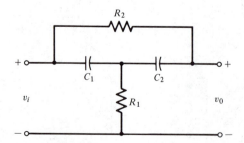

6-16. Obtain a state model for the network shown.

SECTION 6-8

6-17. Obtain a state model for:
(a) $\dddot{y} + 2\ddot{y} + 6\dot{y} + 7y = 4u$
(b) $\ddot{y} + 14\dot{y} + 25y = 25u$
(c) $y^{(4)} + 3\dddot{y} + 7\ddot{y} + 10y = u$

6-18. Obtain a state model for

$$\ddot{y} + 2\dot{y} + 10y = 3u + 4\dot{u}$$

CHAPTER 7

Discrete-Time Signals and Systems

7·1

INTRODUCTION

In the preceding chapters we have been concerned only with the analysis of continuous-time signals and continuous-time systems. We now turn our attention to discrete-time signals and systems.

A discrete-time signal is a signal that is defined by specifying the value of the signal only at discrete times, called *sampling instants*. If the sample values are then quantized and encoded, a digital signal results. A digital signal is formed from a continuous-time (analog) signal through the process of analog-to-digital conversion. In the following section we make a slight detour and study the process of analog-to-digital conversion in some detail so that we will fully understand the relationship between discrete-time signals and digital signals. The error sources must be identified so that all later approximations and assumptions can be appreciated.

In this chapter we consider a discrete-time signal to arise by sampling a continuous-time signal. Quantizing and encoding these resulting sample values yields a digital signal. To start with a continuous-time signal at first appears to be a rather narrow viewpoint since in many situations digital signals occur naturally and there is no continuous-time signal related to the sample values. Output data from a computer is a possible example. However, in taking the view that discrete-time signals arise from continuous-time signals we are better able to relate the theoretical concepts in the following three chapters to the previously developed concepts for continuous-time signals and systems.

An important goal of this chapter is to formulate those tools necessary for the analysis of discrete-time signals and systems. The principal tool will be the z-transform, which will allow us to specify the parameters of digital signals and formulate the input–output relationships of digital systems in both the time domain and the frequency domain. In Chapter 8 attention will be turned to the more interesting task of the *synthesis* of digital systems. That is, we shall design digital systems so that specified tasks will be performed.

7-2

ANALOG-TO-DIGITAL CONVERSION

A block diagram of an analog-to-digital (A/D) converter is shown in Figure 7-1. The first component is a sampler that extracts sample values of the input signal at the sampling times. The output of the sampler is a discrete-time signal but a continuous-amplitude signal, since the sample values assume the same continuous range of values assumed by the input signal $x(t)$. These signals are often referred to as *sampled data signals*. The second component in an A/D converter is a quantizer, which quantizes the continuous range of sample values into a finite number of sample values so that each sample value can be represented by a digital word of finite precision or wordlength. The encoder maps each quantized sample value onto a digital word. Each of these processes is now examined in detail.

Sampling

To sample a continuous-time signal $x(t)$ is to represent $x(t)$ at a discrete number of points, $t = nT$, where T is the sampling period, which is the time between samples, and n is an integer that establishes the time position of each sample. This process is illustrated in Figure 7-2a and b, which shows both a set of samples of a continuous-time signal and a *sampling switch,* which is our initial model of a sampling device.

In order to extract samples of $x(t)$, the sampling switch closes briefly every T seconds. This yields samples that have a value of $x(t)$ when the switch is closed and a value of zero when the switch is open. For the sampling process

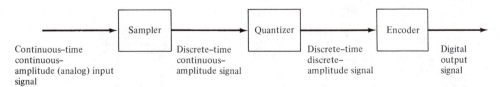

FIGURE 7-1. Analog-to-digital converter.

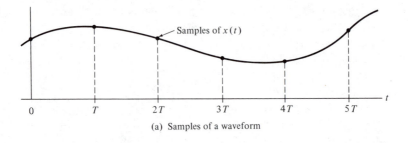

(a) Samples of a waveform

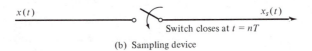

(b) Sampling device

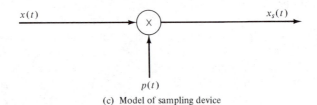

(c) Model of sampling device

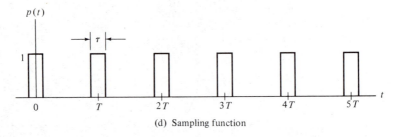

(d) Sampling function

FIGURE 7-2.   The sampling operation.

to be useful, we must be able to show that it is possible to sample in such a way that the signal $x(t)$ can be reconstructed from the samples. This is most easily accomplished in the frequency domain.

First, the sampled $x(t)$, denoted $x_s(t)$, is written

$$x_s(t) = x(t)p(t) \tag{7-1}$$

where $p(t)$, called the *sampling function*, models the action of the sampling switch as shown in Figure 7-2c. The sampling function $p(t)$ is assumed to be the periodic pulse train of period T illustrated in Figure 7-2d. We shall derive the spectrum of $x_s(t)$, and from this spectrum we shall be able to choose appropriate values of T.

The first step in the development of the spectrum of $x_s(t)$ is to recognize that since the sampling function $p(t)$ is periodic, it can be represented by a Fourier series. In other words,

$$p(t) = \sum_{n=-\infty}^{\infty} C_n e^{+jn2\pi f_s t} \tag{7-2}$$

where C_n is the nth Fourier coefficient of $p(t)$ and is given by

$$C_n = \frac{1}{T} \int_{-T/2}^{T/2} p(t) e^{-jn2\pi f_s t} dt \tag{7-3}$$

In (7-2) and (7-3), f_s is the fundamental frequency of $p(t)$, which is the sampling frequency, and is given by

$$f_s = \frac{1}{T} \quad \text{hertz}$$

Since $x_s(t)$ is the product of $x(t)$ and $p(t)$, we have

$$x_s(t) = \sum_{n=-\infty}^{\infty} C_n x(t) e^{+jn2\pi f_s t} \tag{7-4}$$

We can now determine the spectrum of $x_s(t)$, which we shall denote by $X_s(f)$, by taking the Fourier transform of (7-4). The Fourier transform of $x_s(t)$ is defined by

$$X_s(f) = \int_{-\infty}^{\infty} x_s(t) e^{-j2\pi ft} dt$$

which, upon substitution of (7-4) for $x_s(t)$, becomes

$$X_s(f) = \int_{-\infty}^{\infty} \sum_{n=-\infty}^{\infty} C_n x(t) e^{+jn2\pi f_s t} e^{-j2\pi ft} dt$$

Interchanging the order of integration and summation yields

$$X_s(f) = \sum_{n=-\infty}^{\infty} C_n \int_{-\infty}^{\infty} x(t) e^{-j2\pi(f - nf_s)t} dt \tag{7-5}$$

From the definition of the Fourier transform,

$$X(f - nf_s) = \int_{-\infty}^{\infty} x(t) e^{-j2\pi(f - nf_s)t} dt$$

Thus the Fourier transform of the sampled signal can be written

$$X_s(f) = \sum_{n=-\infty}^{\infty} C_n X(f - nf_s) \tag{7-6}$$

which shows that the spectrum of the sampled continuous-time signal, $x(t)$, is composed of the spectrum of $x(t)$ plus the spectrum of $x(t)$ translated to each harmonic of the sampling frequency. Each of the translated spectra is multiplied by a constant, given by the corresponding term in the Fourier series expansion of $p(t)$. This is illustrated in Figure 7-3 for an assumed $X(f)$.

Also shown in Figure 7-3 is the amplitude response of an assumed reconstruction filter. If the sampled signal is filtered by the reconstruction filter, the output

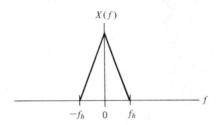

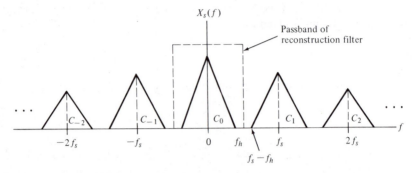

FIGURE 7-3. Spectrum of sampled signal.

of the filter is, in the frequency domain, $C_0X(f)$, and the time-domain signal is $C_0x(t)$.

In Figure 7-3 the assumed $x(t)$ is *bandlimited*. In other words, $X(f)$ is assumed zero for $|f| \geq f_h$. It is clear from Figure 7-3 that if $X(f)$ is to be recoverable from $X_s(f)$, and consequently $x(t)$ from $x_s(t)$, then

$$f_s - f_h \geq f_h$$

or

$$f_s \geq 2f_h \qquad \text{hertz}$$

Thus the minimum sampling frequency is $2f_h$ hertz, where f_h is the highest frequency in $x(t)$.

Thus we have derived the sampling theorem for low-pass *bandlimited* signals.

Sampling Theorem:

A bandlimited signal $x(t)$, having no frequency components above f_h hertz, is *completely* specified by samples that are taken at a *uniform* rate greater than $2f_h$ hertz. In other words, the time between samples is no greater than $1/2f_h$ seconds.

The frequency $2f_h$ is known as the *Nyquist rate*.

Impulse-Train Sampling

The sampling function utilized in the preceding section was general except for the fact that it was assumed to be periodic. In practice, the time during which

$p(t)$ is nonzero—that is, the pulse width—is small compared to the period T. In digital systems, where the sample is in the form of a number whose magnitude represents the value of the signal $x(t)$ at the sampling *instants*, the pulse width of the sampling function is infinitely small. Because an extremely narrow pulse is a situation *modeled* by an impulse and also because of mathematical simplifications that will result, we shall assume $p(t)$ to be composed of an infinite train of impulse functions of period T. Thus

$$p(t) = \sum_{n=-\infty}^{\infty} \delta(t - nT) \tag{7-7}$$

which is the sampling function illustrated in Figure 7-4.

The values of C_n can be computed from (7-3). This yields

$$C_n = \frac{1}{T} \int_{-T/2}^{T/2} \delta(t) e^{-jn2\pi f_s t} dt$$

which is

$$C_n = \frac{1}{T} = f_s$$

by the sifting property of the delta function. Thus C_n is equal to f_s for all n and the expression for the spectrum of the sampled $x(t)$, as given in general by (7-6), becomes

$$X_s(f) = f_s \sum_{n=-\infty}^{\infty} X(f - nf_s) \tag{7-8}$$

The effect of sampling with an impulse train is illustrated in Figure 7-5. It should be noted that the effect is the same as illustrated in Figure 7-3 except that all translated spectra have the same amplitude. Once again $X(f)$ can be obtained from $X_s(f)$, and consequently $x(t)$ from $x_s(t)$, by using a low-pass filter for recontruction of the original continuous-time waveform.

The effect of sampling at too low a rate can also be seen from Figure 7-5. If a signal is sampled below the Nyquist rate,

$$f_s - f_h \leq f_h$$

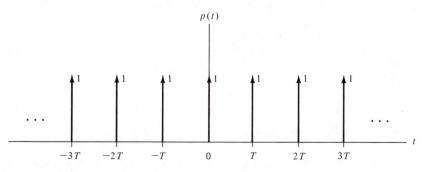

FIGURE 7-4.   Impulse sampling function.

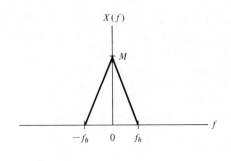

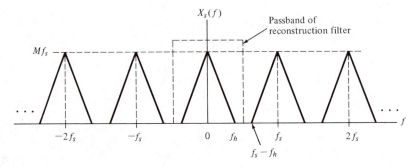

FIGURE 7-5. Spectrum of impulse sampled signal.

and adjacent spectra overlap making it impossible to recover $x(t)$ by filtering. This effect is known as *aliasing* and is illustrated in Figure 7-6 for $X(f)$ assumed real.

In a practical situation it is impossible to sample a signal and reconstruct the original signal from the samples with zero error because no practical signal is strictly bandlimited. However, in all practical signals there is some frequency beyond which the energy is negligible. This frequency is usually taken as the bandwidth. Another source of error arises from the nonexistence of ideal reconstruction filters. This means that some spectral space is required for filtering because of the finite slope of the reconstruction filter beyond the break frequency. All these considerations result in the fact that sampling is usually not performed at the Nyquist rate, but at some significantly higher frequency. Nonbandlimited signals having low-pass spectra are often sampled at approximately 10 times the frequency at which the amplitude spectrum is 3 dB down from its maximum value.

It should be pointed out that the impulse sampling function was chosen in order to have a simple sampling model which yields simple results. One surely does not find true impulse function signal generators as part of a piece of hardware.

Data Reconstruction

We saw in the preceding section that $x(t)$ could be reconstructed from $x_s(t)$ through an ideal low-pass filter having bandwidth greater than f_h but less than

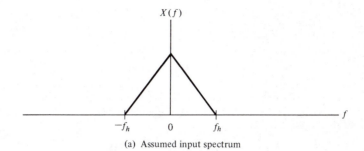

(a) Assumed input spectrum

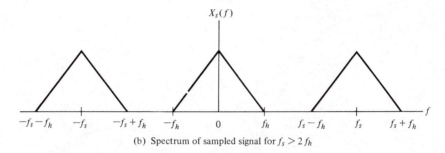

(b) Spectrum of sampled signal for $f_s > 2f_h$

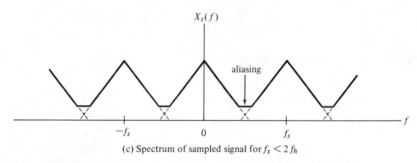

(c) Spectrum of sampled signal for $f_s < 2f_h$

FIGURE 7-6. Illustration of aliasing error for real $X(f)$.

$f_s - f_h$. Also, the amplitude response must be T if the scaling constant induced by the sampler is to be removed. This is illustrated in Figure 7-7a. Assuming the filter bandwidth to be $\frac{1}{2} f_s$, the response to the impulse at $t = 0$, $x(0)\delta(t)$, will be the unit impulse response of the filter weighted by $x(0)$. The unit impulse response of the ideal low-pass filter with gain T is

$$h(t) = T \int_{-f_s/2}^{f_s/2} e^{j2\pi ft} df$$

which is

$$h(t) = \frac{T}{j2\pi t} (e^{j\pi f_s t} - e^{-j\pi f_s t})$$

The expression above can be written

$$h(t) = Tf_s \frac{\sin \pi f_s t}{\pi f_s t} = \text{sinc } f_s t \qquad (7-9)$$

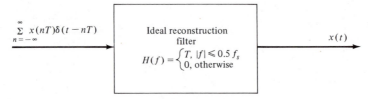

$$\sum_{n=-\infty}^{\infty} x(nT)\delta(t-nT)$$

Ideal reconstruction filter

$$H(f) = \begin{cases} T, & |f| \le 0.5 f_s \\ 0, & \text{otherwise} \end{cases}$$

$x(t)$

(a) Reconstruction filtering

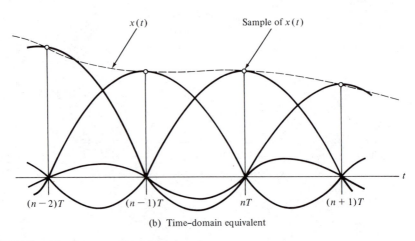

(b) Time–domain equivalent

FIGURE 7-7.   Ideal reconstruction.

so that the response to $x(0)\delta(t)$ is

$$x(0)h(t) = x(0) \text{ sinc } f_s t$$

The response to the nth sample can be obtained from (7-9) by shifting the response to $t = nT$ and weighting the response by the sample value $x(nT)$. Therefore, the response to $x(nT)\,\delta(t - nT)$ is

$$x(nT)h(t - nT) = x(nT) \text{ sinc } f_s(t - nT)$$

The output of the low-pass reconstruction filter is the sum of all the contributions due to the samples applied to the input of the reconstruction filter. Therefore,

$$x(t) = \sum_{n=-\infty}^{\infty} x(nT) \text{ sinc } f_s(t - nT) \tag{7-10}$$

The preceding expression shows that the original data signal can be reconstructed by weighting each sample by a sinc function and summing. This operation is illustrated in Figure 7-7b.

EXAMPLE 7-1
As an example of the sampling and reconstruction operations just studied, we shall consider the simple signal

$$x(t) = 6 \cos 2\pi(5)t$$

sampled at 7 and 14 Hz. Since the highest frequency (the only frequency in this case) is 5 Hz, we shall see the effect of sampling a signal at both a frequency less than and greater than twice the highest frequenccy in $x(t)$. Since

$$X(f) = 3\delta(f - 5) + 3\delta(f + 5)$$

it follows from (7-8) that the spectrum of the sampled $x(t)$ is given by

$$X_s(f) = 3f_s \sum_{n=-\infty}^{\infty} [\delta(f - 5 - nf_s) + \delta(f + 5 - nf_s)]$$

The spectrum of $x(t)$ is shown in Figure 7-8a. The spectrum of the sampled signal is shown in Figure 7-8b for a sampling frequency of 7 Hz and in Figure 7-8c for a sampling frequency of 14 Hz. Both spectra extend from $f = -\infty$ to $f = +\infty$ but are illustrated in Figure 7-8 only in the range $|f| \leq 33$ Hz. The assumed reconstruction filter, $H(f)$, is an ideal low-pass filter with a bandwidth of $0.5f_s$ and an amplitude response of T. Thus the output of the low-pass filter has the spectrum shown in Figure 7-8d for a sampling frequency of 7 Hz and has the spectrum shown in Figure 7-8e for a sampling frequency of 14 Hz.

Figure 7-8d illustrates the effect of aliasing arising from using a sampling frequency less than twice the highest frequency in $x(t)$. The result is a signal, having the proper amplitude but the incorrect frequency. (The student should convince himself that the frequency of the component at the output of the reconstruction filter is always the sampling frequency minus the frequency of the input signal.)

Figure 7-8e illustrates the effect of sampling properly at a frequency greater than twice the highest frequency of $x(t)$. The output of the reconstruction filter is identical to the original signal. ∎

EXAMPLE 7-2

In this example we consider a nonperiodic signal for $x(t)$ so that $X(f)$ has a continuous spectrum. We shall also assume that $X(f)$ is real.

The spectrum of $x(t)$ is shown in Figure 7-9a. The highest frequency is 5 Hz, so that the minimum acceptable sampling frequency is 10 Hz. Once again we shall assume sampling frequencies of 7 and 14 Hz.

Equation (7-8) shows that the spectrum of the sampled signal is obtained by multiplying $X(f)$ by f_s and then reproducing $f_s X(f)$ about dc and all harmonics of the sampling frequency. This is shown in Figure 7-9b for a sampling frequency of 7 Hz and in Figure 7-9c for a sampling frequency of 14 Hz. For a sampling frequency of 7 Hz, overlap of the translated spectra (aliasing) occurs and $X_s(f)$ is found by summing spectra in those regions of spectral overlap. For a sampling frequency of 14 Hz, there is no spectral overlap with translated spectra since the sampling frequency exceeds the minimum value of 10 Hz.

As in the previous examples, the reconstruction filter is assumed to be an ideal low-pass filter with an amplitude response of T and a bandwidth of $0.5 f_s$. The output spectrum of the reconstruction filter is shown in Figure 7-9d for a sampling frequency of 7 Hz and in Figure 7-9e for a sampling frequency of 14 Hz. The impact of aliasing is clear from a comparison of Figure 7-9d and e. ∎

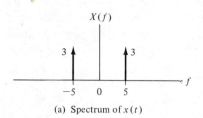

(a) Spectrum of $x(t)$

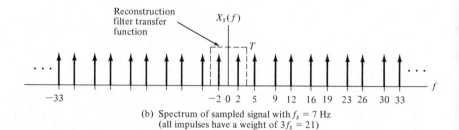

(b) Spectrum of sampled signal with $f_s = 7$ Hz
(all impulses have a weight of $3f_s = 21$)

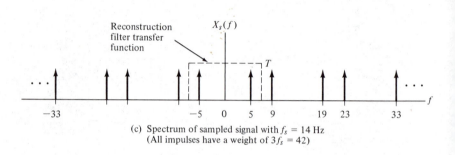

(c) Spectrum of sampled signal with $f_s = 14$ Hz
(All impulses have a weight of $3f_s = 42$)

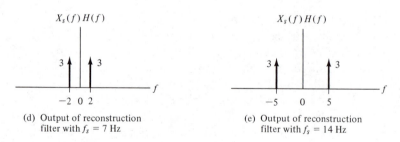

(d) Output of reconstruction filter with $f_s = 7$ Hz

(e) Output of reconstruction filter with $f_s = 14$ Hz

FIGURE 7-8. Sampling a single sinusoid.

Quantizing and Encoding

The process of quantizing and encoding is illustrated in Figure 7-10. To quantize a sample value is to round off the sample value to the nearest of a finite set of permissible values. Encoding is accomplished by representing each of the permissible sample values by a digital word. The number of quantizing levels q and the digital wordlength n are related by

$$q = 2^n \tag{7-11}$$

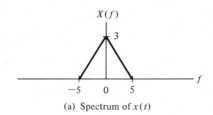

(a) Spectrum of $x(t)$

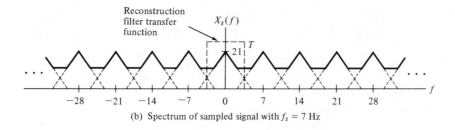

(b) Spectrum of sampled signal with $f_s = 7$ Hz

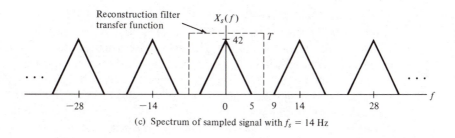

(c) Spectrum of sampled signal with $f_s = 14$ Hz

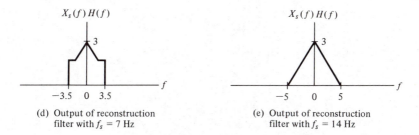

(d) Output of reconstruction filter with $f_s = 7$ Hz

(e) Output of reconstruction filter with $f_s = 14$ Hz

FIGURE 7-9. Sampling a signal having a continuous spectrum.

For example, in Figure 7-10 we have eight quantizing levels, each of which is uniquely specified by a three-bit word.

Since the quantizing level is the only value retained after sample values are quantized, errors are induced in the quantizing process which cannot be removed by additional processing. A quantitative measure of this error is easily derived. Since we would normally "decode" a quantizing level as the center of that level, the maximum error induced by quantizing a sample is $\pm\frac{1}{2}S$, where S is the width of a quantizing level. Assume that there are a large number of quantizing levels, resulting in a small value for S. Then, for *most* quantizing levels,

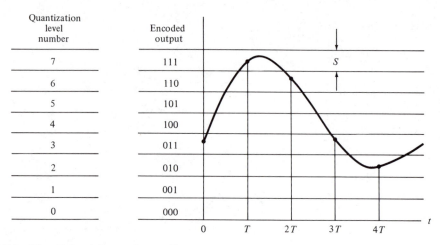

FIGURE 7-10. Quantizing and encoding.

the signal $x(t)$ will be nearly linear within the quantizing level, as shown in Figure 7-11a. For the system of interest it is the samples that are quantized and not the continuous-time signal $x(t)$. However, since sampling and reconstruction can be accomplished without error for bandlimited signals, the quantizer is the only error source in our A/D converter model. Thus we can evaluate the error by simply determining the error resulting from quantizing a continuous-time signal. This error is shown in Figure 7-11b, in which $2t_1$ is the time that the continuous-time signal $x(t)$ remains within the quantizing level. The mean-square error, E, is given by

$$E = \frac{1}{2t_1} \int_{-t_1}^{t_1} \epsilon^2(\alpha) \, d\alpha = \frac{1}{t_1} \int_0^{t_1} \epsilon^2(\alpha) \, d\alpha \qquad (7\text{-}12)$$

Since the error is given by

$$\epsilon(t) = \frac{S}{2t_1} t$$

we have

$$E = \frac{1}{t_1} \int_0^{t_1} \left(\frac{S}{2t_1}\right)^2 \alpha^2 \, d\alpha = \frac{S^2}{12} \qquad (7\text{-}13)$$

which is not a function of t_1. Since E is a mean-square value of a signal, it has dimensions of watts and is therefore interpreted as a noise power.

The quantity of interest at the output of the A/D converter is usually the *signal-to-noise ratio,* which is defined as the ratio of the signal power to the noise power. In order to compute this quantity, we define the A/D converter *dynamic range* D as the range of variation of the A/D converter input signal $x(t)$. Thus

$$D = \max \, [x(t)] - \min \, [x(t)] \qquad (7\text{-}14)$$

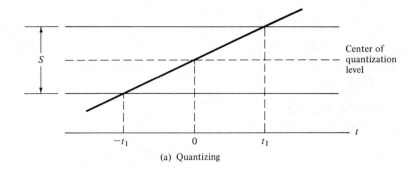

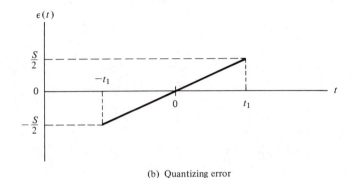

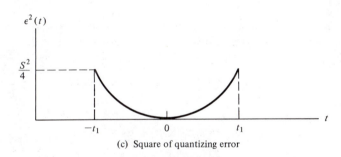

FIGURE 7-11. Calculation of quantizing error.

Since there are $q = 2^n$ quantizing levels, the width of each quantizing level is

$$S = \frac{D}{2^n} = D2^{-n} \qquad (7\text{-}15)$$

which yields

$$E = \frac{D^2}{12} 2^{-2n} \qquad (7\text{-}16)$$

Dividing this quantity into the signal power then yields the signal-to-noise ratio. In calculating the signal power, the assumption is usually made that the signal power at the output of the A/D converter is equal to the signal power at the input of the A/D converter.

It should be pointed out that (7-16) is not always a good approximation even when the number of quantization levels is large. If a signal is at its maximum or minimum values for significant time intervals, (7-16) is not valid since (7-16) ignores the error contribution at the maximum and minimum values. A square wave is a good example of such a signal.

EXAMPLE 7-3

Assume that the sinusoidal signal

$$x(t) = A \cos \omega t$$

is sampled and quantized. The signal power is

$$\overline{x^2} = \frac{1}{T_p} \int_0^{T_p} (A \cos \omega t)^2 \, dt = \frac{A^2}{2}$$

where T_p is the period of the waveform. The dynamic range D is $2A$ for a sinusoidal signal. Thus the noise power becomes

$$E = \frac{4A^2}{12} 2^{-2n} = \frac{A^2}{3} 2^{-2n}$$

The ratio of the signal power to the noise power is

$$\text{SNR} = \frac{A^2/2}{E} = \frac{3}{2} 2^{2n} \tag{7-17}$$

Since a small change in the wordlength, n, results in a large change in the signal-to-noise ratio, it is convenient to express SNR in decibels (dB). This yields

$$10 \log_{10} \text{SNR} = 10 \log_{10} \tfrac{3}{2} + 20n \log_{10} 2$$

or

$$10 \log_{10} \text{SNR} = 1.76 + 6.02n \tag{7-18}$$

Thus the signal-to-noise ratio at the output of the A/D converter increases by approximately 6 dB for each added bit of wordlength. Equation (7-18) is illustrated in Figure 7-12. The importance of using as large a wordlength as possible is obvious. The effect of *not* using the full range of the quantizer is an effective decrease in q and thus in the wordlength, n. Therefore, it is also important that the peak-to-peak value of the signal being processed span the full range (qS) of the quantizer.

Equation (7-18) was derived by assuming a sinusoidal signal. It should be pointed out that a similar result applies to any test signal, since in general, the signal-to-noise ratio increases 6.02 dB for each increment in the wordlength. The bias, 1.76 dB in this case, will change as the waveshape of the test signal is changed (see Problem 7-3). ∎

In the remainder of this book, the assumption will be made that quantizing errors are negligible. Thus no distinction will be made between sampled-data

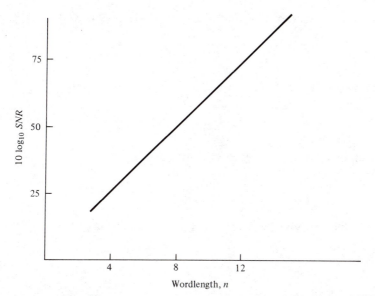

FIGURE 7-12. Quantizing signal-to-noise ratio assuming a sinusoidal signal.

(nonquantized) and digital (quantized) signals. We simply consider the signals to be discrete-time signals.

THE z-TRANSFORM

The z-transform is the basic tool for both the analysis and synthesis of discrete-time systems. In the following sections we develop the basic properties of the z-transform and show how to use the z-transform for the characterization and analysis of discrete-time systems. In Chapter 8 we shall turn our attention to the more interesting synthesis problem.

Definition of the z-Transform

It has been shown that a suitable model of a sampled signal is

$$x_s(t) = \sum_{n=-\infty}^{\infty} x(t)\delta(t - nT) \tag{7-19}$$

Prior to defining the z-transform, we shall make two minor modifications to the preceding expression. First, since $\delta(t - nT)$ is identically zero except at the sampling instants, $t = nT$, $x(t)$ can be replaced by $x(nT)$ if $x(t)$ is continuous at $t = nT$. We shall also assume that

$$x(t) \equiv 0, \quad t < 0$$

which establishes the time reference. Thus (7-19) will be written

$$x_s(t) = \sum_{n=0}^{\infty} x(nT)\delta(t - nT) \qquad (7\text{-}20)$$

Taking the Laplace transform yields

$$X_s(s) = \int_0^{\infty} \sum_{n=0}^{\infty} x(nT)\delta(t - nT)e^{-st}\, dt$$

which is, upon interchanging integration and summation,

$$X_s(s) = \sum_{n=0}^{\infty} x(nT) \int_0^{\infty} \delta(t - nT)e^{-st}\, dt$$

Using the sifting property of the delta function, this integrates to

$$X_s(s) = \sum_{n=0}^{\infty} x(nT)e^{-snT} \qquad (7\text{-}21)$$

Finally, defining the parameter z as

$$z = e^{sT} \qquad (7\text{-}22)$$

yields

$$X(z) = \sum_{n=0}^{\infty} x(nT)z^{-n} \qquad (7\text{-}23)$$

In (7-23) $X(z)$ is known as the z-transform of the sequence of samples, $x(nT)$. Note that the subscript s, which has been used to denote a sampled quantity, has been deleted. This can be done without confusion since only sample values are used to compute the z-transform. Therefore, the subscript s is redundant when the argument of the function is the parameter z.

Within a sequence of samples, a given sample can be represented by an ordered pair of numbers. One number in the ordered pair is used to represent the value of the sample and the other number in the ordered pair is used to specify the occurrence time of the sample. Each term in the summation of (7-23) has this form. The coefficient, $x(nT)$, denotes the sample value and z^{-n} denotes that the sample occurs n sample periods after the $t = 0$ reference. It is also instructive to recognize that e^{sT} is simply the T-second time shift operator discussed in Chapter 4. In (7-21) the sample value $x(nT)$ is multiplied by e^{-snT}, which places that sample value at $t = nT$. Thus the parameter z is simply shorthand notation for the Laplace time shift operator. As an example, $128.7z^{-37}$ denotes a sample, having value 128.7, which occurs 37 sample periods after the $t = 0$ reference.

Recall from Chapter 4 that the Laplace variable s was given by

$$s = \sigma + j\omega$$

where σ was a constant used to ensure convergence of the integral defining the Laplace transform and thus the existence of the transform itself. From (7-22),

$$z = e^{\sigma T}e^{j\omega T} \qquad (7\text{-}24)$$

so that the magnitude of z is given by

$$|z| = e^{\sigma T} \tag{7-25}$$

Thus, the right-half s-plane, $\sigma > 0$, corresponds to $|z| > 1$, while the left-half s-plane, $\sigma < 0$, corresponds to $|z| < 1$. We see that the left-half s-plane maps into the interior of the unit circle in the z-plane and that the right-half s-plane maps outside the unit circle in the z-plane. This mapping of the Laplace variable s into the z-plane through $z = e^{sT}$ is illustrated in Figure 7-13.

Recall from Chapter 4 that stable functions of the Laplace variable s, in the bounded-input bounded-output sense, must not have any poles in the right-half s-plane or on the $j\omega$-axis. Figure 7-13 then yields a similar definition of stability for functions of z. That is, stable functions of z must not have any poles outside or on the unit circle in the z-plane.

It is permissible to take (7-23) as the *definition* of the z-transform of a sequence of sample values. However, relating z to the Laplace variable s allows us to make good use of our previously developed theory of continuous-time signals and systems.

It is certainly possible to develop stability tests for discrete-time systems as was done in Chapter 5 for continuous-time systems. Such a development is

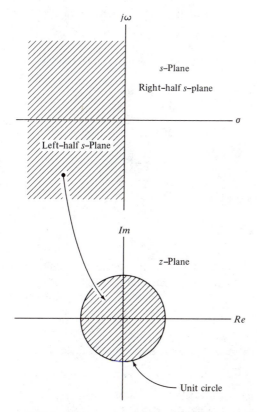

FIGURE 7-13. Mapping defined by $z = e^{sT}$.

beyond the scope of this text, however. An excellent reference is the book by Kuo.†

EXAMPLE 7-4

The unit pulse sequence is defined by the sample values

$$x(nT) = \begin{cases} 1, & n = 0 \\ 0, & n \neq 0 \end{cases} \triangleq \delta(n)$$

and is illustrated in Figure 7-14a. Substituting $x(nT)$ into the defining equation for the z-transform, (7-23), yields

$$X(z) = 1 + 0 \, z^{-1} + 0 \, z^{-2} + \cdots$$

Thus

$$X(z) = 1$$

This is an extremely important result. It will be shown later that the unit pulse sequence plays the same role in discrete-time systems that the unit impulse function plays in continuous-time systems. ∎

EXAMPLE 7-5

The unit step sample sequence is defined by the sample values

$$x(nT) = 1, \qquad n \geq 0$$

and is illustrated in Figure 7-14b. Substitution into (7-23) yields

$$X(z) = \sum_{n=0}^{\infty} z^{-n} \tag{7-26}$$

We know that for $|x| < 1$

$$\sum_{n=0}^{\infty} x^n = \frac{1}{1 - x} \tag{7-27}$$

Thus, with $x = z^{-1}$, we have‡

$$X(z) = \sum_{n=0}^{\infty} z^{-n} = \frac{1}{1 - z^{-1}}, \qquad |z| > 1 \tag{7-28}$$

It should be noted that the unit step sample sequence is defined such that $x(0) = 1$. Therefore, it is not simply a sampled $u(t)$ unless $u(0)$ is defined as 1, since $u(t)$ is not continuous at $t = 0$. The unit step sample sequence is, however, often denoted $u(n)$. ∎

---

†B. C. Kuo, *Digital Control Systems* (New York: Holt, Rinehart and Winston, 1980).

‡The sum $\sum_{n=0}^{\infty} z^{-n}$ converges absolutely to $1/(1 - z^{-1})$ outside the unit circle ($|z| > 1$). The function $1/(1 - z^{-1})$ is well behaved everywhere in the complex z-plane except at the pole $z = 1$. We take (7-28) as the definition of the z-transform of the sampled unit step. This result could have been derived rigorously by other methods beyond the scope of this text.

EXAMPLE 7-6

The unit exponential sequence is defined by the sample values

$$x(nT) = e^{-\alpha nT}, \qquad \alpha > 0, \quad n \geq 0$$

which is illustrated in Figure 7-14c and has the z-transform

$$X(z) = \sum_{n=0}^{\infty} e^{-\alpha nT} z^{-n} = \sum_{n=0}^{\infty} (e^{-\alpha T} z^{-1})^n \tag{7-29}$$

Using (7-27) with $x = e^{-\alpha T} z^{-1}$ yields

$$X(z) = \frac{1}{1 - e^{-\alpha T} z^{-1}}, \qquad |z| > e^{-\alpha T} \tag{7-30}$$

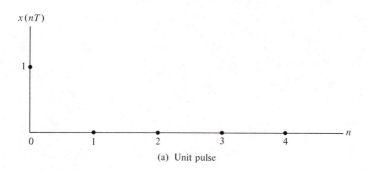

(a) Unit pulse

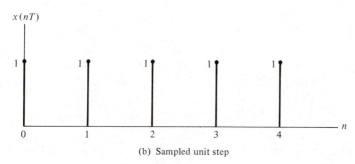

(b) Sampled unit step

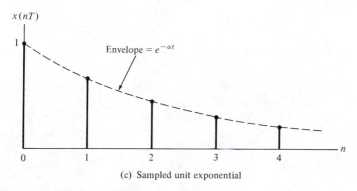

(c) Sampled unit exponential

FIGURE 7-14. Elementary sample sequences.

With the sampling period T and the parameter α fixed, $e^{-\alpha T}$ is a constant. Letting

$$K = e^{-\alpha T}$$

gives the z-transform pair

$$X(z) = \mathscr{L}[K^n] = \frac{1}{1 - Kz^{-1}}, \qquad |z| > K \qquad (7\text{-}31)$$

This is often the most convenient form to use. Equation (7-31) can be written

$$\frac{1}{1 - Kz^{-1}} = \frac{z}{z - K}$$

which shows that $X(z)$ has a zero at $z = 0$ and a pole at $z = K$. Thus only for $|K| < 1$ does K^n go to zero for large n. ∎

A short table of z-transforms is given in Table 7-1. The first three transforms have been derived in the preceding examples. The remaining transform pairs are easily proven (see Problems 7-7 through 7-9).

Linearity

The z-transformation is a linear operation. In other words, if A and B are constants,

$$\sum_{n=0}^{\infty} [Ax_1(nT) + Bx_2(nT)]\, z^{-n} = AX_1(z) + BX_2(z) \qquad (7\text{-}32)$$

where $X_1(z)$ and $X_2(z)$ are the z-transforms of $x_1(nT)$ and $x_2(nT)$, respectively. This is easily seen by recognizing that the left-hand side of (7-32) can be written

$$\sum_{n=0}^{\infty} [Ax_1(nT) + Bx_2(nT)]\, z^{-n} = A \sum_{n=0}^{\infty} x_1(nT)z^{-n} + B \sum_{n=0}^{\infty} x_2(nT)z^{-n}$$

By definition, the two sums on the right-hand side of the expression above are $X_1(z)$ and $X_2(z)$.

Initial Value and Final Value Theorems

The initial value and the final value theorems are useful in that they provide quick insight into the behavior of a time sequence, $x(nT)$, without having to compute the inverse z-transform of $X(z)$. When the inverse z-transform of $X(z)$ is computed, the initial value and final value theorems provide a partial check on the results.

The initial value theorem states that

$$x(0) = \lim_{z \to \infty} X(z) \qquad (7\text{-}33)$$

TABLE 7-1. Short Table of z-Transforms

Transform Pair Number	Continuous-time Function $f(t)$ for $t > 0$	Sample values $f(nT)$ for $n \geq 0$	z-Transform of $f(nT)$
1.	—	$f(nT) = \begin{cases} 1, & n = 0 \\ 0, & n \neq 0 \end{cases} \triangleq \delta(n)$	1
2.	1 (unit step)	1	$\dfrac{1}{1 - z^{-1}}$
3.	e^{-at}	$e^{-anT} = (e^{-\alpha T})n = K^n$	$\dfrac{1}{1 - e^{-\alpha T}z^{-1}} = \dfrac{1}{1 - Kz^{-1}}$
4.	t	nT	$\dfrac{Tz^{-1}}{(1 - z^{-1})^2}$
5.†	te^{-at}	nTe^{-anT}	$\dfrac{Te^{-aT}z^{-1}}{(1 - e^{-aT}z^{-1})^2}$
6.†	$\sin bt$	$\sin bnT$	$\dfrac{(\sin bT)z^{-1}}{1 - 2(\cos bT)z^{-1} + z^{-2}}$
7.†	$\cos bt$	$\cos bnT$	$\dfrac{1 - (\cos bT)z^{-1}}{1 - 2(\cos bT)z^{-1} + z^{-2}}$
8.†	$e^{-at}\sin bt$	$e^{-anT}\sin bnT$	$\dfrac{e^{-aT}(\sin bT)z^{-1}}{1 - 2e^{-aT}(\cos bT)z^{-1} + e^{-2aT}z^{-2}}$
9.†	$e^{-at}\cos bt$	$e^{-anT}\cos bnT$	$\dfrac{1 - e^{-aT}(\cos bT)z^{-1}}{1 - 2e^{-aT}(\cos bT)z^{-1} + e^{-2aT}z^{-2}}$

†In transforms 5 through 9, e^{-aT} and bT can be replaced by constants, K_1 and K_2, respectively, as was done in transform 3. Convergence of the z-transform requires $|K_1| \geq 1$.

This result is easily derived. By definition

$$X(z) = \sum_{n=0}^{\infty} x(nT)z^{-n} = x(0) + \sum_{n=1}^{\infty} x(nT)z^{-n}$$

As $z \to \infty$, the summation on the right vanishes and (7-33) results.

The final value theorem states that

$$x(\infty) = \lim_{z \to 1} (1 - z^{-1})X(z) \tag{7-34}$$

The proof of this theorem is more difficult than the proof of the initial value theorem and we shall not take the time and space to construct the proof here. However, (7-34) can be easily justified. If $X(z)$ has any of its poles outside the unit circle, $X(z)$ corresponds to an unstable function and $x(\infty) = \infty$. If $X(z)$ has all its poles inside the unit circle, $x(nT)$ represents samples of a function that decays with time and $x(\infty) = 0$. If $X(z)$ has *simple* poles on the unit circle but not at $z = 1$, the resulting time response is oscillatory and $x(\infty)$ is not defined, although it is bounded by the limits of the oscillation. By continuing this argument we see that a nonzero steady-state constant value must result from a simple pole in the z-plane at $z = 1$. Thus assume that $X(z)$ has a simple pole at $z = 1$ and decompose $X(z)$ as

$$X(z) = \frac{K}{1 - z^{-1}} + G(z)$$

where all the poles of $G(z)$ lie inside the unit circle. Clearly, by the preceding argument, K is the steady-state value of $X(z)$. Applying (7-34) yields

$$x(\infty) = \lim_{z \to 1} (1 - z^{-1})X(z) = K + \lim_{z \to 1} (1 - z^{-1})G(z) = K$$

which is the expected result.

Several interesting proofs of the final value theorem are given in the literature.[†]

Inverse z-Transform

The z-transform of a sample sequence is written, by definition,

$$X(z) = x(0) + x(T)z^{-1} + x(2T)z^{-2} + \cdots \tag{7-35}$$

If we can manipulate $X(z)$ into this form, the sample values, $x(nT)$, can be determined by inspection. This is easily accomplished by long division when $X(z)$ is expressed as a ratio of polynomials in z. Prior to performing the long division, it is usually most convenient to arrange both the numerator and the denominator in ascending powers of z^{-1}. A simple example will illustrate the method.

[†]As an example of two very different proofs of the final value theorem, the interested student should refer to S. R. Atre, "An Alternate Derivation of Final and Initial Value Theorems in the z Domain," *Proceedings of the IEEE*, Vol. 63, No. 3, March 1975, pp. 537–538, and J. R. Ragazzini and G. F. Franklin, *Sampled-Data Control Systems* (New York: McGraw-Hill, 1958), pp. 61–63.

EXAMPLE 7-7

The inverse of

$$X(z) = \frac{z^2}{(z - 1)(z - 0.2)}$$

will be determined by long division. First $X(z)$ is written as

$$X(z) = \frac{z^2}{z^2 - 1.2z + 0.2}$$

which is, after multiplying numerator and denominator by z^{-2},

$$X(z) = \frac{1}{1 - 1.2z^{-1} + 0.2z^{-2}}$$

The division is illustrated below:

$$
\begin{array}{r}
1 + 1.2z^{-1} + 1.24z^{-2} + 1.248z^{-3} \cdots \\
1 - 1.2z^{-1} + 0.2z^{-2} \overline{\smash{)}\, 1} \\
\underline{1 - 1.2z^{-1} + 0.2z^{-2}} \\
1.2z^{-1} - 0.20z^{-2} \\
\underline{1.2z^{-1} - 1.44z^{-2} + 0.240z^{-3}} \\
1.24z^{-2} - 0.240z^{-3} \\
\underline{1.24z^{-2} - 1.488z^{-3} + 0.248z^{-4}} \\
1.248z^{-3} \cdots
\end{array}
$$

Thus

$$X(z) = 1 + 1.2z^{-1} + 1.24z^{-2} + 1.248z^{-3} + \cdots$$

from which

$$x(0) = 1$$
$$x(T) = 1.2$$
$$x(2T) = 1.24$$
$$x(3T) = 1.248$$

The division can obviously be continued to give as many terms as desired. Although long division is generally the easiest method for inverting a given $X(z)$, the method has the disadvantage of not yielding the general sample value term, $x(nT)$, in closed form.

Note that from the initial value theorem

$$x(0) = \lim_{z \to \infty} X(z) = 1$$

which checks with our previous result. The final value theorem yields

$$x(\infty) = \lim_{z \to 1} \frac{z - 1}{z} X(z) = \lim_{z \to 1} \frac{z}{z - 0.2} = 1.25$$

■

This is not a surprising result considering the sequence $\{x(nT)\}$ found earlier. Although this method of inverse z-transforming a function does not yield the general term in closed form, it is sometimes sufficient. For example, if the time sequence approaches an asymptote in a well-behaved manner, often one only needs to compute the first several values of the time sequence and then compute the asymptote by the final value theorem.

Partial-fraction expansion overcomes the disadvantage of the long-division method in that partial-fraction expansion yields the general term of the time expansion, $x(nT)$. The basic disadvantage of the partial-fraction-expansion method is that the denominator of $X(z)$ must be factored. The technique is exactly the same method as that used to find inverse Laplace transforms by partial-fraction expansion. The idea is to manipulate $X(z)$ into a form that can be inverse z-transformed by using Table 7-1. To use Table 7-1, it is necessary to perform a partial-fraction expansion of $X(z)/z$. An example illustrates the method.

EXAMPLE 7-8

The function to be inverse transformed is

$$X(z) = \frac{1}{1 - 1.2z^{-1} + 0.2z^{-2}} = \frac{z^2}{(z - 1)(z - 0.2)}$$

First, we write

$$\frac{X(z)}{z} = \frac{z}{(z - 1)(z - 0.2)} = \frac{K_1}{z - 1} + \frac{K_2}{z - 0.2}$$

The constants K_1 and K_2 are given by

$$K_1 = \lim_{z \to 1} (z - 1) \frac{X(z)}{z} = 1.25$$

and

$$K_2 = \lim_{z \to 0.2} (z - 0.2) \frac{X(z)}{z} = -0.25$$

Thus

$$X(z) = \frac{1.25z}{z - 1} - \frac{0.25z}{z - 0.2}$$

or

$$X(z) = \frac{1.25}{1 - z^{-1}} - \frac{0.25}{1 - 0.2z^{-1}}$$

We shall find the inverse transform of $X(z)$ by using transform pair 3 in Table 7-1. The first term has $K = 1$, which is a unit step sample sequence; and the second term has $K = 0.2$, which is a sampled exponential. Thus

$$x(nT) = 1.25 - 0.25(0.2)^n, \qquad n \geq 0$$

The values of $x(nT)$ are

$$x(0) = 1.25 - 0.25 (0.2)^0 = 1$$

$$x(T) = 1.25 - 0.25 (0.2)^1 = 1.2$$

$$x(2T) = 1.25 - 0.25 (0.2)^2 = 1.24$$

$$x(3T) = 1.25 - 0.25 (0.2)^3 = 1.248$$

It should be noted that these values agree with the values determined by long division in the preceding example. Also, note that

$$x(\infty) = 1.25$$ ■

EXAMPLE 7-9

Now consider the function

$$Y(z) = \frac{1}{z^2 - 1.2z + 0.2} = \frac{1}{(z - 1)(z - 0.2)}$$

The first step in performing the inverse z-transform is to divide both sides by z. This yields

$$\frac{Y(z)}{z} = \frac{1}{z(z - 1)(z - 0.2)} = \frac{K_1}{z} + \frac{K_2}{z - 1} + \frac{K_3}{z - 0.2}$$

from which

$$K_1 = \lim_{z \to 0} z \frac{Y(z)}{z} = \frac{1}{0.2} = 5$$

$$K_2 = \lim_{z \to 1} (z - 1) \frac{Y(z)}{z} = \frac{1}{0.8} = 1.25$$

and

$$K_3 = \lim_{z \to 0.2} (z - 0.2) \frac{Y(z)}{z} = -6.25$$

Thus

$$Y(z) = 5 + 1.25 \frac{z}{z - 1} - 6.25 \frac{z}{z - 0.2}$$

or

$$Y(z) = 5 + 1.25 \frac{1}{1 - z^{-1}} - 6.25 \frac{1}{1 - 0.2z^{-1}}$$

We can now inverse transform using transform pairs 1 and 3 in Table 7-1. This yields

$$y(nT) = 5\delta(n) + 1.25 - 6.25(0.2)^n, \qquad n \geq 0$$

The student should check this result by using long division. Upon doing so, it will be noted that $y(0) = 0$, $y(T) = 0$, and $y(2T) = 1$. This is obvious from an inspection of $Y(z)$, since the leading term in the division process is z^{-2}. Simple observations of this type are useful in that computational errors are often revealed. ∎

The essential difference between Examples 7-8 and 7-9 is that division by z to form $Y(z)/z$ introduced an additional pole, and therefore an additional term in the partial-fraction expansion, in Example 7-9.

Delay Operator

The delay operation is of fundamental importance in the analysis and synthesis of discrete-time or digital systems, as we shall shortly see. We now show that if the time sequence $\{x(nT)\}$ is delayed by K sample periods, the effect in the z-domain is to multiply $X(z)$ by z^{-K}.

This is easily proved. Since

$$X(z) = \sum_{n=0}^{\infty} x(nT)z^{-n}$$

the z-transform of the sequence $\{x(nT - KT)\}$ is

$$\mathcal{L}[x(nT - KT)] = \sum_{n=0}^{\infty} x(nT - KT)z^{-n}$$

Letting $m = n - K$ yields

$$\mathcal{L}\{x(nT - KT)\} = \sum_{m=-K}^{\infty} x(mT)z^{-m-K}$$

Since $x(mT)$ is assumed zero for $m < 0$, we have

$$\mathcal{L}\{x(nT - KT)\} = \sum_{m=0}^{\infty} x(mT)z^{-m-K}$$

or

$$\mathcal{L}\{x(nT - KT)\} = z^{-K} \sum_{m=0}^{\infty} x(mT)z^{-m} = z^{-K}X(z) \qquad (7\text{-}36)$$

The delay operation is illustrated in Figure 7-15, where it is clear that delay by K sample periods is equivalent to multiplication by z^{-K}.

EXAMPLE 7-10
We shall now use the delay operator and the result of Example 7-8 to work Example 7-9. Note that $Y(z)$ from Example 7-9 can be written

$$Y(z) = \left[\frac{z^2}{z^2 - 1.2z + 0.2} \right] z^{-2}$$

However, the term in brackets is $X(z)$ from Example 7-8. Thus

$$Y(z) = X(z)z^{-2}$$

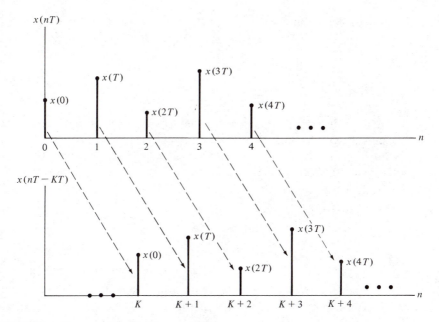

FIGURE 7-15. Illustration of delay operation.

Inverse z-transforming yields

$$y(nT) = x\{(n - 2)T\}$$

Using the result of Example 7-8 yields

$$y(nT) = 1.25 - 0.25(0.2)^{n-2}, \qquad n \geq 2$$

The student should verify that this agrees with the result of Example 7-9. ∎

7-4

DIFFERENCE EQUATIONS AND DISCRETE-TIME SYSTEMS

Just as the differential equation can be used to model a continuous-time system, the difference equation can be used to model a discrete-time system. Consider, for example, a first-order continuous-time system defined by the differential equation

$$\frac{dy}{dt} + ay(t) = bx(t) \tag{7-37}$$

The system output, $y(t)$, can be expressed in the form

$$y(t) = b \int_{-\infty}^{t} x(\alpha) \, d\alpha - a \int_{-\infty}^{t} y(\alpha) \, d\alpha \tag{7-38}$$

This equation tells us that the present value of the system output, $y(t)$, is a function of all previous values of the system input, $b \int_{-\infty}^{t} x(\alpha) \, d\alpha$, and is also a function of all past values of the system output, $\int_{-\infty}^{t} y(\alpha) \, d\alpha$.

Difference Equations

A linear discrete-time system operates in somewhat the same way in that the present system output, $y(nT)$, must be computed using the present input $x(nT)$, past inputs $x(nT - kT)$, and past system outputs $y(nT - kT)$. The structure of such a processor is illustrated in Figure 7-16. The general difference equation for this processor is

$$
\begin{aligned}
y(nT) = & \; L_0 x(nT) + L_1 x(nT - T) + L_2 x(nT - 2T) + \cdots \\
& + L_r x(nT - rT) - K_1 y(nT - T) \\
& - K_2 y(nT - 2T) - \cdots - K_m y(nT - mT)
\end{aligned}
\tag{7-39}
$$

In other words, the processor illustrated in Figure 7-16 generates an output by weighting the present input, the past r inputs, and the past m outputs.

At the present time, we are concerned with the analysis of discrete-time systems. The analysis problem is usually a problem of determining the system output given the system input and a specification of the system. There are many different ways in which the system may be specified, but the most convenient method is to specify the coefficients of the difference equation.

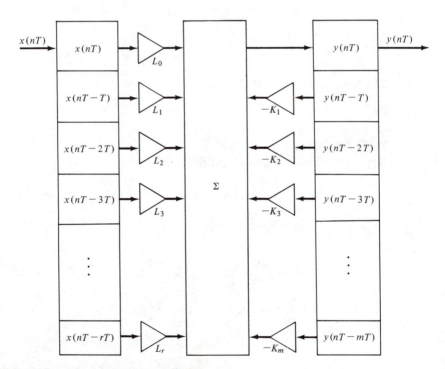

FIGURE 7-16. Linear digital signal processor.

In Chapter 8 we turn our attention to the synthesis of discrete-time systems, in which the problem is to determine the coefficients of the difference equation in order to perform some specified task.

The first step in the analysis of a discrete-time processor is to solve the difference equation. This is accomplished by first writing the difference equation in the form

$$y(nT) + K_1 y(nT - T) + K_2 y(nT - 2T) + \cdots + K_m y(nT - mT)$$

$$= L_0 x(nT) + L_1 x(nT - T) + L_2 x(nT - 2T) + \cdots + L_r x(nT - rT) \tag{7-40}$$

and z-transforming both sides of the difference equation by using the time delay operator

$$\mathscr{Z}[x(nT - kT)] = z^{-k} X(z)$$

developed previously. Application of the time delay operator yields

$$Y(z) + K_1 z^{-1} Y(z) + K_2 z^{-2} Y(z) + \cdots + K_m z^{-m} Y(z)$$

$$= L_0 X(z) + L_1 z^{-1} X(z) + L_2 z^{-2} X(z) + \cdots + L_r z^{-r} X(z)$$

which is

$$Y(z)(1 + K_1 z^{-1} + K_2 z^{-2} + \cdots + K_m z^{-m}) \tag{7-41}$$

$$= X(z)(L_0 + L_1 z^{-1} + L_2 z^{-2} + \cdots + L_r z^{-r})$$

The ratio of $Y(z)$ to $X(z)$ is the transfer function of the discrete-time system, which is often referred to as the *pulse transfer function*. For the processor of Figure 7-16, the pulse transfer function is given by

$$\frac{Y(z)}{X(z)} = H(z) = \frac{L_0 + L_1 z^{-1} + L_2 z^{-2} + \cdots + L_r z^{-r}}{1 + K_1 z^{-1} + K_2 z^{-2} + \cdots + K_m z^{-m}} \tag{7-42}$$

By the definition of the pulse transfer function, we can write

$$Y(z) = H(z)X(z) \tag{7-43}$$

It was shown previously that the z-transform of a unit pulse is unity. Thus if a discrete-time system has a *unit pulse input*, the system output is

$$Y(z) = H(z) \tag{7-44}$$

which can be inverse z-transformed to yield

$$y(nT) = h(nT) \tag{7-45}$$

The time sequence $\{h(nT)\}$, which is the inverse z-transform of the pulse transfer function $H(z)$, is known as the *unit pulse response*. We shall now see that the unit pulse response plays essentially the same role in discrete-time systems that the unit impulse response plays in continuous time systems.

A system described by the difference equation (7-39) in which L_i and K_j are constant for all i and j is known as a *fixed* discrete-time system. These systems are sometimes referred to as *shift invariant* since a time shift of the input time shifts the output without changing the shape of the output. If $h(nT)$ is zero for $n < 0$, the discrete-time system is known as a *causal* system.

Discrete Convolution

The time-domain equivalent of

$$Y(z) = H(z)X(z)$$

is easily determined. Be definition of the z-transform,

$$X(z) = x(0) + x(T)z^{-1} + x(2T)z^{-2} + \cdots$$

and

$$H(z) = h(0) + h(T)z^{-1} + h(2T)z^{-2} + \cdots$$

where $x(iT)$ and $h(iT)$ are the ith values of the system input and unit pulse response, respectively. Combining the preceding three equations, we get

$$Y(z) = [x(0) + x(T)z^{-1} + x(2T)z^{-2} + \cdots] \tag{7-46}$$
$$\times [h(0) + h(T)z^{-1} + h(2T)z^{-2} + \cdots]$$

Performing the multiplication and grouping together terms with the same z^{-n} multiplier yields

$$Y(z) = x(0)h(0) + [x(0)h(T) + x(T)h(0)]z^{-1}$$
$$+ [x(0)h(2T) + x(T)h(T) + x(2T)h(0)]z^{-2}$$
$$+ \cdots \tag{7-47}$$

Note that if a term multiplies z^{-n}, the sum of the arguments of $x(\cdot)$ and $h(\cdot)$ is nT. Thus the general term is

$$[x(0)h(nT) + x(T)h(nT - T) + \cdots$$
$$+ x(nT - T)h(T) + x(nT)h(0)]z^{-n} \tag{7-48}$$

Since

$$Y(z) = y(0) + y(T)z^{-1} + y(2T)z^{-2} + \cdots + y(nT)z^{-n} + \cdots$$

we have

$$y(0) = x(0)h(0)$$
$$y(T) = x(0)h(T) + x(T)h(0)$$
$$y(2T) = x(0)h(2T) + x(T)h(T) + x(2T)h(0)$$

$$\cdot$$
$$\cdot$$
$$\cdot$$

$$y(nT) = x(0)h(nT) + x(T)h(nT - T) + \cdots$$
$$+ x(nT - T)h(T) + x(nT)h(0)$$

so that

$$y(nT) = \sum_{m=0}^{n} x(mT)h(nT - mT) \qquad (7\text{-}49)$$

which is equivalent to

$$y(nT) = \sum_{m=0}^{n} h(mT)x(nT - mT) \qquad (7\text{-}50)$$

This is defined as *discrete convolution*.

EXAMPLE 7-11

We shall use discrete convolution to compute the output of a discrete processor whose input $x(nT)$ and unit pulse response $h(nT)$ are shown in Figure 7-17a. The terms in the summation

$$y(nT) = \sum_{m=0}^{n} x(mT)h(nT - mT)$$

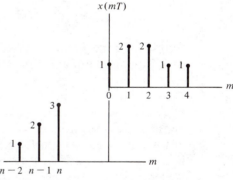

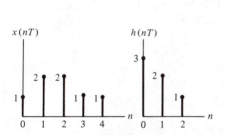

(a) Input and unit pulse response

(b) Functions for computing convolution sum

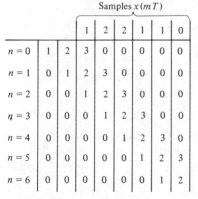

Samples $x(mT)$

		1	2	2	1	1	0
$n = 0$	1	2	3	0	0	0	0	0
$n = 1$	0	1	2	3	0	0	0	0
$n = 2$	0	0	1	2	3	0	0	0
$n = 3$	0	0	0	1	2	3	0	0
$n = 4$	0	0	0	0	1	2	3	0
$n = 5$	0	0	0	0	0	1	2	3
$n = 6$	0	0	0	0	0	0	1	2

(c) Table for evaluating summation

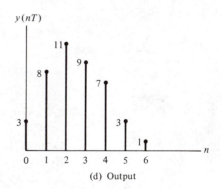

(d) Output

FIGURE 7-17. Discrete convolution.

are shown in Figure 7-17b for $n < 0$. Clearly, $y(nT) \equiv 0$ for $n < 0$ since there is no overlap of the functions $x(mT)$ and $h(nT - mT)$. The table shown in Figure 7-17c is used to evaluate $y(nT)$ for $0 \leq n \leq 6$. The sample values of $x(mT)$ are shown across the top and the samples of $h(nT - mT)$ are shown in the table for $n = 0, 1, 2, 3, 4, 5, 6$. The samples of $h(nT - mT)$ shift to the right 1 unit for each increment of n. The value of the convolution sum is determined by multiplying and summing the two sets of sample values for each value of n. For example,

$$y(0) = 1 \cdot 3 = 3$$

$$y(T) = 1 \cdot 2 + 2 \cdot 3 = 8$$

$$y(2T) = 1 \cdot 1 + 2 \cdot 2 + 2 \cdot 3 = 11$$

$$\cdot$$

$$\cdot$$

$$\cdot$$

$$y(6T) = 1 \cdot 1 = 1$$

For $n > 6$, $y(nT) \equiv 0$ since there is no overlap of $x(mT)$ and $h(nT - mT)$. ∎

Steady-State Frequency Response of a Linear Discrete-Time System

The steady-state frequency response is an important characterization of a linear fixed discrete-time system, just as it was for a continuous-time system. Recall that the steady-state frequency response of a continuous-time system was determined by placing a sinusoidal signal on the input of the system. Since the system is fixed and linear, the output signal is also sinusoidal at the same frequency as the input signal. Thus, in passing through the system, the input signal is subjected only to an amplitude scaling and a phase shift. The ratio of the output signal amplitude to the input signal amplitude is the amplitude response at the input frequency and the output phase minus the input phase is the phase response at the input frequency.

As a parallel to the previous discussion, the sinusoidal steady-state response of a fixed, linear, discrete-time system is determined by placing the complex sampled sinusoid

$$x(nT) = Ae^{j(\omega nT + \theta)} \tag{7-51}$$

on the input of a discrete-time system as illustrated in Figure 7-18. Using (7-50), the output of the system can be written

$$y(nT) = A \sum_{m=0}^{n} h(mT)e^{j[\omega(n-m)T + \theta]} \tag{7-52}$$

The system output can be placed in the form

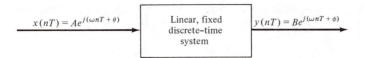

$$x(nT) = Ae^{j(\omega nT + \theta)} \quad \boxed{\begin{array}{c} \text{Linear, fixed} \\ \text{discrete–time} \\ \text{system} \end{array}} \quad y(nT) = Be^{j(\omega nT + \phi)}$$

FIGURE 7-18. Input–output relationship for a discrete-time system.

$$y(nT) = Ae^{j(\omega nT + \theta)} \left[\sum_{m=0}^{n} h(mT)e^{-j\omega mT} \right] \tag{7-53}$$

For n sufficiently large, the terms $h(nT)$ of the unit pulse response are negligible, and we say that the steady-state response has been reached. The term in brackets is then a complex function depending only on the input frequency, ω. Denoting this function by $G(\omega)$, we can write, for n sufficiently large,

$$y(nT) = G(\omega)Ae^{j(\omega nT + \theta)} \tag{7-54}$$

Since we know that the output $y(nT)$ is a sampled sinusoid, we can write

$$Be^{j(\omega nT + \phi)} = G(\omega)Ae^{j(\omega nT + \theta)} \tag{7-55}$$

in which B is the output amplitude and ϕ is the phase of the output. This yields

$$G(\omega) = \frac{Be^{j(\omega nT + \phi)}}{Ae^{j(\omega nT + \theta)}} = \frac{B}{A} e^{j(\phi - \theta)} \tag{7-56}$$

Thus, at frequency ω,

$$|G(\omega)| = \frac{B}{A} \tag{7-57}$$

and

$$\underline{/G(\omega)} = \phi - \theta \tag{7-58}$$

Since $G(\omega)$ can be represented

$$G(\omega) = \sum_{m=0}^{\infty} h(mT)e^{-j\omega mT}$$

and

$$H(z) = \sum_{m=0}^{\infty} h(mT)z^{-m}$$

it follows that the sinusoidal steady-state response of a discrete-time system is given by

$$G(\omega) = H(e^{j\omega T}) \tag{7-59}$$

Thus the steady-state response is easily obtained from the pulse transfer function by simply replacing z in the pulse transfer function by $e^{j\omega T}$.

Of fundamental importance is the fact that the sinusoidal steady-state frequency response of a discrete-time system is periodic in the sampling frequency, ω_s. This is easily seen by writing $H(e^{j\omega T})$ with ω replaced by

$$\omega_a = \omega + k\omega_s \tag{7-60}$$

where k is an integer. This yields

$$H[e^{j(\omega + k\omega_s)T}] = H(e^{j\omega T}e^{jk\omega_s T}) \tag{7-61}$$

Since $\omega_s T = 2\pi$, we have

$$H[e^{j(\omega + k\omega_s)T}] = H(e^{j\omega T}e^{jk2\pi}) = H(e^{j\omega T}) \tag{7-62}$$

which shows that $H(e^{j\omega T})$ is periodic with period ω_s.

Since $H(e^{j\omega T})$ is periodic in the sampling frequency, it is often advantageous to normalize the frequency variable with respect to the sampling frequency. Defining the frequency ratio r as

$$r = \frac{\omega}{\omega_s} \tag{7-63}$$

allows ωT to be replaced by

$$\omega T = r\omega_s T = 2\pi r \tag{7-64}$$

so that the sinusoidal steady-state frequency response is expressed as $H(e^{j2\pi r})$.

The concepts developed in this section are illustrated in the following example.

EXAMPLE 7-12

In this example we determine the amplitude and phase response of the discrete-time system defined by the difference equation

$$y(nT) = x(nT) + x(nT - 2T) \tag{7-65}$$

It follows from the difference equation that the pulse transfer response of the system is

$$H(z) = 1 + z^{-2} \tag{7-66}$$

Thus the sinusoidal steady-state frequency response is

$$H(e^{j\omega T}) = 1 + e^{-j2\omega T}$$

or

$$H(e^{j\omega T}) = (e^{j\omega T} + e^{-j\omega T}) e^{-j\omega T}$$

which can be written

$$H(e^{j\omega T}) = (2 \cos \omega T) e^{-j\omega T} \tag{7-67}$$

In terms of the normalized frequency r, the preceding expression is

$$H(e^{j2\pi r}) = (2 \cos 2\pi r) e^{-j2\pi r} \tag{7-68}$$

The magnitude and phase response of the system are illustrated in Figure 7-19 for r between 0 and 0.5. The π radian discontinuity in the phase response at $r = 0.25$ is due to the sign change of $\cos 2\pi r$ at $r = 0.25$. ■

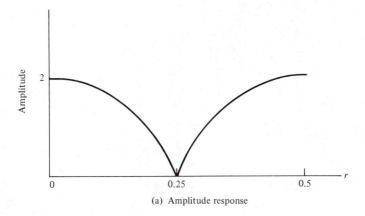

(a) Amplitude response

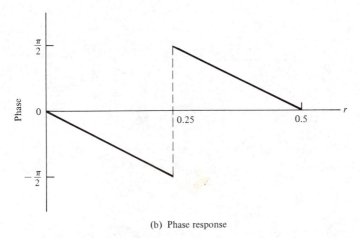

(b) Phase response

FIGURE 7-19. Amplitude and phase response for Example 7-12.

EXAMPLE 7-13

A very important discrete-time system is one that delays an input sample by K sample periods. Such a system is illustrated in Figure 7-20a and as we saw previously, the pulse transfer function is

$$H(z) = z^{-K} \qquad (7\text{-}69)$$

The sinusoidal steady-state frequency response is, in terms of normalized frequency, given by

$$H(e^{j2\pi r}) = e^{-j2\pi Kr} \qquad (7\text{-}70)$$

so that the time delay system has the unity amplitude response and the linear phase response illustrated in Figures 7-20b and c, respectively. The student should note that these results are consistent with the Fourier transform result given in Table 3-2 (entry 2), which states that a time shift in the time domain is equivalent to a linear phase shift in the frequency domain. ∎

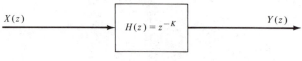

(a) Discrete–time delay system

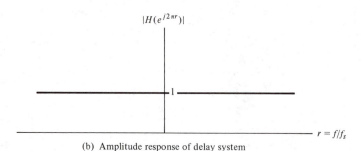

(b) Amplitude response of delay system

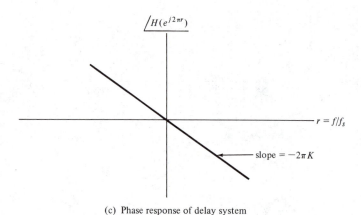

(c) Phase response of delay system

FIGURE 7-20. Discrete-time delay system.

7-5.

EXAMPLE OF A DISCRETE-TIME SYSTEM

As a simple example of a discrete-time system, and as a review of the various techniques that have been studied in this chapter, we shall examine the performance of the system illustrated in Figure 7-21. We shall determine the difference equation, the pulse transfer function, and the sinusoidal steady-state frequency response (both magnitude and phase) of the system.

It follows from the block diagram of the system that the difference equation defining the operation of the system is

$$y(nT) = kx(nT) + \alpha y(nT - T) \tag{7-71}$$

from which we have

$$y(nT) - \alpha y(nT - T) = kx(nT) \tag{7-72}$$

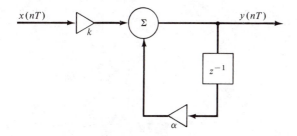

FIGURE 7-21. Example digital processor.

Since a delay of T is equivalent to multiplication by z^{-1}, (7-72) becomes

$$Y(z) - \alpha z^{-1}Y(z) = kX(z)$$

which yields the pulse transfer function

$$H(z) = \frac{Y(z)}{X(z)} = \frac{k}{1 - \alpha z^{-1}} \qquad (7\text{-}73)$$

As we saw in the preceding section, the sinusoidal steady-state response can be determined from $H(z)$ by setting $z = e^{j\omega T}$. For our example this yields

$$H(e^{j\omega T}) = \frac{k}{1 - \alpha e^{-j\omega T}} \qquad (7\text{-}74)$$

or

$$H(e^{j\omega T}) = \frac{k}{1 - \alpha \cos \omega T + j\alpha \sin \omega T} \qquad (7\text{-}75)$$

In terms of the normalized frequency ratio, $r = f/f_s$, this becomes

$$H(e^{j2\pi r}) = \frac{k}{1 - \alpha \cos 2\pi r + j\alpha \sin 2\pi r} \qquad (7\text{-}76)$$

In order to plot the amplitude and phase response of the system, (7-76) is placed in the form

$$H(e^{j2\pi r}) = A(r)e^{j\phi(r)} \qquad (7\text{-}77)$$

in which $A(r)$ and $\phi(r)$ are the amplitude and phase responses, respectively. It follows from (7-77) that the amplitude response $A(r)$ is given by

$$A(r) = \frac{k}{\sqrt{1 + \alpha^2 - 2\alpha \cos 2\pi r}} \qquad (7\text{-}78)$$

and that the phase response, $\phi(r)$ is given by

$$\phi(r) = -\tan^{-1} \frac{\alpha \sin 2\pi r}{1 - \alpha \cos 2\pi r} \qquad (7\text{-}79)$$

Observation of (7-78) shows that the discrete-time system in Figure 7-21 is a low-pass system since $A(r)$ has a maximum at $r = 0$ and decreases with increasing r for r in the range $0 < r < \frac{1}{2}$. If the filter is to have unity gain at $r = 0$,

$$A(0) = \frac{k}{\sqrt{1 - 2\alpha + \alpha^2}} = \frac{k}{1 - \alpha} = 1$$

from which

$$k = 1 - \alpha$$

It should be noted that any desired dc gain can be achieved by an adjustment of k.

The amplitude and phase responses of the system are illustrated in Figures 7-22 and 7-23, respectively, for r in the range $0 \le r \le \frac{1}{2}$. The amplitude response is even and the phase response is odd since they are the magnitude and phase of a Fourier transform. This, together with the fact that the amplitude and phase responses are periodic, since $H(e^{j2\pi r})$ is periodic, fully defines $A(r)$ and $\phi(r)$ for all values of r. The amplitude response is illustrated with a log frequency scale so that the 6 dB per octave slope characteristic of a first-order system can be observed.

*7-6.

INVERSE z-TRANSFORMATION BY THE INVERSION INTEGRAL

In Section 4-5 the technique of determining inverse Laplace transforms by the inversion integral was briefly investigated. We now briefly consider the inversion integral method as applied to z-transforms.

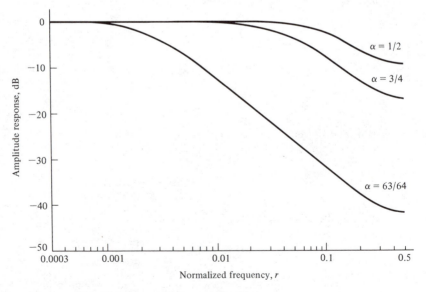

FIGURE 7-22. Amplitude response of example filter.

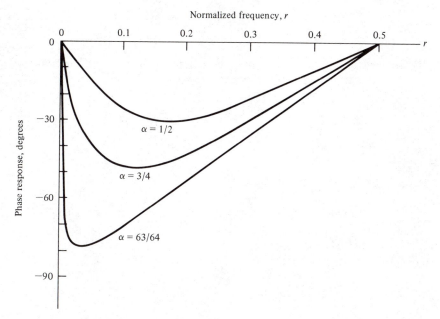

FIGURE 7-23. Phase response of example filter.

The basic inversion integral is easy to derive. By definition, the z-transform of the sample sequence $\{x(nT)\}$ is

$$X(z) = \sum_{m=0}^{\infty} x(mT)z^{-m}$$

Multiplying both sides of the expression above by z^{n-1} and integrating along a path C that lies entirely within the region of convergence of $X(z)$ yields

$$\oint_C X(z)z^{n-1}\,dz = \oint_C \sum_{m=0}^{\infty} x(mT)z^{-m+n-1}\,dz \qquad (7\text{-}80)$$

Assuming that

$$\sum_{m=0}^{\infty} |x(mT)| < \infty$$

in other words, the sequence $\{x(mT)\}$ is absolutely summable, the order of integration and summation can be interchanged. This yields

$$\oint_C X(z)z^{n-1}\,dz = \sum_{m=0}^{\infty} x(mT) \oint_C z^{-m+n-1}\,dz \qquad (7\text{-}81)$$

The integral in the right-hand side can be evaluated through the use of the residue theorem presented in Section 4-5. For $m = n$, z^{-m+n-1} has a first-order pole at $z = 0$ and the residue is one. For $m \neq n$, z^{-m+n-1} has no first-order pole and therefore the residue is zero. Thus

$$\oint_C z^{-m+n-1}\, dz = \begin{cases} 2\pi j, & m = n \\ 0, & m \ne n \end{cases} \qquad (7\text{-}82)$$

and the only nonzero term in the summation on the right-hand side of (7-81) is the term for which $m = n$. Therefore,

$$\oint_C X(z)z^{n-1}\, dz = x(nT)2\pi j$$

which gives

$$x(nT) = \frac{1}{2\pi j} \oint_C X(z)z^{n-1}\, dz \qquad (7\text{-}83)$$

We shall illustrate the inversion integral by means of an example.

EXAMPLE 7-14

The inverse z-transform of

$$X(z) = \frac{1}{1 - Kz^{-1}}, \qquad |Kz^{-1}| < 1 \qquad (7\text{-}84)$$

will be determined by use of the inversion integral. The integral for $x(nT)$ can be written

$$x(nT) = \frac{1}{2\pi j} \oint_C \frac{z^n}{z - K}\, dz \qquad (7\text{-}85)$$

For $n \ge 0$, the integrand has only a single pole at $z = K$. The region of convergence is $|Kz^{-1}| < 1$ or $|z| > 1/K$. Assuming that $K \le 1$, so that the resulting time sequence is bounded, ensures that $X(z)$ converges everywhere outside the unit circle. Thus in (7-85) the path of integration is taken as a circle with $z = 0$ as the center and radius $1 + \epsilon$, in which ϵ is positive. Application of the residue theorem then gives

$$x(nT) = K^n, \qquad n \ge 0 \qquad (7\text{-}86)$$

∎

7-7

SUMMARY

The basic theme of this chapter was the analysis of discrete-time systems. We found that A/D converters consist of three components: a sampler, a quantizer, and an encoder. The sampler extracts from a continuous-time function only those values that occur at the sampling instants. The sampling operation produces a discrete-time signal. In the frequency domain, sampling reproduces the spectrum of the sampled signal about dc and all harmonics of the sampling frequency. Reconstruction, if desired, is accomplished by low-pass filtering. If a signal is to be sampled and reconstructed without error, the sampling frequency must be at least twice the highest frequency of the signal being sampled. The quantizer maps the sample values into a discrete number of levels known as quantization levels. The number of quantization levels is a function of the wordlength of the A/D converter. The wordlength determines quantizing error since quantizing error decreases as the wordlength of the A/D converter increases. The encoder maps each of the quantization levels onto a digital word. The difference between a discrete-time signal and a digital signal is that digital signals result from quantizing and encoding discrete-time signals.

The basic tool for representing a sample sequence and for analyzing a discrete-time system is the z-transform. The z-transform of a sample is of the form Az^{-n}, where A is the value of the sample and n denotes that the sample occurs at time $t = nT$, where T is the time between samples. By using this basic definition, a useful table of z-transforms can be developed. The process of inverse z-transformation is accomplished by long division or by partial-fraction expansion. Partial-fraction expansion is more general in that it gives all terms in the sample sequence in closed form.

Several transform theorems have been developed. A very important theorem is the time delay theorem, which states that a time delay of K sample periods is equivalent to a multiplication by z^{-K} in the z-domain.

Linear, fixed, discrete-time systems are defined by the difference equation of the system or by the pulse transfer function, which results by solving the difference equation using the z-transform. The sinusoidal steady-state transfer function of the system is determined from the pulse transfer function by replacing z by $e^{j\omega T}$.

The analysis of a simple discrete-time system served to bring together the basic ideas of discrete-time analysis. With these ideas firmly in mind, we are now ready to turn our attention from analysis to the more interesting problem of synthesis.

FURTHER READING

The study of the z-transform is usually accompanied by a study of either sampled-data control systems or digital filters. The list of books given in the Further Reading section of Chapter 8 is appropriate for additional study.

PROBLEMS

SECTION 7-2

7-1. The signal

$$x(t) = 6 + 4 \cos 10\pi t + 4 \cos 14\pi t + 2 \cos 20\pi t$$

is sampled at a rate of 30 samples per second. Plot the spectrum of the sampled signal showing all components for $|f| < 80$. Fully explain how $x(t)$ can be reconstructed from the samples.

7-2. The signal having the spectrum shown below is sampled. Plot the spectrum of the sampled signal for the three sampling frequencies $f_s = 30, 40,$ and 60 hertz.

[*Note:* Assume that $X(f)$ is real.] Label all amplitudes and frequencies of interest. Which of the three sampling frequencies are acceptable?

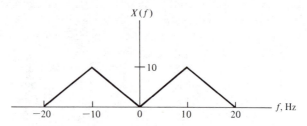

$X(f)$

/.3 An A/D converter has the input signal

$$x(t) = 5 \cos 200\pi t + 7 \cos 600\pi t$$

Determine:
(a) The required dynamic range of the A/D converter.
(b) The output signal-to-noise ratio as a function of the wordlength, n. Express this result in decibels, plot, and compare with the result determined in Example 7-3.

7-4. Repeat Problem 7-3 assuming the signal $x(t)$ shown.

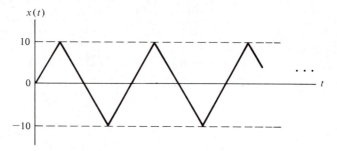

$x(t)$

7-5. An A/D converter has an input signal given by the Fourier series

$$x(t) = \sum_{n=-\infty}^{\infty} X_n e^{jn\omega_0 t}$$

Show that the signal-to-noise ratio at the output of the A/D converter is given by

$$\text{SNR} = \frac{12}{D^2} 2^{2n} \sum_{n=-\infty}^{\infty} |X_n|^2$$

7-6. Assume that the signal $x(t)$ given in Problem 7-1 is digitized with an A/D converter utilizing 10-bit arithmetic. Compute the signal-to-noise ratio at the output of the A/D converter.

SECTION 7-3

7-7. Use z-transform pair 3 in Table 7-1 to establish z-transform pairs 4 and 5. (*Hint:* First write

$$\sum_{n=0}^{\infty} e^{-\alpha nT} z^{-n} = \frac{1}{1 - e^{-\alpha T}z^{-1}}$$

and differentiate both sides with respect to α.)

7-8. Use z-transform pair 3 in Table 7-1 to establish z-transform pairs 6 and 7. (*Hint:* Again first write

$$\sum_{n=0}^{\infty} e^{-\alpha nT} z^{-n} = \frac{1}{1 - e^{-\alpha T}z^{-1}}$$

Let $\alpha = jb$ and equate real and imaginary parts.)

7-9. Use z-transform pair 3 in Table 7-1 to establish z-transform pairs 8 and 9. (*Hint:* The procedure is the same as in Problem 7-8. What should you let α be now?)

7-10. Samples of a time function $x(t)$ are given below. Determine $X(z)$.

$$x(0) = 1$$
$$x(T) = 4.7$$
$$x(2T) = x(3T) = 0$$
$$x(4T) = 0.75$$
$$x(5T) = \sqrt{2}$$
$$x(nT) = 0, \ n \geq 6$$

7-11. The function $x(t) = 9\Pi[(t - 4)/5.2]$ is sampled. Determine the z-transform of the sampled sequence for each of the following sampling frequencies.
(a) $f_s = 1$ Hz
(b) $f_s = 0.5$ Hz
(c) $f_s = 3$ Hz

7-12. Determine the sample values $x(nT)$ corresponding to each of the z-transforms given.
(a) $X(z) = 1 + z^{-1} + 0.5z^{-7}$
(b) $X(z) = 5(1 - z^{-1})(1 + z^{-3})$
(c) $X(z) = 6(1 - 0.25z^{-1})^3$

7-13. Determine $x(0)$, $x(T)$, $x(2T)$, $x(3T)$, and $x(4T)$ for the two functions given.
(a) $X(z) = \dfrac{1 - 4z^{-3}}{1 - 0.7z^{-3}}$
(b) $X(z) = \dfrac{z + z^{-3}}{z + z^{-1}}$

7-14 Determine $x(0)$, $x(T)$, $x(2T)$, and $x(3T)$ for the two functions given.
(a) $X(z) = \dfrac{5 + 4.7z^{-1} - 3z^{-2}}{1 + 2z^{-1} + 3z^{-2} + 4z^{-3}}$

(b) $X(s) = \dfrac{1 + 2z^{-1} + 4z^{-3}}{0.5 + 0.1z^{-1} + 0.25z^{-2} + 0.8z^{-3}}$

7-15. Determine the inverse z-transform of each of the following functions in closed form.

(a) $X(z) = \dfrac{10}{1 - 0.5z^{-1}}$

(b) $X(z) = \dfrac{10z^{-1}}{1 - 0.5z^{-1}}$

(c) $X(z) = \dfrac{10}{1 + 0.25z^{-1} - 0.125z^{-2}}$

(d) $X(z) = \dfrac{10(1 - z^{-1})}{(1 - 0.5z^{-1})(1 - 0.25z^{-1})}$

7-16. Determine the inverse z-transform of each of the following functions in closed form.

(a) $X(z) = \dfrac{1}{1 - 1.5z^{-1} + 0.5z^{-2}}$

(b) $X(z) = \dfrac{1}{1 - 0.5z^{-1} - 0.5z^{-2}}$

(c) $X(z) = \dfrac{z^2}{z^2 - 0.5z + 0.06}$

(d) $X(z) = \dfrac{z^2}{(z + 1)(z^2 + 0.1z - 0.06)}$

7-17. For the function given below, determine $x(nT)$ for all n. Partially check your result by computing $x(0)$, $x(T)$, $x(2T)$, and $x(\infty)$ by an alternative method.

$$X(z) = \frac{5}{(1 - z^{-1})(1 - 0.25z^{-1})(1 - 0.5z^{-1})}$$

7-18. For the functions given below, determine $x(nT)$ for all n. Partially check your result by computing $x(0)$, $x(T)$, $x(2T)$ and $x(\infty)$ by an alternative method.

(a) $X(z) = \dfrac{5}{z(z - 1)}$

(b) $X(z) = \dfrac{1}{(z - 1)(z - 0.5)^2}$

7-19. Determine the inverse z-transform for the functions given below.

(a) $X(z) = \dfrac{z^{-1}}{1 - z^{-1} + z^{-2}}$

(b) $X(z) = \dfrac{1 + z^{-1}}{1 - z^{-1} + z^{-2}}$

(c) $X(z) = \dfrac{1}{1 + z^{-2}}$

7-20.   Determine the inverse z-transform of

$$X(z) = \frac{z^2}{z^2 - 0.5z + 0.6}$$

SECTION 7-4

7-21.   Determine the pulse transfer function, $H(z)$, and the unit pulse response $h(nT)$, for each of the systems defined by the following difference equations.
(a) $y(nT) - y(nT - T) = x(nT)$
(b) $y(nT) - 2y(nT - T) + y(nT - 2T) = x(nT)$
(c) $y(nT) - y(nT - T) + 0.16y(nT - 2T) = x(nT) + 2x(nT - T)$
(d) $y(nT) + 0.6y(nT - T) - 0.16y(nT - 2T) = x(nT) + 2x(nT - T)$
(e) $y(nT) - 0.707y(nT - T) + 0.25y(nT - 2T) = x(nT)$

7-22.   Determine the range of values of the parameter K for which the systems defined by the following difference equations are stable.
(a) $y(nT) - Ky(nT - T) + K^2y(nT - 2T) = x(nT)$
(b) $y(nT) - 2Ky(nT - T) + K^2y(nT - 2T) = x(nT)$
(c) $y(nT) - K^2y(nT - 2T) = x(nT)$

7-23.   Convolve the two functions, $x_1(nT)$ and $x_2(nT)$, illustrated.

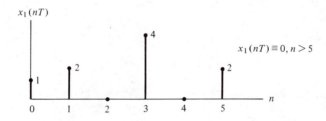

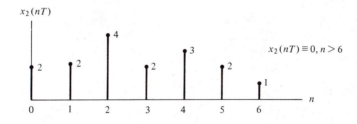

7-24.   A discrete-time system has the input $x(nT)$, and the unit pulse response $h(nT)$, illustrated. Determine and sketch the system output.

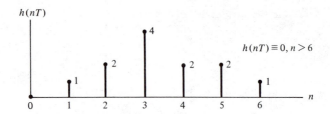

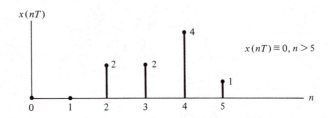

7-25. Determine the amplitude and phase response of the system defined by the difference equation

$$y(nT) = x(nT) - x(nT - 4T)$$

Sketch the amplitude and phase response as a function of normalized frequency $r = f/f_s$ for r in the range $0 \le r \le \frac{1}{2}$.

7-26. Repeat Problem 7-25 for the system defined by the difference equation

$$y(nT) = 2 x(nT) + 4 x(nT - T) + 2 x(nT - 2T)$$

7-27. Repeat Problem 7-25 for the system defined by the difference equation

$$y(nT) + 0.25y(nT - T) = x(nT - T)$$

7-28. A simple discrete-time differentiator is defined by the difference equation

$$y(nT) = \frac{1}{T}[x(nT) - x(nT - T)]$$

Determine the amplitude and phase response of the discrete-time differentiator. Plot the amplitude and phase response as a function of normalized frequency $r = f/f_s$. Compare these responses with the amplitude and phase response of an ideal differentiator.

CHAPTER 8

Analysis and Design of Digital Filters

8-1

INTRODUCTION

Throughout this book we have been concerned with the analysis of continuous-time and discrete-time linear systems. We now turn our attention to a much different type of problem. This is the synthesis problem wherein one is asked not to analyze a given system but is asked to find the system that satisfies a given set of specifications. Typically, the synthesis problem for discrete-time systems is to determine the pulse transfer function $H(z)$ which satisfies the given specifications. Then, as we shall see in the first section of this chapter, a number of different systems (structures or realizations) can be found, all of which have the same pulse transfer function $H(z)$. The system designer then chooses that realization which is most compatible with the requirements.

A number of different techniques which allow us to go from a set of specifications to a pulse transfer function are studied in this chapter. All have their advantages and disadvantages. These will be summarized later. Several synthesis techniques have as their starting point the transfer function of an analog filter or system which satisfies a given set of specifications. Synthesis of the digital filter or system then simply consists of finding the discrete-time equivalent of some given analog system. For those who are not familiar with the terminology and characteristics of analog filters, a brief review is contained in Appendix B.

STRUCTURES OF DIGITAL PROCESSORS

In this section several different techniques for realizing pulse transfer functions are examined.

Direct-Form Realizations

In Chapter 7 we determined the general form of the pulse transfer function of a fixed discrete-time system. The result was (7-42), which can be written as

$$H(z) = \frac{Y(z)}{X(z)} = \frac{\displaystyle\sum_{i=0}^{r} L_i z^{-i}}{1 + \displaystyle\sum_{j=1}^{m} K_j z^{-j}} \tag{8-1}$$

The difference equation corresponding to (8-1) is

$$y(nT) = \sum_{i=0}^{r} L_i x(nT - iT) - \sum_{j=1}^{m} K_j y(nT - jT) \tag{8-2}$$

and is realized by the structure illustrated in Figure 7-16. If the storage of the present and past inputs and storage of past outputs is represented through the use of the z^{-1} delay operator, Figure 8-1 results. The block labeled z^{-1} represents storage of a sample value (a number) for one sample period T. The summing operation illustrated in Figure 7-16 has been broken into two summing junctions, each of which forms one of the summation terms in (8-2). The structure illustrated in Figure 8-1 is known as the Direct Form I realization of the digital signal processor defined by the difference equation (8-2).

Implementation of the processor shown in Figure 8-1 requires $r + m$ unit delays (z^{-1} elements). This requirement can be significantly reduced in many applications by identifying the processor as the cascade of two networks, $H_1(z)$ and $H_2(z)$. The first network, $H_1(z)$, realizes the zeros of the processor and the second network, $H_2(z)$, realizes the poles. Interchanging $H_1(z)$ and $H_2(z)$ results in the structure shown in Figure 8-2. Clearly, the signals at points A_1 and B_1 are equal, as are the signals at A_2 and B_2, and so on down the z^{-1} delay elements. Thus the cascaded z^{-1} elements can be combined to yield the realization illustrated in Figure 8-3, which is known as the Direct Form II realization. Figure 8-3 is drawn for the case in which the number of poles, m, exceeds the number of zeros, r.

Cascade and Parallel Realizations

The Direct Form I realization resulted directly from the general discrete-time difference equation, (8-2). A simple manipulation of the block diagram resulted in the more practical Direct Form II realization. The cascade realization results

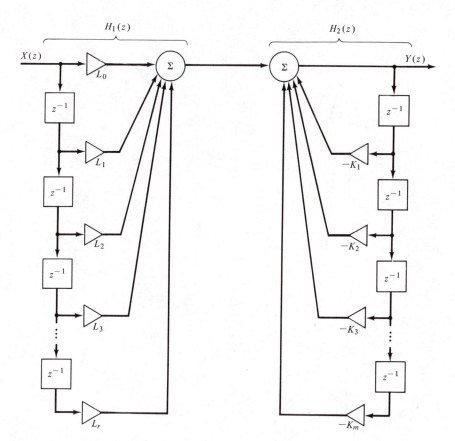

FIGURE 8-1.   Direct Form I realization.

by realizing that since poles or zeros of the pulse transfer function are either real, or occur in complex conjugate pairs, $H(z)$ can be written in the factored form

$$H(z) = Kz^{-M} \frac{\displaystyle\prod_{i=1}^{N_1} (1 - a_i z^{-1}) \prod_{j=1}^{N_2} (1 - b_j^* z^{-1}) (1 - b_j z^{-1})}{\displaystyle\prod_{k=1}^{D_1} (1 - c_k z^{-1}) \prod_{l=1}^{D_2} (1 - d_l^* z^{-1}) (1 - d_l z^{-1})} \qquad (8\text{-}3)$$

In this expression we have N_1 real zeros at values $z = a_i$, N_2 complex-conjugate zero pairs at $z = b_j$ and $z = b_j^*$, D_1 real poles at $z = c_k$, and D_2 complex-conjugate pole pairs at $z = d_l$ and $z = d_l^*$. The real poles and real zeros are typically realized using the Direct Form II realization. For example, a general term of the form

$$H_1(z) = \frac{1 - a_i z^{-1}}{1 - c_k z^{-1}} \qquad (8\text{-}4)$$

is realized using the structure illustrated in Figure 8-4a. The complex-conjugate poles and zeros are realized by pairs. For example, the general term

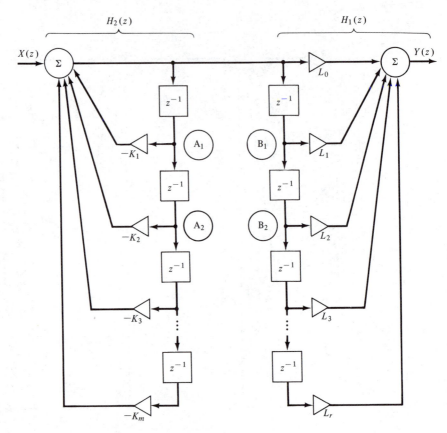

FIGURE 8-2.   Rearrangement of Figure 8-1.

$$H_2(z) = \frac{(1 - b_j z^{-1})(1 - b_j^* z^{-1})}{(1 - d_l z^{-1})(1 - d_l^* z^{-1})} = \frac{1 - (b_j + b_j^*)z^{-1} + b_j b_j^* z^{-2}}{1 - (d_l + d_l^*)z^{-1} + d_l d_l^* z^{-2}} \quad (8\text{-}5)$$

is realized by the structure illustrated in Figure 8-4b. Since $A + A^*$ and AA^* are both real numbers, where A is complex, all multipliers in Figure 8-4b are real numbers. The multiplier Kz^{-M} is included in (8-3) to realize any multiplicative constant and also to allow for the fact that $H(z)$ may have a factor of the form z^{-M}, where $M > 0$.

The parallel form results by expanding $H(z)$ using partial-fraction expansion. The general form of the partial fraction is

$$H(z) = \sum_{i=0}^{M} A_i z^{-i} + \sum_{k=1}^{D_1} B_k \frac{1}{1 - c_k z^{-1}}$$

$$+ \sum_{l=1}^{D_2} C_l \frac{1 - e_l z^{-1}}{(1 - d_l z^{-1})(1 - d_l^* z^{-1})} \quad (8\text{-}6)$$

in which the first summation is included to realize the terms in the partial-fraction expansion that result if $r > m$ in (8-1). The second summation realizes

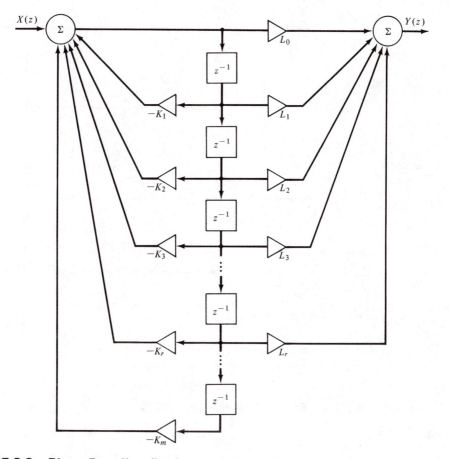

FIGURE 8-3. Direct Form II realization.

the real poles, and the third summation realizes the complex-conjugate pole pairs.

EXAMPLE 8-1

In this example, a cascade and parallel realization of

$$H(z) = \frac{(1 - z^{-1})^3}{(1 - \frac{1}{2} z^{-1}) (1 - \frac{1}{8} z^{-1})} \tag{8-7}$$

will be found.

The cascade realization is relatively simple. For example, $H(z)$ can be written in the form

$$H(z) = \frac{1 - z^{-1}}{1 - \frac{1}{2} z^{-1}} \frac{1 - z^{-1}}{1 - \frac{1}{8} z^{-1}} (1 - z^{-1}) \tag{8-8}$$

which is realized using the structure illustrated in Figure 8-5a.

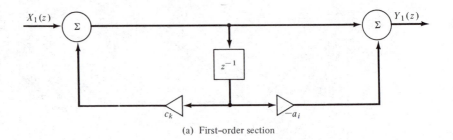

(a) First-order section

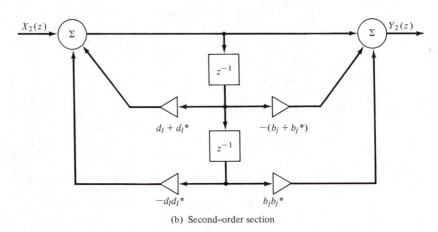

(b) Second-order section

FIGURE 8-4. Basic first- and second-order filters.

The parallel realization is found by first writing $H(z)$ as

$$H(z) = \frac{(z - 1)^3}{z(z - \frac{1}{2})(z - \frac{1}{8})}$$

In order to find the parallel realization, we first determine the partial-fraction expansion of $z^{-1}H(z)$, just as we did to determine inverse z-transforms. This yields

$$\frac{H(z)}{z} = \frac{(z - 1)^3}{z^2(z - \frac{1}{2})(z - \frac{1}{8})} = \frac{A}{z^2} + \frac{B}{z} + \frac{C}{z - \frac{1}{2}} + \frac{D}{z - \frac{1}{8}} \qquad (8\text{-}9)$$

The constants A, B, C, and D are

$$A = \lim_{z \to 0} \frac{(z - 1)^3}{(z - \frac{1}{2})(z - \frac{1}{8})} = -16$$

$$B = \lim_{z \to 0} \frac{d}{dz}\left[\frac{(z - 1)^3}{(z - \frac{1}{2})(z - \frac{1}{8})}\right] = -112$$

$$C = \lim_{z \to 1/2} \frac{(z - 1)^3}{z^2(z - \frac{1}{8})} = -\frac{4}{3}$$

$$D = \lim_{z \to 1/8} \frac{(z - 1)^3}{z^2(z - \frac{1}{2})} = \frac{343}{3}$$

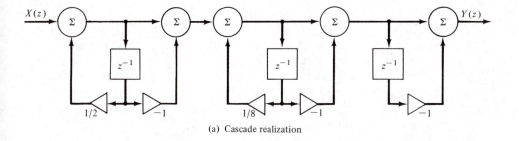

(a) Cascade realization

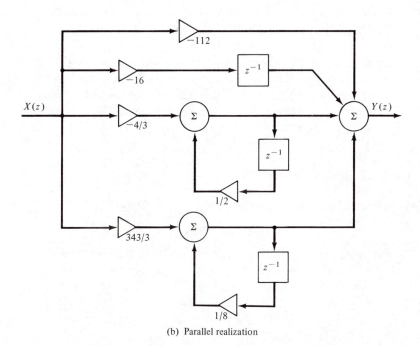

(b) Parallel realization

FIGURE 8-5. Cascade and parallel realizations for Example 8-1.

Thus

$$H(z) = -112 - 16z^{-1} - \frac{4}{3}\frac{1}{1 - \frac{1}{2}z^{-1}} + \frac{343}{3}\frac{1}{1 - \frac{1}{8}z^{-1}} \quad (8\text{-}10)$$

which is realized using the structure illustrated in Figure 8-5b. This example problem is chosen to illustrate the role of the first summation in (8-6). ■

If quantizing errors are neglected, all realizations of the pulse transfer function are equivalent. If quantizing errors are important and must be considered, it is then no longer true that all realizations are equivalent. For example, suppose that the computed value of one of the K_j's is an irrational number. Before the $H(z)$ can be realized, the value of K_j must be truncated or rounded to a value that can be represented by a finite number of bits. This changes the denominator polynomial slightly and as a result *all* poles of the transfer function change by some amount. For high-order systems this quickly becomes a complicated problem. If the original $H(z)$ has poles close to the unit circle, the movement of the poles due to quantization of K_j can actually drive the system unstable.

Consider, however, the corresponding problem if the system is realized using a cascade or parallel combination of first- and second-order sections. In this case, truncating or rounding a coefficient value causes movement in at most two poles. This is a much more tolerable situation and is a basic advantage of cascade or parallel realizations.

DISCRETE-TIME INTEGRATION

As a specific example of a discrete-time approximation to a continuous-time process, we shall now consider the discrete-time integrator. Both rectangular and trapezoidal integration will be considered and we shall see that while both of these techniques are approximations to continuous-time integration, they have distinctly different characteristics.

Rectangular Integration

A discrete-time integrator is defined by

$$y(nT) = y(nT - T) + \Delta y(nT - T, nT) \qquad (8\text{-}11)$$

where $\Delta y(nT - T, nT)$ is the value added to the integral in the time increment $(nT - T, nT)$. The algorithm for rectangular integration is illustrated in Figure 8-6a, in which the value of $\Delta y(nT - T, nT)$ is defined to be

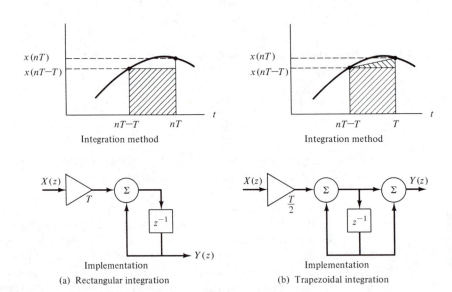

(a) Rectangular integration

(b) Trapezoidal integration

FIGURE 8-6. **Rectangular and trapezoidal integration.**

$$\Delta y(nT - T, nT) \triangleq T x(nT - T) \tag{8-12}$$

which results in the difference equation

$$y(nT) = y(nT - T) + Tx(nT - T) \tag{8-13}$$

The pulse transfer function is determined by z-transforming (8-13), which yields

$$Y(z) = z^{-1}Y(z) + z^{-1}TX(z)$$

Solving for $Y(z)/X(z)$ results in

$$H(z) = \frac{Y(z)}{X(z)} = \frac{z^{-1}T}{1 - z^{-1}} \tag{8-14}$$

The digital processor that accomplishes rectangular integration is also illustrated in Figure 8-6a.

Trapezoidal Integration

The algorithm for trapezoidal integration is illustrated in Figure 8-6b. It follows that $\Delta y(nT - T, nT)$ is

$$\Delta y(nT - T, nT) \triangleq Tx(nT - T) + \tfrac{1}{2} T[x(nT) - x(nT - T)] \tag{8-15}$$

which gives the difference equation

$$y(nT) = y(nT - T) + Tx(nT - T) + \tfrac{1}{2} T[x(nT) - x(nT - T)] \tag{8-16}$$

The z-transform of the difference equation is

$$Y(z) = z^{-1}Y(z) + \frac{T}{2}X(z) + \frac{T}{2}z^{-1}X(z)$$

which yields

$$H(z) = \frac{Y(z)}{X(z)} = \frac{T}{2}\frac{1 + z^{-1}}{1 - z^{-1}} \tag{8-17}$$

for the pulse transfer function. The realization of a trapezoidal integrator is illustrated in Figure 8-6b.

EXAMPLE 8-2

The amplitude and phase characteristics of the rectangular and trapezoidal integrators are easily derived. From (8-14) with $z = e^{j\omega T}$, the transfer function of the rectangular integrator, $H_r(e^{j\omega T})$, is

$$H_r(e^{j\omega T}) = \frac{Te^{-j\omega T}}{1 - e^{-j\omega T}} = \frac{Te^{-j\omega T/2}}{e^{+j\omega T/2} - e^{-j\omega T/2}}$$

or

$$H_r(e^{j\omega T}) = \frac{Te^{-j\omega T/2}}{2j \sin \omega T/2} \tag{8-18}$$

In terms of the normalized frequency $r = f/f_s$, this becomes

$$H_r(e^{j2\pi r}) = \frac{Te^{-j\pi r}}{2j \sin \pi r} \tag{8-19}$$

In the range $0 \le r \le \frac{1}{2}$, $\sin \pi r \ge 0$. Thus the amplitude response $A_r(r)$ of the integrator is given by

$$A_r(r) = \frac{T}{2 \sin \pi r}, \qquad 0 \le r \le \frac{1}{2} \tag{8-20}$$

and the phase, $\phi_r(r)$, is given by

$$\phi_r(r) = -\frac{\pi}{2} - \pi r, \qquad 0 \le r \le \frac{1}{2} \tag{8-21}$$

The sinusoidal steady-state response of the trapezoidal integrator is, from (8-17),

$$H_t(e^{j\omega T}) = \frac{T}{2} \frac{1 + e^{-j\omega T}}{1 - e^{-j\omega T}} \tag{8-22}$$

which can be written

$$H_t(e^{j\omega T}) = \frac{T}{2} \frac{e^{+j\omega T/2} + e^{-j\omega T/2}}{e^{+j\omega T/2} - e^{-j\omega T/2}}$$

or

$$H_t(e^{j\omega T}) = \frac{T}{2} \frac{\cos \omega T/2}{j \sin \omega T/2} \tag{8-23}$$

In terms of normalized frequency, (8-23) is

$$H_t(e^{j2\pi r}) = \frac{T}{2} \frac{\cos \pi r}{j \sin \pi r} \tag{8-24}$$

In the range $0 \le r \le \frac{1}{2}$, both $\cos \pi r$ and $\sin \pi r$ are nonnegative. Thus it follows from (8-24) that the amplitude response of a trapezoidal integrator is given by

$$A_t(r) = \frac{T}{2} \frac{\cos \pi r}{\sin \pi r}, \qquad 0 \le r \le \frac{1}{2} \tag{8-25}$$

and the phase response is

$$\phi_t(r) = -\frac{\pi}{2}, \qquad 0 \le r \le \frac{1}{2} \tag{8-26}$$

The magnitude and phase responses of the rectangular and trapezoidal integrators are given in Figure 8-7 for the case where the sampling period T is equal to unity. Also illustrated are the amplitude and phase responses of the ideal continuous integrator, defined by $H(f) = 1/j2\pi f$. Both the rectangular and trapezoidal integrators exhibit amplitude distortion but neither exhibit phase distortion. The rectangular integrator does, however, delay the input signal by $T/2$ seconds.

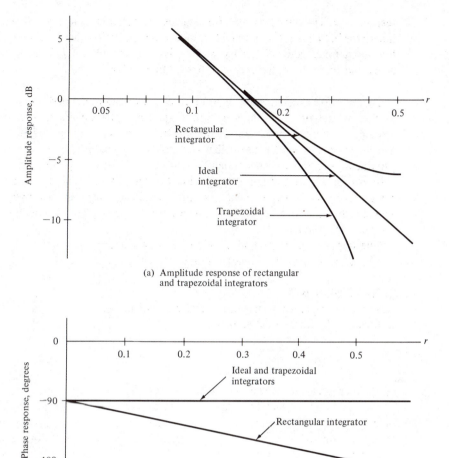

(a) Amplitude response of rectangular
and trapezoidal integrators

(b) Phase response of rectangular and
trapezoidal integrators

FIGURE 8-7. Amplitude and phase response of rectangular and trapezoidal integrators.

It is interesting to consider the form of (8-18) and (8-23) for input frequencies much less than the sampling frequency (i.e., for $\omega T \ll 1$). For this case

$$\frac{\cos \omega T/2}{\sin \omega T/2} \simeq \frac{1}{\sin \omega T/2} \simeq \frac{2}{\omega T}$$

so that (8-18) and (8-23) become

$$H_r\left(e^{j\omega T}\right) \simeq \frac{1}{j\omega} e^{-j\omega T/2}, \qquad \omega T \ll 1$$

and

$$H_t\left(e^{j\omega T}\right) = \frac{1}{j\omega}, \qquad \omega T \ll 1$$

respectively. Thus, for low input frequencies, the rectangular and the trapezoidal integrator both closely approximate an ideal integrator. The basic difference for low input frequencies is that the rectangular integrator has a $T/2$ second time delay and the trapezoidal integrator has no time delay. ∎

Example 8-2 illustrates the importance of investigating the performance of a system by studying both the time- and frequency-domain characteristics of the system. A simple comparison of the integration methods illustrated in Figure 8-6 leads us to the conclusion that trapezoidal integration yields a better approximation to the area under a curve. However, a comparison of the frequency responses in Figure 8-7 provides additional insight. We see that the rectangular integrator provides less attenuation of high frequencies than an ideal integrator, whereas the trapezoidal integrator provides more high frequency attenuation than an ideal integrator.

The trapezoidal integration algorithm will gain additional significance when the bilinear z-transform is studied in the following section.

EXAMPLE 8-3

In this example we determine a digital equivalent (one of many) for the system defined by the differential equation

$$\frac{dy}{dt} = x(t) - \beta y(t) \tag{8-27}$$

The block diagram of the system represented by the differential equation is shown in Figure 8-8a. The only block in the system having memory is the integrator. A simple digital equivalent is derived by substituting a trapezoidal integrator for the analog integrator. Using (8-18) results in the block diagram illustrated in Figure 8-8b.

The transfer function of the system illustrated in Figure 8-8b is easily derived. From the block diagram it follows that

$$Y(z) = \frac{T}{2} \frac{1 + z^{-1}}{1 - z^{-1}} [X(z) - \beta Y(z)]$$

which reduces to

$$H(z) = \frac{Y(z)}{X(z)} = \frac{T(1 + z^{-1})}{(\beta T + 2) + (\beta T - 2)z^{-1}} \tag{8-28}$$

We shall return to this problem in a later section in order to derive this same result by a different method. ∎

8-4

INFINITE IMPULSE RESPONSE FILTER DESIGN

We now turn our attention to a fundamental problem, the problem of determining the required coefficients of a linear difference equation in order that a specified task be performed. This *is* the synthesis problem. The resulting structures are

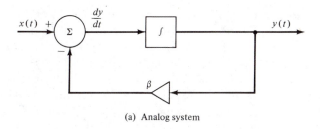

(a) Analog system

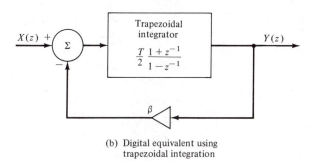

(b) Digital equivalent using
trapezoidal integration

FIGURE 8-8. Analog and digital equivalents using trapezoidal integration.

referred to as *digital filters* and are classified as either *infinite-duration* unit pulse response (IIR) filters or *finite-duration* unit pulse response (FIR) filters, depending on the form of the unit pulse response of the system. We will see that IIR filters are usually implemented using structures having feedback (recursive structures) and FIR filters usually are implemented using structures having no feedback (nonrecursive structures), but this is not a necessary restriction. In this section we examine the synthesis of IIR filters, and in the following section we examine the synthesis of FIR filters.

In designing IIR filters, the usual starting point will be an analog (continuous-time) system transfer function, $H_a(s)$. The problem will be to determine a discrete-time system, in other words, to determine $H(z)$, which in some sense approximates the performance of the analog system. The process of deriving a digital filter from an analog prototype can be performed using either time-domain or frequency-domain techniques and we shall see that satisfactory correspondence between the analog and digital systems in the time domain does not imply that satisfactory correspondence exists in the frequency domain.

Synthesis in the Time Domain—Invariant Design

The concept of time-domain invariance is a simple one. A digital filter is equivalent to an analog filter, in the time-domain-invariance sense, if equivalent inputs yield equivalent outputs. The simplest, as well as the most often used, invariant design is the impulse-invariant digital filter. After a quick look at the impulse-invariant filter we shall consider the general invariant synthesis technique.

Impulse-Invariant Design. The synthesis technique for an impulse-invariant digital filter is illustrated in Figure 8-9. Assume an impulse function signal source. The output of the analog filter will be the unit impulse response $h_a(t)$. Sampling this impulse response yields the sample values $h_a(nT)$.

We now consider a second signal path in which the analog filter and sampler are replaced by a sampler and a digital filter. The discrete-time equivalent of a unit impulse function is a unit pulse (the weight of the impulse denotes the sample value). Therefore, the input to the digital filter is a unit pulse, and the digital filter output is the unit pulse response of the digital filter. If the parameters of the digital filter are adjusted so that this unit pulse response is identical to the previously specified $h_a(nT)$, the analog filter and sampler are equivalent to the sampler and digital filter. With this condition we say that the digital filter is the *impulse-invariant* equivalent of the analog filter.

To illustrate the impulse-invariant synthesis technique, assume that the transfer function of an analog filter has m distinct real poles such that the partial-fraction expension of $H_a(s)$ has the form

$$H_a(s) = \sum_{i=1}^{m} \frac{K_i}{s + s_i} \tag{8-29}$$

in which $s = -s_i$ are the pole locations and K_i is the residue of the pole at s_i. Taking the inverse Laplace transform yields

$$h_a(t) = \sum_{i=1}^{m} K_i e^{-s_i t}, \qquad t \geq 0 \tag{8-30}$$

for the unit impulse response of the analog filter. The z-transform of the unit impulse response is

$$H_1(z) = \sum_{n=0}^{\infty} h_a(nT) z^{-n}$$

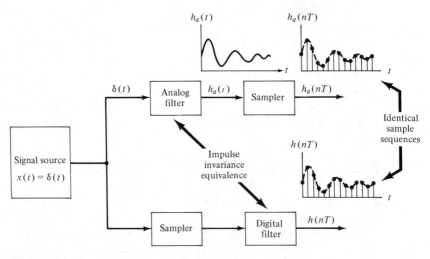

FIGURE 8-9. The concept of impulse invariance.

which, upon substitution of (8-30), yields

$$H_1(z) = \sum_{n=0}^{\infty} \sum_{i=1}^{m} K_i e^{-s_i nT} z^{-1})^n \qquad (8\text{-}31)$$

Interchanging the order of summation allows (8-31) to be written

$$H_1(z) = \sum_{i=1}^{m} K_i \sum_{n=0}^{\infty} (e^{-s_i T} z^{-1})^n$$

which is, after summing on n,

$$H_1(z) = \sum_{i=1}^{m} \frac{K_i}{1 - e^{-s_i T} z^{-1}} \qquad (8\text{-}32)$$

The filter described in (8-32) has a unit pulse response equivalent to the sampled impulse response of the analog filter from which it was derived. We shall later see, however, that the amplitude response of the digital filter will be scaled by f_s due to the sampling operation. Therefore, scaling the amplitude response of the digital filter to approximate the amplitude response of the analog filter requires multiplication of $H_1(z)$ by $T = 1/f_s$. Thus the pulse transfer function of the impulse invariant digital filter equivalent to (8-29) is

$$H(z) = T \sum_{i=1}^{m} \frac{K_i}{1 - e^{-s_i T} z^{-1}} \qquad (8\text{-}33)$$

It should be noted that this synthesis technique yields the parallel realization illustrated in Figure 8-10.

Now that the impulse-invariant design technique has been illustrated, we examine the general concept of invariant design. Then we shall consider several example problems.

General Time-Invariant Synthesis.

The general concept of time-domain invariance is illustrated in Figure 8-11. The input to the analog filter is $x(t)$ and the input to the digital filter is $x(nT)$, the sampled version of $x(t)$. With these equivalent inputs applied to the analog and digital filters, the filter coefficients that determine $H(z)$ are adjusted until the sampled output of the analog filter corresponds to the output of the digital filter.

The required $H(z)$ is determined by writing

$$y(t) = \mathcal{L}^{-1}[H_a(s)X_a(s)]$$

where $\mathcal{L}^{-1}$ denotes the inverse Laplace transform. The output samples of the digital filter are defined to be

$$y(nT) = [\mathcal{L}^{-1}[H_a(s)X_a(s)]]\big|_{t=nT} \qquad (8\text{-}34)$$

The z-transform of this quantity yields the z-domain output of the digital filter. This gives

$$Y(z) = H(z)X(z) = G\mathcal{Z}\{[\mathcal{L}^{-1}[H_a(s)X_a(s)]]\big|_{t=nT}\} \qquad (8\text{-}35)$$

where $\mathcal{Z}$ denotes the z-transform. The constant, G, has been included in order

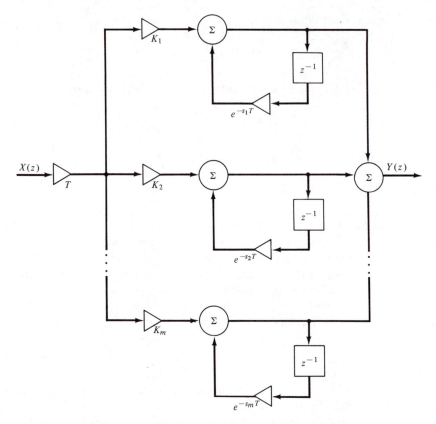

FIGURE 8-10. Impulse-invariant digital filter equivalent to (8-29).

to give similar frequency responses for both the analog and digital filters. Solving (8-35) for $H(z)$ yields the general synthesis equation

$$H(z) = \frac{G}{X(z)} \mathcal{Z}\{[\mathcal{L}^{-1}[H_a(s)X_a(s)]]|_{t=nT}\} \qquad (8\text{-}36)$$

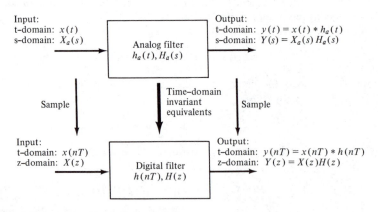

FIGURE 8-11. Time-domain invariance.

The impulse-invariant filter is perhaps the most popular digital filter synthesized using a time-domain invariant synthesis technique. For impulse invariance, (8-36) is used with

$$X(z) = X_a(s) = 1$$

and

$$G = T$$

In order to understand the role of G in this example, recall that the sampling operation, assuming the impulse train sampling function, multiplies the spectrum of the sampled signal by $C_n = 1/T = f_s$ [see (7-8)]. Multiplication by T is required if the spectrum, which in this case is the transfer function of the filter, is to be scaled so that the analog and digital filters have approximately equivalent amplitude responses. Thus the synthesis equation is

$$H(z) = T\mathscr{Z}\{[\mathscr{L}^{-1}[H_a(s)]]|_{t=nT}\} \tag{8-37}$$

An example illustrates the method.

EXAMPLE 8-4

The impulse-invariant equivalent of

$$H_a(s) = \frac{0.5\,(s\,+\,4)}{(s\,+\,1)(s\,+\,2)} \tag{8-38}$$

will now be found. By partial-fraction expansion $H_a(s)$ can be written

$$H_a(s) = \frac{1.5}{s\,+\,1} - \frac{1}{s\,+\,2}$$

The unit impulse response of the analog filter is, for $t \geq 0$,

$$h_a(t) = 1.5e^{-t} - e^{-2t}$$

Thus the unit pulse response of the desired digital filter is, for $n \geq 0$,

$$h_a(nT) = 1.5e^{-nT} - e^{-2nT}$$

which upon z-transforming and multiplying by T yields

$$H(z) = \frac{1.5T}{1\,-\,e^{-T}z^{-1}} - \frac{T}{1\,-\,e^{-2T}z^{-1}} \tag{8-39}$$

The parallel realization of the digital filter is illustrated in Figure 8-12.

It is interesting to compare the frequency response of the two filters. For the analog filter

$$H_a(j\omega) = \frac{0.5(4\,+\,j\omega)}{(1\,+\,j\omega)(2\,+\,j\omega)} \tag{8-40}$$

while for the digital filter

$$H(e^{j\omega T}) = \frac{1.5T}{1\,-\,e^{-T}e^{-j\omega T}} - \frac{T}{1\,-\,e^{-2T}e^{-j\omega T}} \tag{8-41}$$

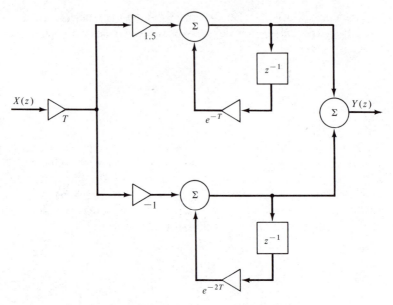

FIGURE 8-12. Impulse-invariant filter for Example 8-4.

It follows that dc responses of the analog and digital filters are given by

$$H_a(0) = 1 \qquad (8\text{-}42)$$

and

$$H(e^{j0}) = H(1) = \frac{1.5T}{1 - e^{-T}} - \frac{T}{1 - e^{-2T}} \qquad (8\text{-}43)$$

respectively. Thus the dc responses are different due to aliasing at dc. However, if the sampling frequency is high, T is small, which yields

$$e^{-T} \simeq 1 - T$$

and

$$e^{-2T} \simeq 1 - 2T$$

which results in

$$H(1) \simeq \frac{1.5T}{1 - (1 - T)} - \frac{T}{1 - (1 - 2T)} = 1 \qquad (8\text{-}44)$$

We shall illustrate this concept using a numerical example. Let the sampling frequency be 20 rad/s so that

$$T = \frac{2\pi}{20} = 0.31416 \text{ s}$$

$$e^{-T} = 0.7304$$

and

$$e^{-2T} = 0.5335$$

With these values, the dc gain of the impulse invariant digital filter is, from (8-43)

$$H(1) = 1.0745$$

which is slightly larger than the gain of the analog filter, as expected, due to the aliasing at dc. For a large sampling frequency, T is small and the approximation $e^{-x} \simeq 1 - x$ becomes valid. For sufficiently large sampling frequency, the aliasing effect becomes negligible and the dc gain is unity as indicated by (8-44).

Figure 8-13 illustrates the amplitude response of both the analog and digital filters. From (8-41) the amplitude response can be written as

$$|H(e^{j2\pi r})| = \frac{\pi}{10} \sqrt{\frac{0.25488 - 0.06983 \cos 2\pi r}{2.74925 - 3.51275 \cos 2\pi r + 0.77932 \cos 4\pi r}} \tag{8-45}$$

Also, the frequency response of the analog filter is

$$|H_a(j\omega)| = 0.5 \sqrt{\frac{16 + \omega^2}{4 + 5\omega^2 + \omega^4}} \tag{8-46}$$

As expected, the amplitude response of the digital filter is a very good approximation to the analog filter for small values of r. The phase responses are illustrated in Figure 8-14.

In some applications it is desirable that both filters have the same dc gain.

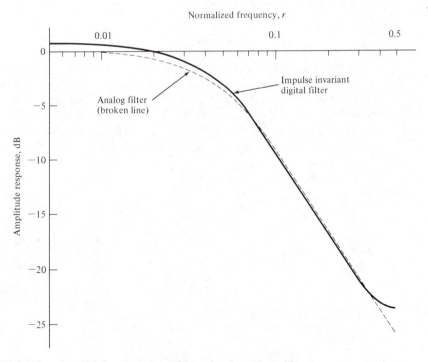

FIGURE 8-13. Amplitude response of impulse-invariant filter.

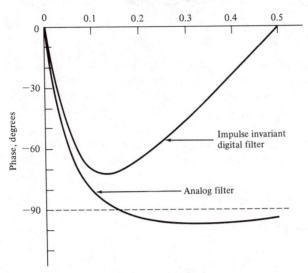

FIGURE 8-14. Phase response of impulse-invariant filter.

Note that this could have been accomplished in this example by multiplying by

$$G = \frac{T}{1.0745}$$

in (8-45) instead of multiplying by

$$G = T = \frac{\pi}{10}$$

∎

Another popular digital filter synthesis procedure using a time-domain criterion is the step invariance synthesis procedure, which is a direct application of (8-36). The step invariant filter is derived by placing a unit step on the input of an analog filter and a sampled unit step on the input to a digital filter. The pulse transfer function of the digital filter, $H(z)$, is adjusted until the output of the digital filter represents samples of the output of the analog filter. Thus, in (8-36), we let

$$G = 1$$

$$X_a(s) = \frac{1}{s}$$

and

$$X(z) = \frac{1}{1 - z^{-1}}$$

This yields

$$H(z) = (1 - z^{-1})\mathscr{Z}\left\{ \mathscr{L}^{-1}\left[\frac{1}{s}H_a(s)\right]\Big|_{t=nT}\right\} \tag{8-47}$$

EXAMPLE 8-5

The step-invariant equivalent of

$$H_a(s) = \frac{0.5(s + 4)}{(s + 1)(s + 2)}$$

will now be found. Since

$$\frac{1}{s} H_a(s) = \frac{0.5(s + 4)}{s(s + 1)(s + 2)} = \frac{1}{s} - \frac{1.5}{s + 1} + \frac{0.5}{s + 2} \qquad (8\text{-}48)$$

we have

$$\mathcal{L}^{-1}\left[\frac{1}{s} H_a(s)\right]\Big|_{t=nT} = 1 - 1.5e^{-nT} + 0.5e^{-2nT} \qquad (8\text{-}49)$$

so that

$$H(z) = (1 - z^{-1})\left(\frac{1}{1 - z^{-1}} - \frac{1.5}{1 - e^{-T}z^{-1}} + \frac{0.5}{1 - e^{-2T}z^{-1}}\right)$$

or

$$H(z) = 1 - 1.5\frac{1 - z^{-1}}{1 - e^{-T}z^{-1}} + 0.5\frac{1 - z^{-1}}{1 - e^{-2T}z^{-1}} \qquad (8\text{-}50)$$

In order to compare this result with the result of Example 8-4, let

$$T = 0.31416 \text{ s}$$

This yields

$$H(z) = \frac{0.17115z^{-1} - 0.04538z^{-2}}{1 - 1.2639z^{-1} + 0.3897z^{-2}} \qquad (8\text{-}51)$$

Figure 8-15 compares the amplitude responses of the step invariant and impulse-invariant digital filters. ∎

Design in the Frequency Domain—The Bilinear z-transform

The time-domain invariance method of filter synthesis was simple to apply, but for many applications the effects of aliasing preclude its use. We now examine a synthesis technique that overcomes the aliasing problem and is even easier to apply than the time-domain invariance method since it is not necessary to inverse transform $H_a(s)$. This synthesis method is known as the *bilinear z-transform* method.

To avoid the effects of aliasing, the transfer function of the analog filter must be bandlimited to the range

$$-\tfrac{1}{2}f_s \leq f \leq \tfrac{1}{2}f_s$$

Since typical analog transfer functions do not satisfy this property, they must first be modified using a nonlinear transformation such that they are bandlimited.

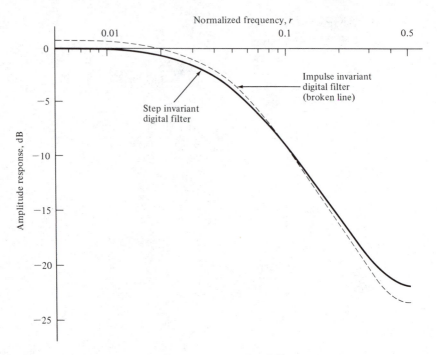

FIGURE 8-15. Amplitude response of step-invariant filter.

The technique is to transform the entire complex s-plane into the s_1-plane such that the entire $j\omega$-axis in the s-plane is mapped into the region

$$-\tfrac{1}{2} f_s \le f_1 \le \tfrac{1}{2} f_s$$

in the s_1-plane. Although there are many transformations that will accomplish this, we seek one that is easy to apply and yields a pulse transfer function for the digital filter directly as a ratio of polynomials in z^{-1}.

A transformation that satisfies these requirements is

$$\omega = C \tan \frac{\omega_1}{\tfrac{1}{2}\omega_s} \frac{\pi}{2} = C \tan \frac{\omega_1 T}{2} \tag{8-52}$$

and is illustrated in Figure 8-16. It can be seen that $\omega = \infty$ corresponds to $\omega_1 = \tfrac{1}{2} \omega_s$. If the constant C is properly chosen, the correspondence $\omega = \omega_1$ can be established at any desired frequency. If $\omega = \omega_1 = \omega_r$, we have, upon solving for C,

$$C = \omega_r \cot \frac{\omega_r T}{2} \tag{8-53}$$

If ω_r is chosen small, such that

$$\frac{\omega_r T}{2} \ll 1$$

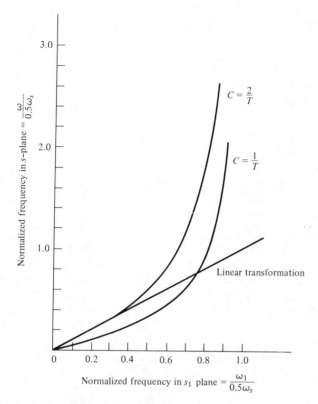

FIGURE 8-16. Bilinear z-transformation frequency mapping.

$\cot x \simeq 1/x$ and the constant C is

$$C = \omega_r \frac{2}{\omega_r T} = \frac{2}{T} \tag{8-54}$$

The relationship between s and s_1 is easily determined from (8-52). First, recall that $\tan x$ can be written

$$\tan x = \frac{\sin x}{\cos x} = -j\frac{e^{jx} - e^{-jx}}{e^{jx} + e^{-jx}} \tag{8-55}$$

Letting $s = j\omega$ and $s_1 = j\omega_1$ in (8-52) then yields

$$-js = -jC\frac{e^{+s_1T/2} - e^{-s_1T/2}}{e^{+s_1T/2} + e^{-s_1T/2}}$$

which can be written

$$s = C\frac{e^{s_1T/2} - e^{-s_1T/2}}{e^{s_1T/2} + e^{-s_1T/2}} = C\tanh\frac{s_1T}{2} \tag{8-56}$$

The digital filter is determined from $H_a(s_1)$ by letting

$$z = e^{s_1T}$$

With this substitution, (8-56) becomes

$$s = C \frac{1 - e^{-s_1 T}}{1 + e^{-s_1 T}} = C \frac{1 - z^{-1}}{1 + z^{-1}} \tag{8-57}$$

Thus the digital filter $H(z)$ is determined from the analog filter $H_a(s)$ by simply making the substitution

$$C \frac{1 - z^{-1}}{1 + z^{-1}}$$

for s in $H_a(s)$. Since $H_a(s)$ is expressed as a ratio of polynomials in s, the resulting $H(z)$ is a ratio of polynomials in z. One only needs to determine the appropriate value of the scaling constant C in order to apply the bilinear z-transform synthesis method.

The effect of transforming the complex s-plane into the s_1-plane is illustrated in Figure 8-17. Note that since $H_a(s)$ is not bandlimited, aliasing occurs when $H_a(s)$ is digitized. However, since $H_a(s_1)$ is bandlimited, no aliasing occurs. It is also important to note that $H_a(s_1)$ is not explicitly determined as part of the synthesis process. The complex s-plane is transformed into the s_1-plane and $H_a(s_1)$ is digitized to form $H(z)$ all in one step by application of (8-57).

EXAMPLE 8-6

A second-order low-pass Butterworth filter, having a 3-dB bandwidth of ω_c radians per second, is given by the system function

$$H_a(s) = \frac{\omega_c^2}{s^2 + \sqrt{2}\,\omega_c s + \omega_c^2} \tag{8-58}$$

as shown in Appendix B. In this example the bilinear z-transform digital filter equivalent to (8-58) will be determined.

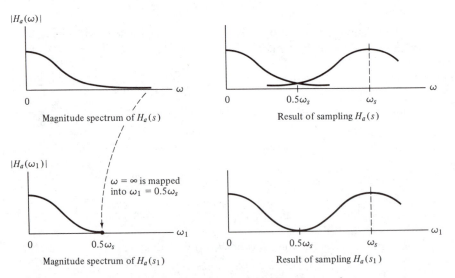

FIGURE 8-17. Effect of mapping the s-plane into the s_1-plane.

We wish the digital filter also to have a 3-dB bandwidth of ω_c, so that

$$\omega_r = \omega_c$$

yielding

$$C = \omega_c \cot \frac{\omega_c T}{2} = \omega_c \cot \frac{\pi \omega_c}{\omega_s} \tag{8-59}$$

Thus, for s in (8-58) we substitute

$$\omega_c \cot \frac{\pi \omega_c}{\omega_s} \frac{1 - z^{-1}}{1 + z^{-1}}$$

This substitution is simplified by defining

$$m = \cot \frac{\pi \omega_c}{\omega_s} = \cot \frac{\pi f_c}{f_s} \tag{8-60}$$

so that

$$s = \omega_c m \frac{1 - z^{-1}}{1 + z^{-1}} \tag{8-61}$$

This yields, in (8-58), the pulse transfer function

$$H(z) = \frac{\omega_c^2}{\left(\omega_c m \dfrac{1 - z^{-1}}{1 + z^{-1}}\right)^2 + \sqrt{2}\omega_c \left(\omega_c m \dfrac{1 - z^{-1}}{1 + z^{-1}}\right) + \omega_c^2} \tag{8-62}$$

It is important to note that ω_c cancels in (8-62). This illustrates the fact that the pulse transfer function is only a function of the ratio of ω_c to ω_s through the parameter m as defined in (8-60). Multiplying the numerator and denominator of (8-62) by $(1 + z^{-1})^2$ and grouping like terms yields

$$H(z) = \frac{1 + 2z^{-1} + z^{-2}}{(m^2 + \sqrt{2}m + 1) + 2(1 - m^2)z^{-1} + (m^2 - \sqrt{2}m + 1)z^{-2}} \tag{8-63}$$

For the case where $f_c = 500$ Hz and $f_s = 2000$ Hz,

$$m = \cot \frac{\pi}{4} = 1$$

and the pulse transfer function becomes

$$H(z) = \frac{1 + 2z^{-1} + z^{-2}}{3.414214 + 0.585786z^{-2}} \tag{8-64}$$

Dividing numerator and denominator by 3.414214 yields the pulse transfer function in standard form:

$$H(z) = \frac{0.292893 + 0.585786z^{-1} + 0.292893z^{-2}}{1 + 0.171573z^{-2}} \tag{8-65}$$

The amplitude response is determined by substituting $e^{j2\pi r}$ for z in (8-65) and taking the magnitude of the resulting expression. This was done and the result is illustrated in Figure 8-18. Also illustrated in Figure 8-18 is the amplitude response of the analog second-order Butterworth filter given by

$$|H_a(f)| = \frac{1}{\sqrt{1 + (f/f_c)^4}} \tag{8-66}$$

Since for the case analyzed

$$f_c = \tfrac{1}{4} f_s$$

f/f_c in (8-66) reduces to

$$\frac{f}{f_c} = \frac{4f}{f_s} = 4r$$

Thus, $f = 4rf_c$, and

$$|H_a(4rf_c)| = \frac{1}{\sqrt{1 + 256r^4}} \tag{8-67}$$

The analog and digital filter responses illustrated in Figure 8-18 compare poorly because the 3-dB break frequency of the digital filter is relatively close to half the sampling frequency. The amplitude responses of bilinear z-transform digital filters derived from a lowpass Butterworth prototype was determined for $n = 1, 2, 3, 4,$ and 5. The results are shown in Figure 8-19 for $f_c/f_s = 0.01$

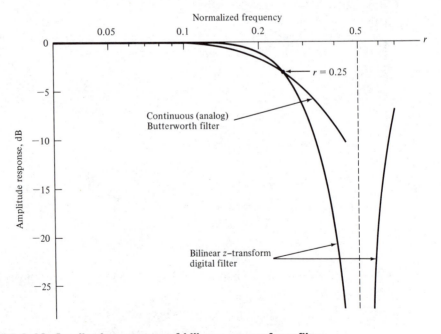

FIGURE 8-18. Amplitude response of bilinear z-transform filter.

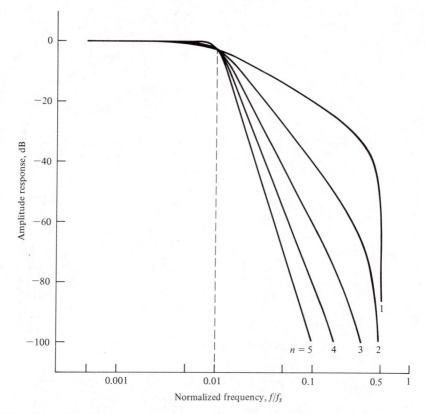

FIGURE 8-19. Digital Butterworth filter responses with $f_c/f_s = 0.01$.

and in Figure 8-20 for $f_c/f_s = 0.05$. It can be seen that the nonlinear effects of the bilinear z-transformation are negligible for $r < 0.1$.

Thus for input frequencies less than $0.1f_s$, the amplitude response of the digital filter closely approximates the amplitude response of the analog Butterworth prototype. ∎

EXAMPLE 8-7

An analog integrator is defined by

$$H_a(s) = \frac{1}{s}$$

so that the bilinear z-transform integrator is

$$H(z) = \frac{1}{s}\bigg|_{s=C[(1-z^{-1})/(1+z^{-1})]} = \frac{1}{C}\frac{1+z^{-1}}{1-z^{-1}}$$

If C is set equal to the limiting value of $2/T$, we have

$$H(z) = \frac{T}{2}\frac{1+z^{-1}}{1-z^{-1}}$$

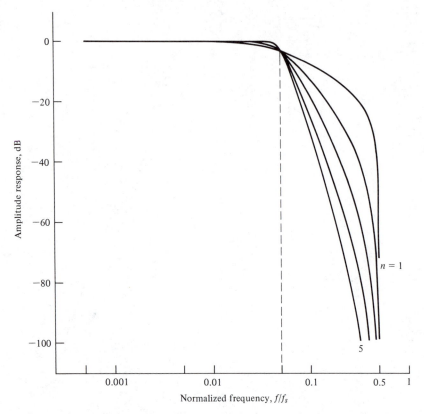

FIGURE 8-20.   Digital Butterworth filter responses with $f_c/f_s = 0.05$.

Comparison of this expression with (8-17) illustrates the equivalence of trape-zoidal integration and the bilinear z-transformation.

As an additional illustration of this equivalence, let us return to Example 8-3, in which we determined the transfer function of the digital system based on the differential equation

$$\frac{dy}{dt} = x(t) - \beta y(t)$$

Laplace transforming this equation yields

$$sY(s) = X(s) - \beta Y(s)$$

or

$$H_a(s) = \frac{Y(s)}{X(s)} = \frac{1}{s + \beta}$$

With s replaced by

$$\frac{2}{T}\frac{1 - z^{-1}}{1 + z^{-1}}$$

we have

$$H(z) = \frac{T(1 + z^{-1})}{(\beta T + 2) + (\beta T - 2)z^{-1}}$$

which is exactly the same result achieved by replacing the continuous-time integrator by the trapezoidal integrator. ∎

8-5

FINITE IMPULSE RESPONSE FILTER DESIGN

In Section 8-4, methods for designing digital filters having an infinite duration unit pulse response were studied. We found that, by starting with an analog filter, the design techniques were easy to apply. We now turn our attention to the design of finite duration unit pulse response filters. The design technique to be used now differs from these previously considered in that our starting point will not be an analog filter.

At the present time, the most popular tool for the design of FIR digital filters is to employ one of the several computer-aided design algorithms currently available (see the references). However, to illustrate the basic characteristics of FIR filters, we now consider FIR digital filters designed using the Fourier series and windowing method.

The design technique that will be used for FIR digital filters originates with the observation that since the sinusoidal steady-state transfer function of a digital filter is periodic in the sampling frequency, it can be expanded in a Fourier series. Thus

$$H(z)\big|_{z=e^{j\omega T}} = H(e^{j\omega T}) = \sum_{n=-\infty}^{\infty} h(nT)e^{-jn\omega T} \qquad (8\text{-}68)$$

in which $h(nT)$ represents the terms of the unit pulse response. In writing (8-68) we allowed the filter to be noncausal by starting the summation at $n = -\infty$ rather than $n = 0$. It will be seen later that a noncausal filter can be modified to yield a causal filter.

The transfer function can be decomposed into real and imaginary components by writing (8-68) as

$$H(e^{j\omega T}) = \sum_{n=-\infty}^{\infty} h(nT) \cos n\omega T - j \sum_{n=-\infty}^{\infty} h(nT) \sin n\omega T \qquad (8\text{-}69)$$

which is

$$H(e^{j2\pi fT}) = H_r(f) + jH_i(f)$$

where the real and imaginary parts of the transfer function are given by

$$H_r(f) = \sum_{n=-\infty}^{\infty} h(nT) \cos 2\pi nfT \qquad (8\text{-}70)$$

and

$$H_i(f) = -\sum_{n=-\infty}^{\infty} h(nT) \sin 2\pi nfT \qquad (8\text{-}71)$$

The two preceding expressions illustrate that $H_r(f)$ is an even function of frequency and $H_i(f)$ is an odd function of frequency. We see that if $h(nT)$ is an even sequence, the imaginary part of the transfer function, $H_i(f)$, will be zero. [The even sequence, $h(nT)$, multiplied by the odd sequence, $\sin 2\pi nfT$, will yield an odd sequence. An odd sequence summed over symmetric limits yields zero.] In a similar manner, if $h(nT)$ is an odd sequence, the real part of the transfer function, $H_r(f)$, will be zero. Thus an even unit pulse response yields a real transfer function and an odd unit pulse response yields an imaginary transfer function. Recall that a real transfer function has 0 or $\pm\pi$ radians phase shift, while an imaginary transfer function has $\pm\pi/2$ radians phase shift, as illustrated in Figure 8-21. Therefore, by making the unit pulse response either even or odd, we can generate a transfer function that is either real or imaginary.

In designing digital filters we usually have interest in one of the two following situations:

1. *Filtering*. For filtering applications the function of interest is usually the amplitude response of the filter. In other words, some portion of the input signal spectrum is to be attenuated and some portion of the input signal spectrum is to be passed to the output with no attenuation. This should be accomplished without phase distortion. Thus, ideally, we realize the amplitude response by using only a real transfer function. In other words, let

$$H(e^{j2\pi fT}) = H_r(f)$$

with

$$H_i(f) \equiv 0$$

2. *Filtering Plus Quadrature Phase Shift*. These applications include integrators, differentiators, and Hilbert transform devices. For all of these applications

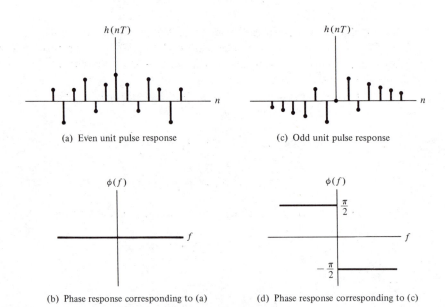

(a) Even unit pulse response   (c) Odd unit pulse response

(b) Phase response corresponding to (a)   (d) Phase response corresponding to (c)

FIGURE 8-21.   Even and odd unit pulse responses.

the desired transfer function is imaginary. Thus our design procedure will be to realize the required amplitude response by using only $H_i(f)$. In other words, let

$$H(e^{j2\pi fT}) = jH_i(f)$$

with

$$H_r(f) \equiv 0$$

Design Equations

Since $H(e^{j2\pi fT})$ is periodic in the sampling frequency, both $H_r(f)$ and $H_i(f)$ are periodic in the sampling frequency. Thus both $H_r(f)$ and $H_i(f)$ can be expanded in a Fourier series. Since the real part of the transfer function, $H_r(f)$, is an even function of frequency, its Fourier series will be of the form

$$H_r(f) = a_0 + \sum_{n=1}^{\infty} a_n \cos 2\pi nfT \tag{8-72}$$

The Fourier coefficients a_n are given by

$$a_n = \frac{2}{f_s} \int_{-f_s/2}^{f_s/2} H_r(f) \cos 2\pi nfT \, df, \qquad n \neq 0 \tag{8-73}$$

and the a_0 term is given by

$$a_0 = \frac{1}{f_s} \int_{-f_s/2}^{f_s/2} H_r(f) \, df \tag{8-74}$$

In like manner, the imaginary part of the transfer function, which is an odd function of frequency, can be expanded in the Fourier series

$$H_i(f) = \sum_{n=1}^{\infty} b_n \sin 2\pi nfT \tag{8-75}$$

The Fourier coefficients b_n are given by

$$b_n = \frac{2}{f_s} \int_{-f_s/2}^{f_s/2} H_i(f) \sin 2\pi nfT \, df \tag{8-76}$$

Since an odd function has zero average value, $b_0 \equiv 0$.

At this point the Fourier coefficients a_n and b_n must be related to the unit pulse response of the filter. First we consider the case in which $H_i(f) \equiv 0$. Equation (8-72) can then be written

$$H(e^{j2\pi fT}) = H_r(f) = a_0 + \sum_{n=1}^{\infty} a_n \cos 2\pi nfT \tag{8-77}$$

$$= a_0 + \sum_{n=1}^{\infty} \frac{a_n}{2} (z^n + z^{-n}) \Big|_{z=e^{j2\pi fT}}$$

The transfer function can be also written

$$H(e^{j2\pi fT}) = \sum_{n=-\infty}^{\infty} h(nT)z^{-n}\bigg|_{z=e^{j2\pi fT}} \tag{8-78}$$

$$= h(0) + \sum_{n=1}^{\infty} \{h(-nT)z^n + h(nT)z^{-n}\}\bigg|_{z=e^{j2\pi fT}}$$

The terms of the unit pulse response $h(nT)$ can easily be related to the Fourier coefficients of the frequency response by comparing (8-77) and (8-78). This yields

$$h(0) = a_0$$

$$\left.\begin{aligned} h(-nT) &= \tfrac{1}{2}a_n \\[2em] h(nT) &= \tfrac{1}{2}a_n \end{aligned}\right\} \quad n > 0 \tag{8-79}$$

For the case in which $H_r(f) \equiv 0$, we have, from (8-75),

$$H(e^{j2\pi fT}) = jH_i(f) = j\sum_{n=1}^{\infty} b_n \sin 2\pi nfT \tag{8-80}$$

$$= \sum_{n=1}^{\infty} \frac{b_n}{2}(z^n - z^{-n})\bigg|_{z=e^{j2\pi fT}}$$

Comparing this to (8-78), we have the design equations

$$\left.\begin{aligned} h(-nT) &= \frac{b_n}{2} \\[2em] h(nT) &= -\frac{b_n}{2} \end{aligned}\right\} \quad n > 0 \tag{8-81}$$

Note that (8-79) and (8-81) are even and odd unit pulse responses, respectively.

Causal Filters—Windowing

Before the design equations developed in the preceding section can be used to implement a realizable digital filter, we must modify the pulse transfer function

$$H(z) = \sum_{n=-\infty}^{\infty} h(nT)z^{-n}$$

so that the filter is causal. To do this, the unit pulse response is first truncated to $2M + 1$ terms. This yields the new pulse transfer function

$$H_1(z) = \sum_{n=-M}^{M} h(nT)z^{-n} \tag{8-82}$$

Next, $H_1(z)$ is multiplied by z^{-M}, which results in the causal pulse transfer function

$$H_2(z) = \sum_{n=-M}^{M} h(nT)z^{-(n+M)} = \sum_{k=0}^{2M} h(kT - MT)z^{-k} \qquad (8\text{-}83)$$

The weighting coefficients in the general FIR pulse transfer function

$$H_2(z) = \sum_{k=0}^{2M} L_k z^{-k} \qquad (8\text{-}84)$$

are then given by

$$L_k = h(kT - MT) \qquad (8\text{-}85)$$

The resulting digital filter, and its unit pulse response, are shown in Figure 8-22a and b, respectively. Note that the unit pulse response is identically equal to zero after $2M$ sample periods. Thus the filter is indeed an FIR filter with a unit pulse response duration of $2MT$ seconds. The implementation of the filter is simply the feedforward network, $H_1(z)$, in Figure 8-1.

In obtaining (8-84), two distinct operations were used. First the unit pulse response was truncated to $2M + 1$ terms. The resulting finite-duration unit pulse response was then shifted to the right (delayed) M sample periods in order to obtain a causal digital filter.

Of these two operations, the truncation of the unit pulse response has by far the more serious effect. Truncating the unit pulse response is equivalent to multiplying the unit pulse response $h(nT)$ by a rectangular window $w_r(nT)$ defined by

$$w_r(nT) = \begin{cases} 1, & |n| \leq M \\ 0, & |n| > M \end{cases} \qquad (8\text{-}86)$$

Thus the steady-state frequency response corresponding to the truncated unit pulse response (8-84) is the desired steady-state frequency response, $H(e^{j\omega T})$, convolved with the Fourier transform of the rectangular window function

$$W_r(f) = 2MT \text{ sinc } 2fMT \qquad (8\text{-}87)$$

This, as we shall see in the examples to follow, induces ripples on the desired steady-state frequency response. As M becomes large, $W(f)$ becomes more like an impulse function in the frequency domain. Then, convolution of $H(e^{j\omega T})$ with $W_r(f)$ is essentially convolution with an impulse function, which has no effect on $H(e^{j\omega T})$.

The effect of the truncation of the unit pulse response is reduced by using a window function which still truncates the unit pulse response to $2M + 1$ terms but which does not have the large discontinuity at the end points, $n = \pm M$. An example of such a window is the Hamming window defined by

$$w_h(nT) = 0.54 + 0.46 \cos \frac{\pi n}{M}, \qquad -M \leq n \leq M \qquad (8\text{-}88)$$

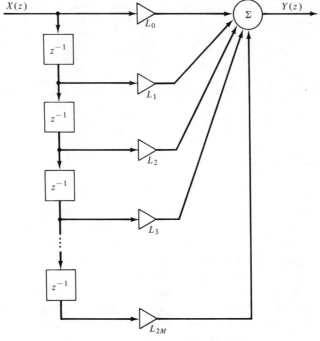

(a) Implementation of FIR filter

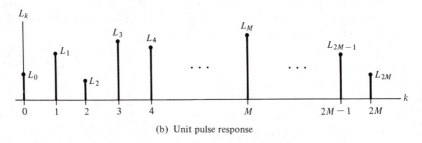

(b) Unit pulse response

FIGURE 8-22.   FIR filter and unit pulse response.

and

$$w_h(nT) = 0, \qquad |n| > M \qquad (8\text{-}89)$$

The procedure is to simply replace $h(nT)$ by the product $h(nT)w_h(nT)$ in (8-83). The examples to follow illustrate both the method of applying a window and the effect of the window on the amplitude response.

The second operation, multiplying $H_1(z)$ by z^{-M} to obtain the causal pulse transfer function $H_2(z)$, simply time delays the unit pulse response M sample periods or MT seconds. This is equivalent to cascading the nonrealizable system with a system having a sinusoidal steady-state transfer function

$$z^{-M}\Big|_{z=e^{j\omega T}} = e^{-j\omega MT} = e^{-j2\pi f MT} \qquad (8\text{-}90)$$

The amplitude response is unaffected and the phase response is changed only by a linear function of frequency. No distortion of *any kind* results from the multiplication by z^{-M}. Thus the digital filter illustrated in Figure 8-22 will have a phase response similar to one of the three phase responses shown in Figure 8-23. Variations of these phase responses occur if $H_r(f)$ or $H_i(f)$ changes sign.

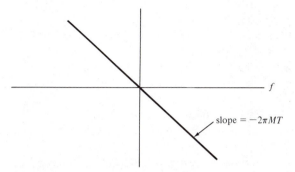

(a) Phase response of FIR digital filter with $H_r(f) > 0$ for $f > 0$

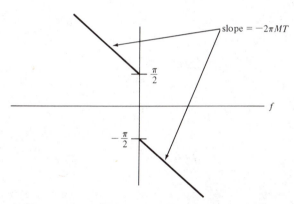

(b) Phase response of FIR digital filter with $H_i(f) < 0$ for $f > 0$

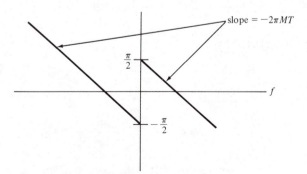

(c) Phase response of FIR digital filter with $H_i(f) > 0$ for $f > 0$

FIGURE 8-23. FIR digital filter phase responses.

Design Procedure

In summary, the procedure for designing an FIR digital filter consists of the following steps:

1. Decide whether $H_r(f)$ or $H_i(f)$ is to be set equal to zero. For filtering applications we typically set $H_i(f) = 0$. For integrators, differentiators, and Hilbert transformers we set $H_r(f) = 0$.
2. Expand $H_r(f)$ or $H_i(f)$ in a Fourier series.
3. The unit pulse response is determined from the Fourier coefficients using (8-79) or (8-81).
4. Decide on an appropriate number of weighting coefficients for the filter, in order to determine the point at which the unit pulse response is to be truncated.
5. Apply a window function to the unit pulse response.
6. Check the resulting amplitude response to ensure that the resulting sinusoidal steady-state response is satisfactory. If not, the value of M may have to be increased or one of the many different window functions described in the literature may be tried.†

EXAMPLE 8-8

In this example an FIR differentiator is to be designed. The ideal differentiator is defined by the transfer function

$$H(f) = j2\pi f \tag{8-91}$$

and therefore we shall let

$$H_i(f) = 2\pi f \tag{8-92}$$

with

$$H_r(f) = 0 \tag{8-93}$$

Figure 8-24 illustrates $H_i(f)$ together with the resulting amplitude and phase responses of the ideal differentiator. From (8-76) and (8-92) we have

$$b_n = \frac{2}{f_s} \int_{-f_s/2}^{f_s/2} 2\pi f \sin 2\pi n f T \, df \tag{8-94}$$

which, with the change of variable

$$x = 2\pi n f T$$

becomes

$$b_n = \frac{1}{\pi f_s (nT)^2} \int_{-n\pi}^{n\pi} x \sin x \, dx \tag{8-95}$$

---

†See F. J. Harris, "On the Use of Windows for Harmonic Analysis with the Discrete Fourier Transform," *Proceedings of the IEEE*, Vol. 66, No. 1, January 1978, pp. 51–83, for a complete review of the many window functions which have been found useful. Window functions will be discussed again in Chapter 9.

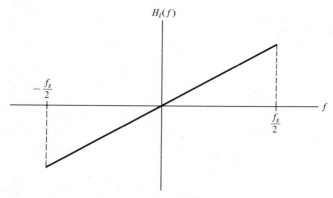

(a) $H_i(f)$ such that $H(e^{j2\pi fT}) = j\,H_i(f)$

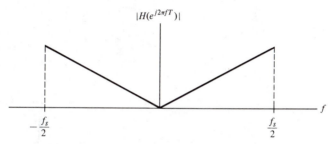

(b) Amplitude response of ideal differentiator

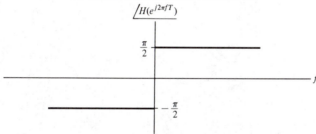

(c) Phase response of ideal differentiator

FIGURE 8-24. Ideal differentiator characteristics.

which reduces to

$$b_n = \frac{2}{\pi n^2 T} \int_0^{n\pi} x \sin x \, dx \qquad (8\text{-}96)$$

Evaluating the integral yields

$$b_n = \frac{2}{\pi n^2 T} (\sin x - x \cos x) \Big|_0^{n\pi} = -\frac{2}{nT} \cos n\pi \qquad (8\text{-}97)$$

or

$$b_n = -\frac{2}{nT}(-1)^n \qquad (8\text{-}98)$$

Note that the b_n's must be scaled by the sampling period, T. For $T = 1$ we have (recall that $b_0 = 0$)

$$b_n = -\frac{2}{n}(-1)^n \qquad (8\text{-}99)$$

Thus

$$\left.\begin{aligned} h(-nT) &= \frac{1}{n}(-1)^{n+1} \\[2ex] h(nT) &= \frac{1}{n}(-1)^n \end{aligned}\right\} n > 0 \qquad (8\text{-}100)$$

Figure 8-25 illustrates the performance of the differentiator for the case where 15 weights are used. In other words, $h(nT) = 0$, $|n| \geq 8$. A Hamming window for this case is given by

$$w_h(nT) = 0.54 + 0.46 \cos\frac{n\pi}{7}, \qquad |n| \leq 7 \qquad (8\text{-}101)$$

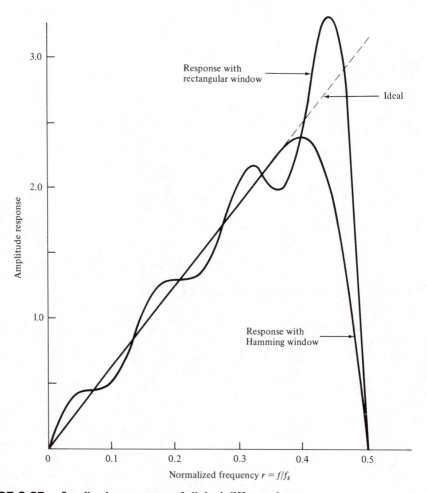

FIGURE 8-25. Amplitude response of digital differentiator.

TABLE 8-1. Filter Weights for Transversal Differentiator

n	Unit Pulse Response with Rectangular Window, $h(nT)$	Hamming Window Function, $w_h(nT)$	Unit Pulse Response with Hamming Window, $w_h(nT)h(nT)$
-7	$+0.142857$	0.08	$+0.011429$
-6	-0.166667	0.125554	-0.020926
-5	$+0.2$	0.253195	$+0.050639$
-4	-0.25	0.437640	-0.109410
-3	$+0.333333$	0.642360	$+0.214120$
-2	-0.5	0.826805	-0.413403
-1	$+1.0$	0.954446	$+0.954446$
0	0	1.0	0
1	-1.0	0.954446	-0.954446
2	$+0.5$	0.826805	$+0.413403$
3	-0.333333	0.642360	-0.214120
4	$+0.25$	0.437640	$+0.109410$
5	-0.2	0.253195	-0.050639
6	$+0.166667$	0.125554	$+0.020926$
7	-0.142857	0.08	-0.011429

Table 8-1 illustrates the terms of the unit pulse response for the FIR differentiator with a rectangular window and with a Hamming window. The phase response is shown in Figure 8-26. ∎

EXAMPLE 8-9

As another example of an FIR design we approximate an ideal low-pass filter using 17 weights. Assuming that the filter bandwidth is $0.15f_s$, we have

$$H_r(f) = 1, \quad |f| < 0.15f_s$$

$$H_r(f) = 0, \quad |f| > 0.15f_s$$

and

$$H_i(f) = 0$$

Thus (8-73) becomes

$$a_n = \frac{2}{f_s} \int_{-0.15f_s}^{0.15f_s} \cos 2\pi nfT \, df \tag{8-102}$$

and (8-74) becomes

$$a_0 = \frac{1}{f_s} \int_{-0.15f_s}^{0.15f_s} df = 0.3 \tag{8-103}$$

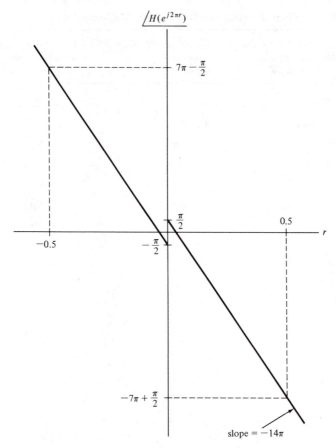

FIGURE 8-26. Phase response for **M = 7** FIR differentiator.

Equation (8-102) yields

$$a_n = \frac{2}{\pi n} \int_0^{0.3\pi n} \cos x \, dx = \frac{2}{\pi n} \sin 0.3\pi n \tag{8-104}$$

Thus, from (8-79),

$$h(0) = 0.3 \tag{8-105}$$

$$h(\pm nT) = \frac{1}{n\pi} \sin 0.3\pi n, \qquad n > 0 \tag{8-106}$$

The Hamming window for this application is defined by

$$w_h(nT) = 0.54 + 0.46 \cos\left(\frac{n\pi}{8}\right), \qquad |n| \leq 8 \tag{8-107}$$

Figure 8-27 illustrates the amplitude response with and without the Hamming window. In drawing Figure 8-27 we have allowed the amplitude response to be negative to avoid having to draw the phase. Table 8-2 gives the terms in the unit pulse response with a rectangular window and with a Hamming window. ∎

TABLE 8-2. Filter Weights for Transversal Low-pass Filter ($f_c = 0.15f_s$)

n	Unit Pulse Response with Rectangular Window, $h(nT)$	Hamming Window Function, $w_h(nT)$	Unit Pulse Response with Hamming Window, $w_h(nT)h(nT)$
-8	0.037841	0.08	0.003027
-7	0.014052	0.115015	0.001616
-6	-0.031183	0.214731	-0.006696
-5	-0.063662	0.363966	-0.023171
-4	-0.046774	0.54	-0.025258
-3	0.032788	0.716034	0.023477
-2	0.151365	0.865269	0.130972
-1	0.257518	0.964985	0.248501
0	0.3	1.0	0.3
1	0.257518	0.964985	0.248501
2	0.151365	0.865269	0.130972
3	0.032788	0.716034	0.023477
4	-0.046774	0.54	-0.025258
5	-0.063662	0.363966	-0.023171
6	-0.031183	0.214731	-0.006696
7	0.014052	0.115015	0.001616
8	0.037841	0.08	0.003027

EXAMPLE 8-10

As our last example of an FIR digital filter design, consider a 90° phase shifter. The desired sinusoidal steady-state transfer function is

$$H(e^{j\omega T}) = \begin{cases} -j1, & \omega > 0 \\ +j1, & \omega < 0 \end{cases} \tag{8-108}$$

so that $H(e^{j\omega T})$ is imaginary. Thus

$$H_i(f) = \begin{cases} -1, & \omega > 0 \\ +1, & \omega < 0 \end{cases} \tag{8-109}$$

as shown in Figure 8-28.

Application of (8-76) yields

$$b_n = \frac{2}{f_s} \int_{-f_s/2}^{0} (1) \sin 2\pi nfT \, df + \frac{2}{f_s} \int_{0}^{f_s/2} (-1) \sin 2\pi nfT \, df \tag{8-110}$$

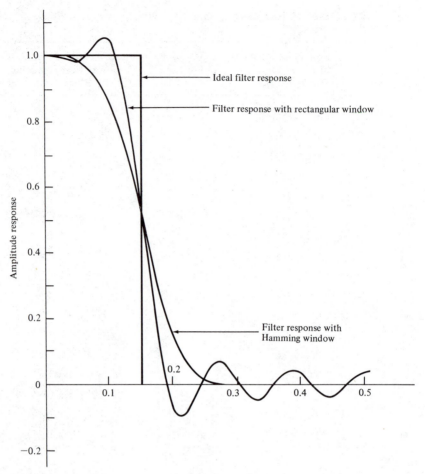

FIGURE 8-27. **Amplitude responnse of digital low-pass filter. (The negative portions are shown negative for convenience.)**

which gives

$$b_n = -\frac{4}{f_s}\int_0^{f_s/2} \sin 2\pi n f T \, df = -\frac{2}{n\pi}\int_0^{n\pi} \sin x \, dx \qquad (8\text{-}111)$$

This yields

$$b_n = \begin{cases} 0, & n \text{ even} \\ -\dfrac{4}{\pi n}, & n \text{ odd} \end{cases} \qquad (8\text{-}112)$$

from which the unit pulse response is obtained using (8-81). The result is

$$h(nT) = 0, \qquad n \text{ even}$$

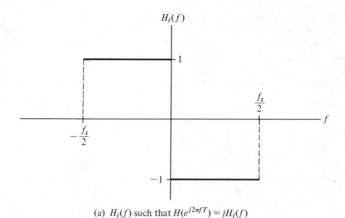

(a) $H_i(f)$ such that $H(e^{j2\pi fT}) = jH_i(f)$

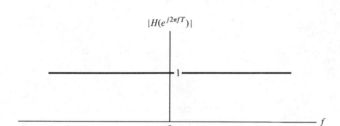

(b) Amplitude response of ideal 90 degree phase shifter

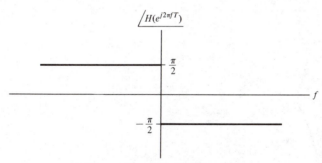

(c) Phase response of ideal 90° phase shifter

FIGURE 8-28. Ideal 90° phase shifter characteristics.

$$
\left.
\begin{aligned}
h(-nT) &= -\frac{2}{\pi n} \\
\\
h(nT) &= \frac{2}{\pi n}
\end{aligned}
\right\} \quad n \text{ odd}
\tag{8-113}
$$

The amplitude response is shown in Figure 8-29 for $M = 7$. Both responses with and without a Hamming window are illustrated. The Hamming window for this case is given by (8-101). The filter weights are given in Table 8-3. ∎

TABLE 8-3. Filter Weights for Transversal 90° Phase Shifter

n	Unit Pulse Response with Rectangular Window, $h(nT)$	Hamming Window Function, $w_h(nT)$	Unit Pulse Response with Hamming Window, $w_h(nT)h(nT)$
−7	−0.090946	0.08	−0.007276
−6	0	0.125554	0
−5	−0.127324	0.253195	−0.032238
−4	0	0.437640	0
−3	−0.212207	0.642360	−0.136313
−2	0	0.826805	0
−1	−0.636620	0.954446	−0.607619
0	0	1.0	0
1	0.636620	0.954446	0.607619
2	0	0.826805	0
3	0.212207	0.642360	0.136313
4	0	0.437640	0
5	0.127324	0.253195	0.032238
6	0	0.125554	0
7	0.090946	0.08	0.007276

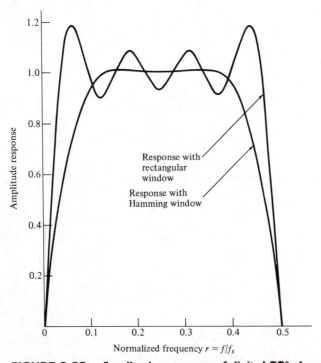

FIGURE 8-29. Amplitude response of digital 90° phase shifter.

Systems that have unity amplitude response and 90° phase shifts are sometimes referred to as *Hilbert transformers* since the output is the Hilbert transform of the input. These filters are useful in the implementation of single-sideband modulators, and have many other applications as well.

8-6

SUMMARY

In this chapter various filter designs have been investigated. As both a summary and an attempt to place the topics of this chapter in better perspective, we now pause to compare these filter types. Perhaps the best way to accomplish this is to list the advantages and disadvantages of both classes of filters.

First, let us look at IIR (infinite-duration impulse response) digital filters. The main advantages of IIR filters are:

1. The design techniques for IIR filters are easy to apply and since the design follows directly from an analog prototype, the engineer usually has some feel for the performance of a given filter in a given application.
2. Typically, the hardware requirements for an IIR filter are less than the hardware requirements for a comparable FIR filter. However, with today's LSI techniques, hardware considerations are becoming less important.

The main advantages of FIR (finite-duration impulse response) filters are:

1. FIR filters can be designed which have perfectly linear phase.
2. Since FIR filters have no feedback, they have no poles, and are therefore always stable.
3. The fast Fourier transform algorithm, which is considered in the following chapter, gives the filter designer a very simple tool for determining the filter weights.

FURTHER READING

W. A. STANLEY, *Digital Signal Processing*. Reston, Va.: Reston, 1975.
   This book is an excellent introduction to discrete-time signals and systems written at about the same level as this book. Since Stanley's book is devoted entirely to digital signal processing, he covers several topics not found in this book. An example is a section on the synthesis of bandpass digital filters.
A. V. OPPENHEIM and R. W. SCHAFER, *Digital Signal Processing*. Englewood Cliffs, N.J.: Prentice-Hall, 1975.
L. R. RABINER and B. GOLD, *Theory and Application of Digital Signal Processing*. Englewood Cliffs, N.J.: Prentice-Hall, 1975.
   Both of these books are important contributions and together they give a thorough introduction to digital signal processing. Included are such topics as two-dimensional processing and the effects of finite-wordlength arithmetic.
   One important tool not mentioned in this chapter is computer-aided design of digital filters. Techniques exist for both IIR and FIR filters. Chapter 5 of Oppenheim and Schafer contains a brief review of the techniques and algorithms available. The Parks–McClellan algorithm is extremely useful for the design of FIR filters and is

extensively covered in the literature. The Parks–McClellan algorithm is covered in Chapter 3 of Rabiner and Gold.

L. R. RABINER and C. M. RADER, eds., *Digital Signal Processing*. New York: IEEE Press, 1972.

Selected Papers in Digital Signal Processing II. New York: IEEE Press, 1976.

These two volumes contain many key papers and offer insight into many diverse applications. They are essential to anyone actively working in the field and are highly recommended to the interested and ambitious student.

Programs for Digital Signal Processing. New York: IEEE Press, 1980.

This book contains a large number of carefully documented computer programs, including design programs for both IIR and FIR filters. The Parks–McClellan algorithm is among those contained in this package.

PROBLEMS

SECTION 8-2

8-1. Determine and sketch the Direct Form I and Direct Form II realization for each of the pulse transfer functions given below:

(a) $H(z) = \dfrac{3 + 7.5z^{-1}}{4 + 4z^{-1} + 3z^{-2}}$

(b) $H(z) = \dfrac{1}{(1 - 0.1z^{-1})(1 - 0.2z^{-1})}$

(c) $H(z) = \dfrac{z^{-2} + 4z^{-3}}{(2 - z^{-1})(3 + z^{-1})(1 - z^{-2})}$

8-2. Determine and sketch the Direct Form I and Direct Form II realizations for the pulse transfer function

$$H(z) = \frac{3}{1 + 0.1z^{-1}} + \frac{1 + 5z^{-2}}{(1 - z^{-1})(1 + 0.5z^{-1})}$$

8-3. Determine the pulse transfer function of the system shown.

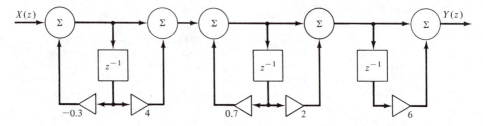

8-4. Determine the Direct Form II realization for each of the pulse transfer functions given below.

(a) $H(z) = \left[\dfrac{z^{-1}(1 - z^{-3})}{(1 - z^{-1})(1 + 0.2z^{-2})}\right]^2$

(b) $H(z) = \left[\dfrac{z^{-1}}{1 - 0.1z^{-1}} + \dfrac{3}{1 - 0.2z^{-1}}\right]^2$

8-5.   For the system shown determine the pulse transfer function and determine the Direct Form II realization.

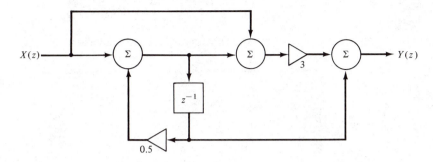

8-6.   Realize the system shown using only one z^{-1} element.

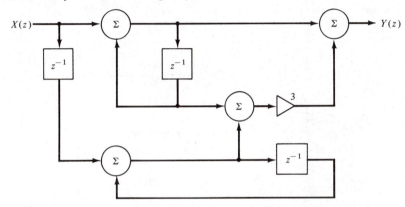

8-7.   Determine a cascade and parallel realization, using only first-order structures, for each of the following pulse transfer functions.

(a) $H(z) = \dfrac{1 + z^{-1}}{1 + 0.5z^{-1} - 0.5z^{-2}}$

(b) $H(z) = \dfrac{z^{-2}}{1 - 1.5z^{-1} + 0.5z^{-2}}$

(c) $H(z) = \dfrac{0.5z^{-1} + 4z^{-2} + 6z^{-3}}{(1 - z^{-1})(1 - 0.2z^{-1})(1 - 0.5z^{-1})}$

8-8.   Determine a cascade and parallel realization, using sections having the minimum possible order, for

$$H(z) = \frac{z^{-2} + 2z^{-3}}{(1 - z^{-1})^2(1 + 0.5z^{-1})}$$

8-9.   Determine a parallel realization for

$$H(z) = \frac{z^{-2}}{(1 - 0.1z^{-1})^2}$$

8-10. Determine a parallel realization for

$$H(z) = \frac{z^{-3}}{(1 - 0.1z^{-1})^2}$$

8-11. Determine the pulse transfer function for the structure shown.

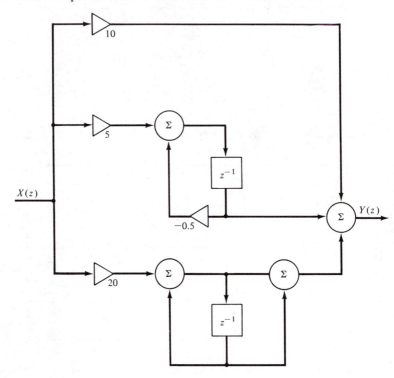

SECTION 8-3

8-12. The two-term running average filter is defined by the difference equation

$$y(nT) = \tfrac{1}{2} [x(nT) + x(nT - T)]$$

Determine the pulse transfer function. Determine and plot the amplitude response and the phase response as a function of r in the range of $0 \le r \le 0.5$. In what sense is the two-term running average filter an integrator?

8-13. Repeat Problem 8-12 for the three-term running average filter, defined by

$$y(nT) = \tfrac{1}{3} [x(nT) + x(nT - T) + x(nT - 2T)]$$

8-14. Determine and plot the unit pulse response of both a rectangular and trapezoidal integrator. Compare with the unit impulse response of an ideal analog integrator. Comment on your results.

8-15. Assume that an ideal analog integrator has input

$$x(t) = Ae^{-\alpha t}$$

Determine and plot the output. Now assume that samples of $x(t)$ defined by

$$x(nT) = A(e^{-\alpha T})^n$$

are applied to both a rectangular and trapezoidal integrator. Determine and plot the output sequence, $y(nT)$, for both integrators and compare with the sampled output of an ideal integrator. Comment on the results.

8-16. A first approximation to a digital differentiator is a system that generates an output equal to the slope of the line connecting the sample points $x(nT)$ and $x(nT - T)$.
   (a) Show that the defining difference equation is

$$y(nT) = \frac{1}{T}\{x(nT) - x(nT - T)\}$$

   (b) Determine the amplitude response and phase response of the system. Compare with an ideal differentiator.

SECTION 8-4

8-17. Write a FORTRAN computer program for determining the amplitude response (in decibels) and phase response (in degrees) of a second order IIR digital filter. The input to the computer program should be the coefficient set of the pulse transfer function $(L_0, L_1, L_2, K_1, K_2)$. The amplitude and phase response are to be determined at 51 values of r in the range $0 \le r \le 0.5$ (Include $r = 0$ and $r = 0.5$). After the program is developed, test it by verifying the results of Example 8-6.

8-18. Determine the pulse transfer function of both an impulse-invariant and step-invariant integrator. Determine and plot both the amplitude and phase responses of each design in the range $0 \le r \le 0.5$. Compare your results with the results of Example 8-2.

8-19. Design a digital integrator by using the bilinear z-transform synthesis technique. Show that the result is trapezoidal integration.

8-20. Design a digital differentiator by using the step-invariant method. Plot the amplitude response and phase response of your design. Compare your results with the differentiator designed in Problem 8-16.

8-21. Determine the pulse transfer function, using impulse-invariant synthesis, corresponding to the following analog transfer functions (leave the sampling period, T, arbitrary).

   (a) $H_a(s) = \dfrac{5}{(s + 1)(s + 5)}$

   (b) $H_a(s) = \dfrac{1}{s(s + 1)(s + 5)}$

   (c) $H_a(s) = \dfrac{50}{(s + 1)(s + 5)(s + 10)}$

8-22. Determine the impulse-invariant pulse transfer function corresponding to the analog transfer function

$$H_a(s) = \frac{5}{(s + 1)^2(s + 5)}$$

Sketch the block diagram of the parallel realization of the filter.

8-23. Determine the pulse transfer function, using step invariant synthesis, corresponding to the following analog transfer functions (leave the sampling period, T, arbitrary).

(a) $H_a(s) = \dfrac{5}{s + 5}$

(b) $H_a(s) = \dfrac{5}{(s + 1)(s + 5)}$

(c) $H_a(s) = \dfrac{5}{s^2 (s + 5)}$

8-24. Determine the impulse-invariant equivalent of the analog transfer function

$$H_a(s) = \frac{4}{(s + 2)^2}$$

assuming a sampling frequency of 10 Hz.

8-25. Repeat Problem 8-24 using step-invariant synthesis.

8-26. Design a digital differentiator using the bilinear z-transform synthesis technique. Plot the amplitude and phase response of both the digital filter and the ideal analog filter. Tell why this is not an effective design even for low input frequencies.

8-27. Determine $H(z)$, using the bilinear z-transform synthesis technique, for the analog transfer function

$$H_a(s) = \frac{8}{(s + 2)(s + 4)}$$

Assume a sampling frequency of 10 Hz and express your result as a ratio of polynomials in z^{-1}. Consider the following two cases:
(a) The digital filter is to have a response closely approximating the analog filter for very low frequencies.
(b) The digital filter and the analog filter are to have the same response at a frequency of 2 Hz.

8-28. Determine $H(z)$, using the bilinear z-transform synthesis technique corresponding to the analog transfer function

$$H_a(s) = \frac{10}{s^2 + 7s + 10}$$

Both the analog and digital filters are to have the same magnitude and phase response at 2 Hz. Assume a sampling frequency of 10 Hz.

8-29. Summarize the advantages, disadvantages, and the characteristics of the sinusoidal steady-state frequency response of the impulse-invariant, step-invariant, and bilinear z-transform digital filters.

8-30. Using the bilinear z-transform synthesis technique, determine the pulse transfer function corresponding to

$$H_a(s) = \frac{0.5(s + 4)}{(s + 1)(s + 2)}$$

Both the analog filter and the digital filter are to have the same response at $\omega = 3$ rad/s. Assume a sampling frequency of $10/\pi$ Hz. Plot the amplitude response of the resulting filter as a function of r in the range $0 \le r \le 0.5$. Compare with the results of Examples 8-4 and 8-5.

8-31. Consider the analog system shown.

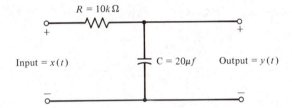

(a) For a sampling frequency of 10 Hz, determine the pulse transfer function for the impulse-invariant, step-invariant, and bilinear z-transform digital filter corresponding to the given analog system. For the bilinear z-transform filter, match the response of the digital filter to the response of the analog filter at $\omega = 5$ rad/s.

(b) Determine and plot the amplitude response (in decibels) of all three filters found in part (a). Also plot the amplitude response of the analog filter. Compare the responses.

(c) Repeat part (b) for the phase responses.

8-32. A third-order Butterworth filter having a 3-dB break frequency of 1 rad/s is given by

$$H_a(s) = \frac{1}{s^3 + 2s^2 + 2s + 1}$$

Assuming a sampling frequency of 5 Hz, determine the pulse transfer function of the bilinear z-transform digital filter equivalent to the given $H_a(s)$. Both the analog and digital filters are to have the same response at $\omega = 1$ rad/s.

8-33. Repeat Problem 8-32 assuming that the digital filter is to have a response closely approximating the analog filter response at very low frequencies.

SECTION 8-5

8-34. Write an equation giving the filter weights, L_k, for the FIR filter defined by the desired amplitude response given. Assume that a Hamming window is to be used and let $M = 10$.

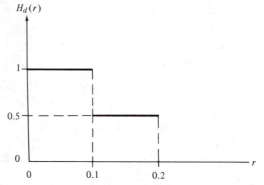

$H_d(r)$

8-35. Plot the amplitude and phase response of the filter derived in Problem 8-34.

8-36. For the system shown, plot the amplitude and phase response as a function of the normalized frequency, r, in the range $0 \le r \le 0.5$.

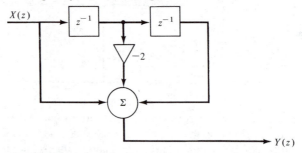

8-37. Plot the amplitude and phase response of the FIR digital filter shown.

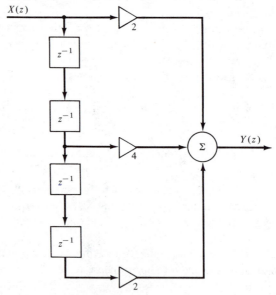

8-38. Repeat Problem 8-37 for the system shown.

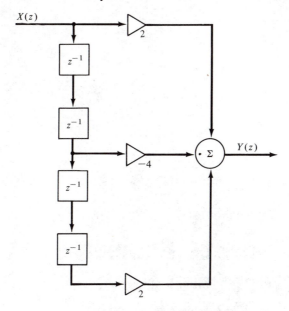

CHAPTER 9

The Discrete Fourier Transform and Fast Fourier Transform Algorithms

9-1

INTRODUCTION

In Chapter 3 we considered the representation of signals by Fourier series. Fourier representations for signals proved useful in both providing spectral information about a signal as well as a frequency-domain tool for systems analysis. In this chapter we consider the approximation of the exponential Fourier series and the integral for the Fourier coefficients developed in Chapter 3 by sums of finite length in order that they can be numerically computed. This discrete-sum spectral representation for a finite-length sequence is referred to, appropriately, as the *discrete Fourier transform* (DFT). It is very often true that the desired number of terms in a DFT sum is large—perhaps 1000 to 10,000 in number. Efficient algorithms for the computation of the DFT are desirable, and we consider the development of such algorithms in this chapter. Computationally efficient algorithms for computing the DFT are known as *fast Fourier transform* (FFT) algorithms, and are very often referred to collectively as the FFT.

9-2

THE DISCRETE FOURIER TRANSFORM

The exponential Fourier series of a periodic signal of period T, which was developed in Chapter 3, is

$$x(t) = \sum_{n=-\infty}^{\infty} X_n e^{j2\pi n f_0 t} \qquad (9\text{-}1)$$

where the coefficients are computed by the formula

$$X_n = \frac{1}{T} \int_T x(t) e^{-j2\pi n f_0 t} \, dt \qquad (9\text{-}2)$$

with $f_0 = 1/T$ and $\int_T(\cdot) \, dt$ denoting integration over any period.

To obtain finite-sum approximations for (9-1) and (9-2), consider the analog periodic signal $x(t)$ shown in Figure 9-1a and its sampled version $x_s(k \, \Delta t)$ shown in Figure 9-1b. Using $x_s(k\Delta t)$, we can approximate the integral for X_n, (9-2), by the sum

$$
\begin{aligned}
X_n &= \frac{1}{T} \sum_{k=0}^{N-1} x_s(k\Delta t) e^{-j2\pi n k \Delta t/T} \frac{T}{N} \\
&= \frac{1}{N} \sum_{k=0}^{N-1} x_s(k\Delta t) e^{-j2\pi n k/N}, \qquad n = 0, 1, \dots N - 1 \qquad (9\text{-}3)
\end{aligned}
$$

A finite series approximation for $x(t)$ is obtained by truncating the series for $x(t)$ given by (9-1) to N terms, and substituting $t = k \, \Delta t$ and $\Delta t = T/N$.

With this as justification, we *define* the *direct discrete Fourier transform (DFT)* of the time series $x_0, x_1, \dots, x_{N-1}$ as

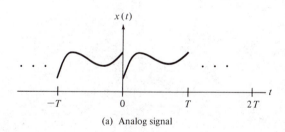

(a) Analog signal

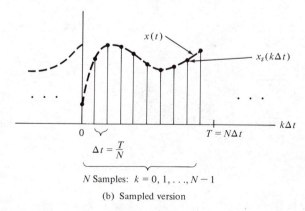

(b) Sampled version

FIGURE 9-1. Approximation of a signal by sample values.

$$X_n = \sum_{k=0}^{N-1} x_k e^{-j(2\pi/N)nk}, \qquad n = 0, 1, \ldots, N-1 \qquad (9\text{-}4)$$

where the factor $1/N$ in (9-3) has been dropped for convenience. The time series is obtained from $X_0, X_1, \ldots, X_{N-1}$ by the inverse relationship

$$x_k = \frac{1}{N} \sum_{n=0}^{N-1} X_n e^{j(2\pi/N)nk}, \qquad k = 0, 1, \ldots, N-1 \qquad (9\text{-}5)$$

where a factor $1/N$ has been inserted to compensate for its omission in (9-4). Equation (9-5) defines the *inverse DFT* operation.

We must show that (9-5) with (9-4) substituted produces an identity. Making this substitution and summing over n first, we obtain

$$x_k \stackrel{?}{=} \sum_{l=0}^{N-1} \frac{1}{N} x_l \sum_{n=0}^{N-1} e^{j(2\pi n/N)(k-l)} \qquad (9\text{-}6)$$

where the $\stackrel{?}{=}$ means that the equality is being tested.

The inside sum over n can be written as

$$\Sigma_n \stackrel{\triangle}{=} \sum_{n=0}^{N-1} [e^{j2\pi(k-l)/N}]^n \qquad (9\text{-}7)$$

Now the sum of a geometric series is

$$S_N \stackrel{\triangle}{=} \sum_{n=0}^{N-1} x^n = \frac{1 - x^N}{1 - x}$$

Letting $x = \exp[j2\pi(k-l)/N]$, we find that

$$\Sigma_N = \frac{1 - \exp[j2\pi(k-l)]}{1 - \exp[j2\pi(k-l)/N]} = 0, \qquad k \neq l$$

which follows because $\exp[j2\pi(k-l)] = 1$. If $k = l$, the result reduces to the indeterminate form $0/0$, so we must sum (9-7) directly for this special case. With $k = l$, the exponential in (9-7) is unity and we obtain

$$\Sigma_N = N, \qquad k = l$$

Summarizing, we have shown that

$$\sum_{n=0}^{N-1} e^{j(2\pi n/N)(k-l)} = \begin{cases} N, & k = l \\ 0, & k \neq l \end{cases} \stackrel{\triangle}{=} N\delta_{kl} \qquad (9\text{-}8)$$

where $\delta_{kl} = 1, k = l$, and $\delta_{kl} = 0, k \neq l$, is called the *Kronecker delta function*. Thus (9-6) becomes

$$x_k \stackrel{?}{=} \frac{1}{N} \sum_{l=0}^{N-1} x_l(N\delta_{kl}) = x_k \qquad (9\text{-}9)$$

which shows that (9-4) and (9-5) are indeed inverse operations and therefore constitute a valid transform pair.

Before finding a fast computational technique for evaluating (9-4) and (9-5), we note that they are really equivalent operations. From (9-5) it follows that the complex conjugates of the time samples, x_k^*, are given by

$$x_k^* = \frac{1}{N} \sum_{n=0}^{N-1} X_n^* e^{-j(2\pi/N)nk}, \qquad k = 0, 1, \ldots, N - 1$$

so that

$$N x_k^* = \sum_{n=0}^{N-1} X_n^* e^{-j(2\pi/N)nk}, \qquad k = 0, 1, \ldots, N - 1 \qquad (9\text{-}10)$$

Comparison of (9-10) and (9-4) show that algorithms used to calculate the forward transform can also be used for the calculation of the inverse transform if the frequency samples X_n are conjugated prior to performing the computation. Since the result yields $N x_k^*$ rather than x_k, it must be conjugated and divided by N. Often, however, the time samples are real and the conjugation of the result can be omitted. Thus once we have developed a computationally efficient algorithm for calculating the sum given by (9-4), we can use the same algorithm for computing the inverse DFT.

THE DFT COMPARED WITH THE EXPONENTIAL FOURIER SERIES

To emphasize the difference between the Fourier series and the DFT, we note that the Fourier series represents an analog periodic signal that lasts for $-\infty < t < \infty$, and the resulting line spectrum generally covers $-\infty < f < \infty$. The DFT represents a finite number of sample values in the finite observation interval $0 \le kT/N < T$, and the resulting line spectrum is limited to $0 \le n/T < N/T$. Of course, (9-4) and (9-5) repeat each N points due to the periodicity of $e^{\pm j2\pi kn/N}$.

We now illustrate by example the effect of sampling and finite observation interval by examining in some detail the approximation of a Fourier transform by a DFT. Specifically, the results from Fourier transform theory that we wish to make use of are the theorems repeated below for convenience.

1. Transform of the ideal sampling waveform (3-82):

$$y_s(t) \triangleq \sum_{m=-\infty}^{\infty} \delta(t - mT_s) \leftrightarrow f_s \sum_{n=-\infty}^{\infty} \delta(f - nf_s), \qquad f_s = \frac{1}{T_s}$$

$$(9\text{-}11)$$

2. Transform of a rectangular pulse (Example 3-12 and the time delay theorem):

$$\Pi\left(\frac{t - t_0}{T}\right) \leftrightarrow T \text{ sinc } Tf \exp\left(-j2\pi t_0 f\right) \qquad (9\text{-}12)$$

3. The convolution theorem of Fourier tranforms:

$$x_1(t) * x_2(t) \leftrightarrow X_1(f)X_2(f) \qquad (9\text{-}13)$$

4. The multiplication theorem of Fourier transforms:

$$x_1(t)x_2(t) \leftrightarrow X_1(f) * X_2(f) \qquad (9\text{-}14)$$

EXAMPLE 9-1

As a specific example, consider the exponential signal

$$x(t) = \exp\left(\frac{-|t|}{\tau}\right)$$

with Fourier transform

$$X(f) = \frac{2\tau}{1 + (2\pi f\tau)^2}$$

These are sketched in Figure 9-2a.

Since we are considering discrete-time signals, we multiply $x(t)$ by the ideal sampling waveform, $y_s(t)$, to produce the sampled exponential signal

$$x_s(t) = y_s(t) \exp\left(\frac{-|t|}{\tau}\right)$$

By the multiplication theorem, the Fourier transform of $x_s(t)$ is the convolution of $\mathcal{F}[y_s(t)] = f_s \sum_{n=-\infty}^{\infty} \delta(f - nf_s)$ and $X(f)$. The result of this convolution is

$$X_s(f) = 2\tau f_s \sum_{n=-\infty}^{\infty} \{1 + [2\pi\tau(f - nf_s)]^2\}^{-1}$$

The sampled signal $x_s(t)$ and its spectrum $X_s(f)$ are shown in Figure 9-2b for the case where $f_s = 1$.

In calculating a DFT only a T-second section of the sampled signal is observed. In effect, $x_s(t)$ is multiplied by a window function $\Pi(t/T)$. In the frequency domain, this corresponds to convolution of $X_s(f)$ with the Fourier transform of the window function. Denoting the Fourier transform of the windowed and sampled signal by $X_{sw}(f)$, we may write it as

$$X_{sw}(f) = 2\tau f_s \sum_{n=-\infty}^{\infty} \{1 + [2\pi\tau(f - nf_s)]^2\}^{-1} * T \text{ sinc } Tf$$

The sampled, windowed signal and its spectrum are shown in Figure 9-2c.

Finally, the result of a DFT operation effectively samples the spectrum $X_{sw}(f)$ at a discrete set of frequencies separated by the reciprocal of the observation interval (window duration), $1/T$, resulting in the sampled spectrum $X_{sp}(f)$. This corresponds to convolution in the time domain with a sequence of δ-functions since

$$T \sum_{m=-\infty}^{\infty} \delta(t - mT) \leftrightarrow \sum_{n=-\infty}^{\infty} \delta(f - nT^{-1})$$

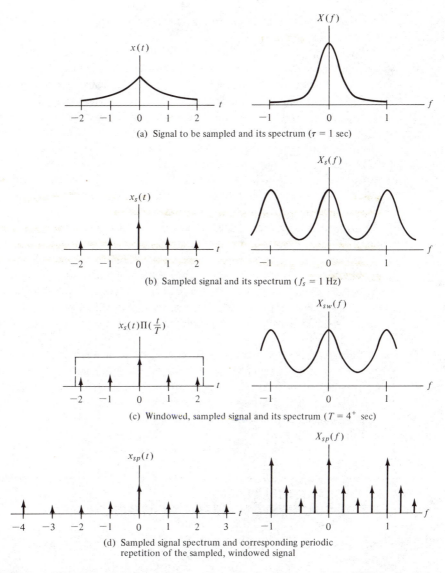

(a) Signal to be sampled and its spectrum ($\tau = 1$ sec)

(b) Sampled signal and its spectrum ($f_s = 1$ Hz)

(c) Windowed, sampled signal and its spectrum ($T = 4^+$ sec)

(d) Sampled signal spectrum and corresponding periodic
repetition of the sampled, windowed signal

FIGURE 9-2.   Signals and spectra illustrating the derivation of the DFT from the Fourier transform.

This produces a periodic sampled sequence, $x_{sp}(t)$, in the time domain. The resultant sampled signal and its spectrum are shown in Figure 9-2d.   ■

Over the years, a descriptive terminology has come into common use to denote the effects illustrated in Figure 9-2. Because of sampling, the spectrum of the analog signal is repeated about integer multiples of the sampling frequency to produce the spectrum of the sampled signal. The resultant spectrum of the sampled signal therefore consists of a superposition of the analog signal spectrum and its translates in the sampling frequency. The overlapping of the analog signal spectrum with its translates, which results in the spectrum shown in Figure 9-2b, is called *aliasing*.

Windowing the samples of the signal in the time domain corresponds to convolution of the sampled-signal spectrum with the Fourier transform of the window that produces the spectrum of Figure 9-2c. Thus the value of the spectrum for the windowed, sampled signal at a given frequency really is the result of integrating over all spectral components of the windowed spectrum after weighting by the Fourier transform of the rectangular window. Thus a large spectral component in the periodic spectrum shown in Figure 9-2b, say at $f = 1$ Hz, can have a large influence on the spectrum at some other frequency, say $f = 0.5$ Hz, of the windowed signal. This effect is referred to as the *leakage effect*, which is descriptive of the fact that spectral energy "leaks" from one frequency to another as a result of the windowing.

The final effect to be discussed is referred to is the *picket-fence effect*. It simply refers to the fact that the spectrum of the sampled, windowed signal is viewed at a discrete set of frequencies. In effect, the spectrum is viewed through a picket fence. Figure 9-2d illustrates a rather fortunate state of affairs in that the peaks and valleys of the spectrum $X_{sw}(f)$ are viewed through the picket fence of impulses. Had there been a large spectral component or a spectral null at, say, $f = 0.375$ Hz, its presence would not be apparent from the spectral samples shown in Figure 9-2d.

With all these effects being present any time an analog signal is sampled and the DFT of a finite sequence of samples taken, it is indeed surprising that the DFT of the samples comes anywhere close to approximating the spectrum of the original signal. Various measures can be taken to minimize the detrimental effects of aliasing, leakage, and the picket-fence effect. These cures are outlined in Table 9-1.

9-4

INTRODUCTION TO FFT (FAST FOURIER TRANSFORM) ALGORITHMS†

FFT algorithms are efficient means of computing the sum given in (9-4) which we now write as‡

$$X(n) = \sum_{k=0}^{N-1} x(k)W_N^{kn}, \qquad n = 0, 1, \ldots, N - 1 \qquad (9\text{-}15)$$

where

$$W_N = \exp\left[\frac{-j2\pi}{N}\right] \qquad (9\text{-}16)$$

---

†The material in this section gives an introductory discussion of the FFT algorithm which follows that given in G. D. Bergland, "A Guided Tour of the Fast Fourier Transform," *IEEE Spectrum*, Vol. 6, July 1969, p. 41–52. Section 9.5 gives a mathematical derivation of two FFT algorithms.

‡In order to simplify notation we now replace the frequency samples X_n by $X(n)$ and the time smples $x_s(k \Delta t)$ by $x(k)$.

TABLE 9-1 Minimizing DFT Error Effects

Problem	Possible Cure
1. Excessive errors due to aliasing	a. Increase the sampling rate
	b. Prefilter the signal to minimize high-frequency spectral content
2. Spectral distortion due to leakage	a. Increase the window width by increasing the number of DFT points
	b. Use window functions that have Fourier transforms with low sidelobes (discussed in Section 9-9)
	c. If large periodic components are present in the signal, eliminate these before windowing by filtering
3. Picket-fence effect results in important spectral components being missed	a. Increase the number of DFT points while holding the sampling rate fixed. This places "pickets" closer together
	b. If the data length is limited, fill in additional DFT points with zeros (called *zero padding*)

Generally, two approaches can be taken to the development of the FFT. These are decimation in time and decimation in frequency. For N a power of 2, the decimation in time algorithm results by representing n and k as binary numbers and separating the sum over k into separate sums over the binary coefficients. For example, for $N = 4$, we represent k and n as†

$$k = 2k_1 + k_0$$

$$n = 2n_1 + n_0$$

where

$$k = \{(k_1 k_0)\} = \{00, 01, 10, 11\}$$

and

$$n = \{(n_1 n_0)\} = \{00, 01, 10, 11\}$$

If the original time-domain samples, $x(k)$, after representing k in binary form, are denoted $x_0(k_1 k_0)$, (9-15) becomes

$$X(n_1 n_0) = \sum_{k_0=0}^{1} \sum_{k_1=0}^{1} x_0(k_1 k_0) W_4^{(2n_1+n_0)(2k_1+k_0)}$$

$$= \sum_{k_0=0}^{1} \left[\sum_{k_1=0}^{1} x_0(k_1 k_0) W_4^{2n_0 k_1} \right] W_4^{(2n_1+n_0)k_0} \qquad (9\text{-}17)$$

†The values that k and n may assume are written as binary numbers; $k = 11$ is shorthand notation for $k = (11)_2 = 1 \times 2^1 + 1 \times 2^0 = (3)_{10}$.

since $W_4^{4n_1k_1} = 1$. The sum in brackets will be a function of n_0 and k_0 since k_1 is the index of summation. Thus let

$$x_1(n_0k_0) = \sum_{k_1=0}^{1} x_0(k_1k_0)W_4^{2n_0k_1} \qquad (9\text{-}18)$$

which yields

$$X(n_1n_0) = \sum_{k_0=0}^{1} x_1(n_0k_0)W_4^{(2n_1+n_0)k_0} = x_2(n_0n_1) \qquad (9\text{-}19)$$

Note that in our notation, $x_j(\cdot)$, the value of j is 0, 1, or 2. The value of j is incremented by one each time a sum over a binary index is evaluated. Each incrementation of j is known as a *recursion* or *stage*. Thus the evaluation of an $N = 4$ point DFT requires two recursions.

It is often convenient to represent the two summations required for the evaluation of (9-17) in matrix form. Writing out (9-18) for $n_0 = 0$, 1 and $k_1 = 0$, 1 illustrates that x_0 and x_1 are related by the matrix operation

$$\begin{bmatrix} x_1(00) \\ x_1(01) \\ x_1(10) \\ x_1(11) \end{bmatrix} = \begin{bmatrix} 1 & 0 & W_4^0 & 0 \\ 0 & 1 & 0 & W_4^0 \\ 1 & 0 & W_4^2 & 0 \\ 0 & 1 & 0 & W_4^2 \end{bmatrix} \begin{bmatrix} x_0(00) \\ x_0(01) \\ x_0(10) \\ x_0(11) \end{bmatrix} \qquad (9\text{-}20)$$

In a similar manner, $x_2(\cdot)$ is derived from $x_1(\cdot)$ by the matrix operation

$$\begin{bmatrix} x_2(00) \\ x_2(01) \\ x_2(10) \\ x_2(11) \end{bmatrix} = \begin{bmatrix} 1 & W_4^0 & 0 & 0 \\ 1 & W_4^2 & 0 & 0 \\ 0 & 0 & 1 & W_4^1 \\ 0 & 0 & 1 & W_4^3 \end{bmatrix} \begin{bmatrix} x_1(00) \\ x_1(01) \\ x_1(10) \\ x_1(11) \end{bmatrix} \qquad (9\text{-}21)$$

As shown by (9-19), the frequency samples, $X(n_1n_0)$ are obtained from $x_2(n_0n_1)$ by reversing the order of the binary coefficients n_1 and n_0. This is the original Cooley–Tukey formulation of the FFT algorithm.

The flow graph shown in Figure 9-3a illustrates (9-18) and (9-19) diagrammatically. The dashed lines indicate the reversal of n_0 and n_1 that must take place in order to get $X(n)$ in sequential order at the output. By rearranging this flow graph, we can reorder the input in bit-reversed order to get natural order at the output. This results in the flow graph shown in Figure 9-3b.

Note that a pair of nodes, referred to as *dual nodes*, are involved in a basic computation of the form

$$X_{m+1}(p) = X_m(p) + W_N^s X_m(q) \qquad (9\text{-}22)$$

$$X_{m+1}(q) = X_m(p) - W_N^s X_m(q)$$

to go from the mth recursion to the $(m + 1)$st recursion. This is called a *butterfly* computation. The integers p and q denote the sample numbers involved in the computation and s is an integer power of W_N.

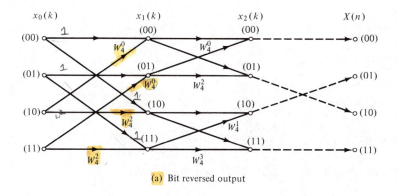

(a) Bit reversed output

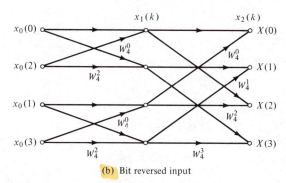

(b) Bit reversed input

FIGURE 9-3. Flow diagrams for a four-point FFT.

Another form of the FFT algorithm can be obtained by writing (9-15) with $N = 4$ as

$$X(n_1 n_0) = \sum_{k_0=0}^{1} \left[\sum_{k_1=0}^{1} x_0(k_1 k_0) W_4^{2n_0 k_1} W_4^{n_0 k_0} \right] W_4^{2n_1 k_0} \qquad (9\text{-}23)$$

With $x_1(\cdot)$ and $x_2(\cdot)$ identified as

$$x_1(n_0 k_0) = \sum_{k_1=0}^{1} x_0(k_1 k_0) W_4^{(2n_0 k_1 + n_0 k_0)}, \qquad n_0, k_0 = 0, 1$$

$$x_2(n_0 n_1) = \sum_{k_0=0}^{1} x_1(n_0 k_0) W_4^{2n_1 k_0}, \qquad n_0, n_1 = 0, 1$$

respectively, and

$$X(n_1 n_0) = x_2(n_0 n_1)$$

the algorithm corresponding to the signal flow graph illustrated in Figure 9-4 results. This algorithm is known as the *Sande–Tukey algorithm.*
    The butterfly computation is now of the form

$$X_{m+1}(p) = X_m(p) + X_m(q)$$
$$X_{m+1}(q) = [X_m(p) - X_m(q)] W_N^s \qquad (9\text{-}24)$$

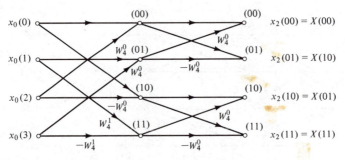

FIGURE 9-4   Flow graph for Sande–Tukey Algorithm.

In either case, each FFT recursion requires $N/2$ butterfly computations. The integer s varies with p, q, and m in a manner that depends on the FFT algorithm used. There are $\log_2 N$ recursions per FFT for a total of $(N/2) \log_2 N$ butterflies per FFT.

For future reference, the flow graph for the Cooley–Tukey algorithm, with $N = 8$, is shown in the Figure 9-5. This flow graph corresponds to a set of equations of the form of (9-18) generalized to the case of three nested sums. A considerable computational savings is facilitated by noting the simplifications in computing powers of W_N and in applying the series in sequence. Even more savings in computation can be achieved by noting the symmetry properties of W_N^m as shown in Table 9-2 for $N = 8$. The fact that $W_N^m = -W_N^{(m + N/2)}$ has been used in constructing the butterflies for the flow graph of Figure 9-5 [also see (9-22)].

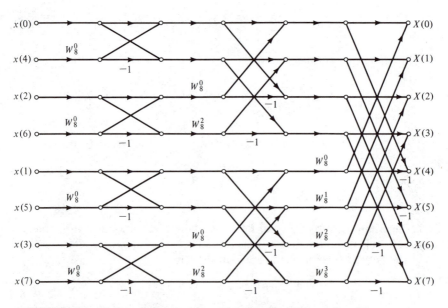

FIGURE 9-5.   Flow graph for eight-point Cooley–Tukey algorithm.

TABLE 9-2. Symmetry Properties of W_N^m for $N = 8$

$m =$	W_8^m	
0	1	
1	$\dfrac{1 - j}{\sqrt{2}}$	
2	$-j$	
3	$-\dfrac{(1 + j)}{\sqrt{2}}$	
4	-1	$= -W^0$
5	$-\dfrac{(1 - j)}{\sqrt{2}}$	$= -W^1$
6	j	$= -W^2$
7	$\dfrac{1 + j}{\sqrt{2}}$	$= -W^3$

The procedure used in evaluating a spectrum by means of an eight-point FFT will now be illustrated by example.

EXAMPLE 9-2

Consider the pulse signal with sample values

$$x(000) = x(001) = x(010) = 1$$

$$x(011) = x(100) = x(101) = 0$$

$$x(110) = x(111) = 1$$

This signal is shown in Figure 9-6. The FFT is developed according to the equations

$$x_1(n_0 k_1 k_0) = \sum_{k_2=0}^{1} x(k_2 k_1 k_0) W_8^{4n_0 k_2} \qquad (9\text{-}25a)$$

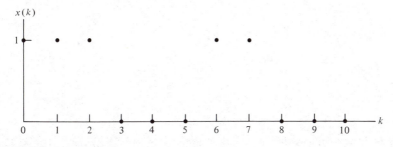

FIGURE 9-6. Sample-data signal for Example 9-2.

$$x_2(n_0 n_1 k_0) = \sum_{k_1=0}^{1} x_1(n_0 k_1 k_0) W_8^{2(2n_1 + n_0)k_1} \qquad (9\text{-}25b)$$

and

$$X(n) = x_3(n_0 n_1 n_2) = \sum_{k_0=0}^{1} x_2(n_0 n_1 k_0) W_8^{(4n_2 + 2n_1 + n_0)k_0} \qquad (9\text{-}25c)$$

The calculation of $x(000)$, $x(001)$, and $x(010)$ is illustrated in Figure 9-7. These figures illustrate how the flow graph shown in Figure 9-5 develops. In these figures, the numbers at the arrow tails are multiplied by the numbers along the respective arrow and summed with numbers computed from other arrows converging on a given point. By continuing the procedure illustrated in Figure 9-7, the entire flow graph shown in Figure 9-5 could be developed. ∎

MATHEMATICAL DERIVATION OF THE FFT

In this section we provide a more mathematical derivation of the decimation-in-time and decimation-in-frequency FFT algorithms discussed in Section 9-4.

Decimation in Time

Consider the DFT sum of (9-15) with the summation carried out separately over the even- and odd-indexed terms in the sum. Letting $k = 2r$ for the even-

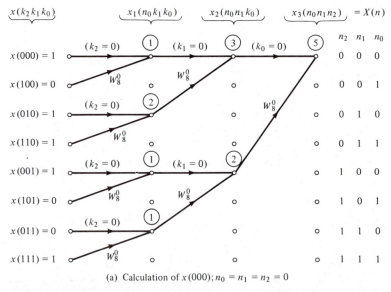

(a) Calculation of $x(000)$; $n_0 = n_1 = n_2 = 0$

FIGURE 9-7. Computations involved in computing $x(000)$, $x(001)$, and $x(010)$ for the signal shown in Figure 9-6 (note that all node indices are bit reversed).

THE DISCRETE FOURIER TRANSFORM AND
FAST FOURIER TRANSFORM ALGORITHMS

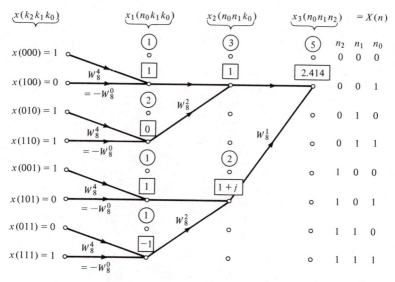

(b) Calculation of $x(001)$; $n_0 = 1$, $n_1 = n_2 = 0$

FIGURE 9-7b.

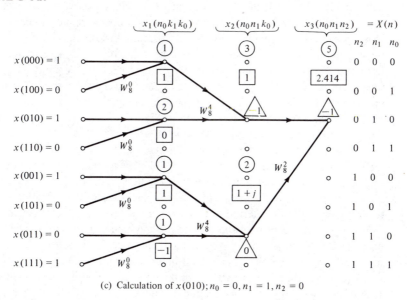

(c) Calculation of $x(010)$; $n_0 = 0$, $n_1 = 1$, $n_2 = 0$

FIGURE 9-7c.

indexed terms and $k = 2r + 1$ for the odd-indexed terms, we may write the DFT of $x(k)$ as

$$X(n) = \sum_{r=0}^{(N/2)-1} x(2r)\, W_N^{2rn} + \sum_{r=0}^{(N/2)-1} x(2r + 1) W_N^{(2r + 1)n}$$

$$= \sum_{r=0}^{(N/2)-1} x(2r) W_N^{2rn} + W_N^n \sum_{r=0}^{(N/2)-1} x(2r + 1) W_N^{2rn}$$

(9-26)

Now

$$W_N^{2rn} = e^{-j(2\pi/N)(2rn)} = e^{-j(2\pi/(N/2))(rn)} = W_{N/2}^{rn} \tag{9-27}$$

which allows (9-26) to be written as

$$X(n) = \sum_{r=0}^{(N/2)-1} x(2r)\, W_{N/2}^{rn} + W_N^n \sum_{r=0}^{(N/2)-1} x(2r+1)W_{N/2}^{rn}$$
$$= G(n) + W_N^n\, H(n), \qquad n = 0, 1, \ldots, N-1 \tag{9-28}$$

where

$$G(n) = \sum_{r=0}^{(N/2)-1} x(2r)W_{N/2}^{rn} \tag{9-29}$$

and

$$H(n) = \sum_{r=0}^{(N/2)-1} x(2r+1)\, W_{N/2}^{rn} \tag{9-30}$$

are $(N/2)$-point DFTs of the even- and odd-indexed points of the sequence $x(k)$, respectively. Although the frequency index n ranges over N values from 0 to $N-1$, only $N/2$ values of $G(n)$ and $H(n)$ need to be computed since $G(n)$ and $H(n)$ are periodic in n with period $N/2$. Assuming that an algorithm is available for computing an $(N/2)$-point DFT, we may combine the $(N/2)$-point DFTs, $G(n)$ and $H(n)$, in accordance with (9-28) to provide the N-point DFT of the time sequence $x(k)$. This is illustrated in Figure 9-8a. Furthermore, we can break up each of the $(N/2)$-point DFTs shown in Figure 9-8a into two $(N/4)$-point DFTs, which results in the flow graph of Figure 9-8b. Note that in combining the sums for $G(n)$ and $H(n)$, the relation $W_{N/2}^r = W_N^{2r}$ is used.

The decimation procedure can be continued until a series of $N/2$ two-point DFTs results for the first stage of the N-point DFT computation. The flow graph that results for $N = 8$ is precisely that of Figure 9-9. Since this flow graph was arrived at by separating the input samples into successively smaller sets, the resulting algorithm is referred to as a *decimation-in-time* (DIT) algorithm.

We note that the bit-reversed ordering of the input samples results because at each stage of decimation the sequence of input samples is separated into even- and odd-indexed samples. Each even-indexed sample for the first decimation has a least significant bit (LSB) of zero, while each odd-indexed point has an LSB of one. On the second decimation it is the second-most LSB that determines whether a sequence number is included in the even- or odd-indexed DFT, and so on until the last decimation takes place whereupon it is the most significant bits of the sequence-member indices that determine what DFT sum the sequence members are associated with. What has just been described, therefore, is an ordering of the input samples in bit-reversed order.

Decimation in Frequency

We now consider the DFT sum (9-15) with the sum carried out over the first half and the last half of the input samples separately. This results in the sums

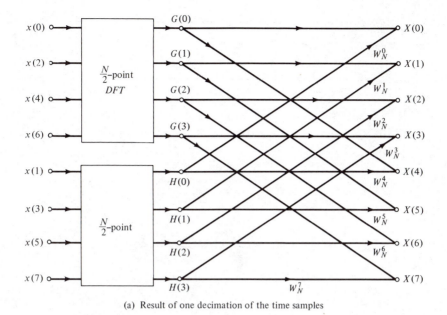

(a) Result of one decimation of the time samples

(b) Result of applying two decimations

FIGURE 9-8. Flow graphs showing the decimation-in-time decomposition of an N-point DFT computation (N = 8).

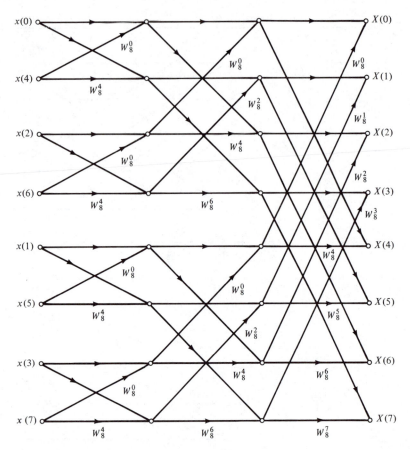

FIGURE 9-9. Complete flow graph for an FFT developed by applying decimation in time ($N = 8$).

$$X(n) = \sum_{k=0}^{(N/2)-1} x(k)W_N^{kn} + \sum_{k=N/2}^{N-1} x(k)W_N^{kn}$$

$$= \sum_{k=0}^{(N/2)-1} x(k)W_N^{kn} + W_N^{(N/2)n} \sum_{m=0}^{(N/2)-1} x\left(m + \frac{N}{2}\right)W_N^{mn} \qquad (9\text{-}31)$$

where the substitution $m = k - N/2$ has been made to obtain the second sum of the second equation above. We may use the fact that $W_N^{(N/2)n} = (-1)^n$ to combine the two sums in (9-31) to obtain

$$X(n) = \sum_{k=0}^{(N/2)-1}\left[x(k) + (-1)^n x\left(k + \frac{N}{2}\right)\right]W_N^{nk} \qquad (9\text{-}32)$$

We now consider n even and odd separately. Letting $n = 2r$ and $n = 2r + 1$, respectively, to indicate the even- and odd-indexed output points, we have

$$X(2r) = \sum_{k=0}^{(N/2)-1}\left[x(k) + x\left(k + \frac{N}{2}\right)\right]W_N^{2rk} \qquad (9\text{-}33)$$

and

$$X(2r + 1) = \sum_{k=0}^{(N/2)-1} \left[x(k) - x\left(k + \frac{N}{2}\right) \right] W_N^k W_N^{2rk},\qquad (9\text{-}34)$$

$$r = 0, 1, \ldots, \frac{N}{2} - 1$$

Since $W_N^{2rk} = W_{N/2}^{rk}$, as noted previously, we recognize (9-33) and (9-34) as $(N/2)$-point DFTs. The flow graph for obtaining $X(n)$ in this fashion is shown in Figure 9-10. This process can be used to replace each $(N/2)$-point DFT by two $(N/4)$-point DFTs, and so on until only two points are left in each DFT. The resulting flow graph, shown in Figure 9-11, is precisely the reverse of Figure 9-9, with all arrows reversed and input and output interchanged. Since the output, or frequency, points were subdivided to obtain this algorithm, it is referred to as a *decimation-in-frequency* (DIF) FFT algorithm.

9-6

PROPERTIES OF THE DFT

As in the case of the Fourier transform, several useful properties can be proved for the DFT. The proofs of these properties may be carried out in a similar fashion to the proofs of the analogous properties of the Fourier transform by using the defining relationships for the DFT given by (9-4) and (9-5). Proofs of the properties stated below are left to the problems. The following notation will

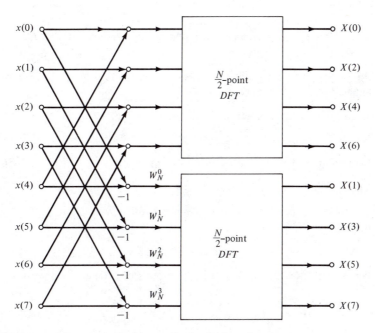

FIGURE 9-10. Flow graph after a single decimation.

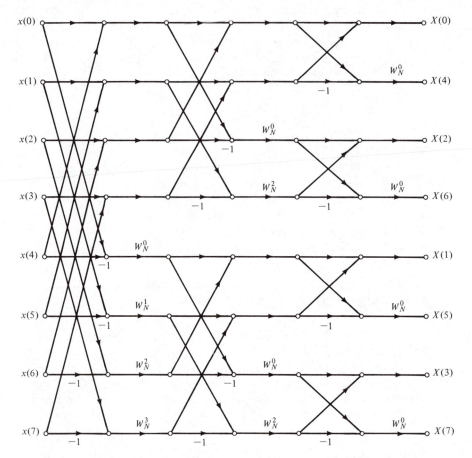

FIGURE 9-11. Flow graph pertaining to decimation in frequency FFT ($N = 8$).

be employed in stating the properties of the DFT, where all sequences are assumed periodic with period N:

Discrete-time sequences are denoted as $x(k)$ and $y(k)$.

Their DFTs are denoted as $X(n)$ and $Y(n)$.

N is the length of the sequence or size of the DFT.

A and B are arbitrary constants.

The subscript e denotes a sequence that is even about the point $(N - 1)/2$ for N even and $N/2$ for N odd.

The subscript o denotes a sequence that is odd.

Any real sequence can be expressed in terms of its even and odd parts according to

$$x(k) = x_e(k) + x_0(k)$$

$$= \tfrac{1}{2} [x(k) + x(N - k)] + \tfrac{1}{2} [x(k) - x(N - k)]$$

(9-35)

The subscript r denotes a real sequence (or the real part of a complex sequence).

The subscript i similarly denotes the imaginary part of a complex sequence.

A double-headed arrow ($\leftrightarrow$) denotes a discrete-Fourier transform pair: for example, $x(k) \leftrightarrow X(n)$.

With this notation, the following properties of the DFT may be stated.

1. Linearity:

$$Ax(k) + By(k) \leftrightarrow AX(n) + BY(n)$$

2. Time shift:

$$x(k - m) \leftrightarrow X(n) \exp(-j2\pi nm/N) = X(n)W_N^{nm}$$

3. Frequency shift:

$$x(k) \exp(j2\pi km/N) \leftrightarrow X(n - m)$$

4. Duality:

$$N^{-1}X(k) \leftrightarrow x(-n)$$

5. Circular convolution:†

$$\sum_{m=0}^{N-1} x(m)y(k - m) \triangleq x(k) \ \textcircled{N} \ y(k) \leftrightarrow X(n)Y(n)$$

6. Multiplication:

$$x(k)y(k) \leftrightarrow N^{-1} \sum_{m=0}^{N-1} X(m)Y(n - m) \triangleq N^{-1}X(n) \ \textcircled{N} \ Y(n)$$

7. Parseval's theorem:

$$\sum_{k=0}^{N-1} |x(k)|^2 = N^{-1} \sum_{n=0}^{N-1} |X(n)|^2$$

8. Transforms of even, real functions:

$$x_{er}(k) \leftrightarrow X_{er}(n)$$

(The DFT of an even, real sequence is even and real.)

9. Transforms of odd, real functions:

$$x_{or}(k) \leftrightarrow jX_{oi}(m)$$

(The DFT of an odd, real sequence is odd and imaginary.)

10. Assume that $x(k)$ and $y(k)$ are the real and imaginary parts, respectively, of a complex sequence $z(k)$; that is,

$$z(k) = x(k) + jy(k)$$

---

†The circular convolution operation can be viewed as placing the N samples of $\{x(k)\}$ on the circumference of a cylinder and the N samples of $\{y(k)\}$ in reverse order on another cylinder, sliding one cylinder inside the other, while multiplying coincident samples, and adding the products.

Then

$$z(k) \leftrightarrow Z(n) = X(n) + jY(n)$$

where

$$x(k) \leftrightarrow X(n)$$

and

$$y(k) \leftrightarrow Y(n)$$

Note that $X(n)$ and $Y(n)$ are, in general, complex. Thus we cannot simply take real and imaginary parts of $Z(n)$ to obtain the separate DFTs of $x(n)$ and $y(n)$. However, using properties 8 and 9, we can show that

$$Z_r(n) = X_{er}(n) - Y_{oi}(n)$$

and

$$Z_i(n) = X_{oi}(n) + Y_{er}(n)$$

where $X_{er}(n)$ is the even, real part of $X(n)$
$X_{oi}(n)$ is the odd, imaginary part of $X(n)$
$Y_{er}(n)$ is the even, real part of $Y(n)$
$Y_{oi}(n)$ is the odd, imaginary part of $Y(n)$

Writing $X(n)$ in terms of its even and odd parts, we have

$x(k) \leftrightarrow X(n) =$ even part of real part of $z(n) +$
j[odd part of imaginary part of $Z(n)$]

$y(k) \leftrightarrow Y(n) = -$[odd part of real part of $Z(n)$] $+$
j[even part of imaginary part of $Z(n)$]

This property is convenient for simultaneously obtaining the DFT of two real sequences and then separating their DFTs. Since most FFT algorithms can provide the DFT of a complex sequence, it is possible to save computation time in this fashion when a real sequence is to be transformed.

It is important to think of the properties listed above as applying to periodic sequences. To illustrate the impact of this, we consider two examples, one involving property 2 and the other involving property 5.

EXAMPLE 9-3
Consider the DFT of

$$x(k) = \delta(k)$$

which is

$$X(n) = \sum_{k=0}^{N-1} \delta(k)W_N^{nk} = 1, \qquad n = 0, 1, \ldots, N - 1$$

Using the time shift property we find that

$$\text{DFT}[x(k - k_0)] = \exp\left(\frac{-j2\pi nk_0}{N}\right) = W_N^{nk_0}$$

Writing out the DFT sum for $x(k - k_0) = \delta(k - k_0)$, we have

$$\text{DFT}[\delta(k - k_0)] = \sum_{k=0}^{N-1} \delta(k - k_0)W_N^{nk} = W_N^{nk_0}$$

which is the same as obtained using the time shift property. Note that the DFT is periodic with period N. ∎

EXAMPLE 9-4

Consider the circular convolution of

$$x_1(k) = x_2(k) = 1, \qquad 0 \le k \le N - 1$$

which is denoted $x_{3c}(k)$. The DFT of each of these sequences is

$$X_1(n) = X_2(n) = \sum_{k=0}^{N-1} W_N^{nk}$$

$$= \begin{cases} N, & n = 0 \\ 0, & n = 1, 2, \ldots, N - 1 \end{cases}$$

where use has been made of (9-8). The *circular convolution* of $x_1(k)$ and $x_2(k)$ is the inverse DFT of

$$X_3(n) = X_1(n)X_2(n) = \begin{cases} N^2, & n = 0 \\ 0, & n = 1, 2, \ldots, N - 1 \end{cases}$$

which is

$$x_{3C}(k) = \frac{1}{N} \sum_{n=0}^{N-1} N^2 \delta(n)W_N^{-nk} = N, \qquad k = 0, 1, \ldots, N - 1$$

This is illustrated in Figure 9-12a for $N = 4$.

The *linear convolution*, denoted $x_{3L}(k)$, of the two sequences $x_1(k)$ and $x_2(k)$ is the triangular envelope sequence given by

$$x_{3L}(k) = \begin{cases} k + 1, & 0 \le k \le N \\ 2N - k - 1, & N \le k \le 2N \\ 0, & \text{otherwise} \end{cases}$$

as illustrated in Figure 9-12b. The student can readily verify this result by considering $\{x_1(k)\}$ and $\{x_2(k)\}$ as infinite sequences which are zero for $n < 0$ and $n > N$.

The reason for the difference in the circular and linear convolution results is that in the former case aliasing takes place as a result of the sequences $x_1(k)$ and $x_2(k)$ being treated as periodic sequences because of the DFT operation. To avoid this aliasing, $x_1(k)$ and $x_2(k)$ each can be augmented with N points that are zero to obtain the sequences

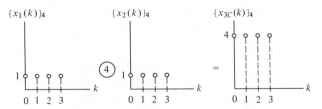

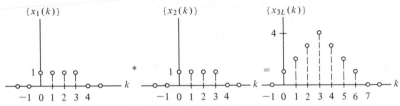

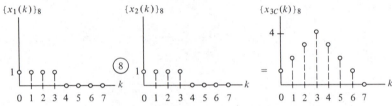

(a) The circular convolution of two four–point sequences whose members are unity is itself a four–point sequence whose members are equal because of aliasing

(b) The linear convolution of two sequences consisting of four equal–value samples preceded and followed by zeros is a triangular envelope sequence

(c) The use of circular convolution to calculate the linear convolution of two sequences requires zero padding with the same number of zeros as non–zero sequence members

FIGURE 9-12. Comparison of circular and linear convolution using the sequences of Example 9-4. In the figures, $\{\cdot\}_m$ indicates a sequence of length m.

$$x_1(k) = x_2(k) = \begin{cases} 1, & 0 \le n \le N - 1 \\ 0, & N \le n \le 2N - 1 \end{cases}$$

The resulting circular convolution is then a triangular-envelope sequence, as illustrated in Figure 9-12c. Any fewer than $N - 1$ zeros would result in aliasing.

■

9-7

COMPARISON OF THE NUMBER OF COMPUTATIONS BETWEEN THE FFT AND DFT

By examining (9-4), we see that $(N - 1)$ complex multiplications and $(N - 1)$ complex additions are required to compute each output point for the DFT (the first term in the sum involves exp $(j0) = 1$ and therefore does not require a multiplication). Thus, to compute N output points, we require $N(N-1)$ complex multiplications and the same number of complex additions. Now each complex multiplication is of the form

$$(A + jB)(C + jD) = (AC - BD) + j(BC + AD) \qquad (9\text{-}36)$$

and therefore requires four real multiplications and two real additions. Hence computation of all the output points for an N-point DFT requires $4N(N-1)$ real multiplications and $4N(N-1)$ real additions.

To compare the number of computations required for the FFT with the number required for the DFT, we note that an N-point FFT, where N is a power of 2, requires $\log_2 N$ recursions which are each composed of $N/2$ butterflies. Regardless of whether the Cooley–Tukey or Sande–Tukey algorithm is used, each butterfly involves one complex multiplication and two complex additions, or four real multiplications and six real additions, Thus each FFT requires $2N \log_2 N$ real multiplications and $3N \log_2 N$ real additions. Actually, the number of multiplications is less than this because several of the multiplications by W^r are really multiplications by ± 1 or $\pm j$, which are not multiplications at all.

We can get a rough comparison of the speed advantage of an FFT over a DFT by computing the number of multiplications for each since these are usually more time consuming than additions. The comparison is given in Table 9-3.

9-8

APPLICATIONS OF THE FFT

Several applications of the FFT are discussed in this section. These include filtering, convolution, power spectrum estimation, system identification and deconvolution.

Filtering

The example given below illustrates the application of the FFT to filtering.

TABLE 9-3. Comparison of the Number of Real Multiplications for the DFT and FFT

Number of Points	DFT	FFT	Speed Factor
2	8	4	2
4	48	16	3
8	224	48	5
16	960	128	8
32	3968	320	12
64	16,128	768	21
128	65,024	1,792	36
256	261,120	4,096	64
512	1,046,528	9,216	114
1024	4,190,208	20,480	205
2048	16,769,024	45,056	372
4096	67,092,480	98,304	683

EXAMPLE 9-5

We first find the eight-point FFT of the sum of two cosine sequences

$$x(k) = \cos \frac{\pi}{4} k + \cos \frac{\pi}{2} k \qquad (9\text{-}37)$$

which is sketched in Figure 9-13. The numbers corresponding to a DIT computation of the FFT for this signal are given in Figure 9-14. Next, we assume that all frequency components of the spectrum above $n = 1$ are eliminated [note that this entails setting $x(n) = 0$, $2 \le n \le 6$]. The inverse FFT of this result is computed as shown in Figure 9-15. The result, when scaled by $N = 8$, is a cosine sequence corresponding to the first term of the input sequence (9-37). This series of operations therefore corresponds to low-pass filtering.

This example illustrates the implementation of filtering by using the FFT to digitally carry out the operations expressed by (3-85). In implementing digital filters by this method, care must be exercised to invoke the proper symmetry conditions on the transfer function, $H(n)$, as expressed by (3-87) and (3-88). In the discrete-time signal case, however, the center of symmetry is the point $n = N/2$. ∎

Another problem that arises is leakage, which can be combatted by proper window selection. This is discussed in more detail in Section 9-9.

A problem related to window selection is aliasing in the time domain. When an ideal filter, such as in Example 9-5, is implemented in the frequency domain, a $(\sin x)/x$ impulse response as expressed by (3-94) is implied in the time domain. In such cases, the impulse response does not go to zero, which results in the tails of the $(\sin x)/x$ impulse response folding back into the N-point region included in the inverse sum (9-5). The result is aliasing in the time domain caused by undersampling the filter transfer function in the frequency domain. A cure for this problem is to choose a filter with an impulse response that dies out or can be truncated so that at least half of its terms are essentially zero, as illustrated by Example 9-4.

Convolution

When convolutions are implemented by means of the FFT, both the signals being convolved are treated as periodic. This was illustrated by Example 9-4. The result is a circular, or cyclical, convolution which is entirely different from the noncyclical convolution desired in order to avoid aliasing.

The problem is easily avoided in computing the cyclical convolution of two N-point sequences, say $\{g(k)\}$ and $\{h(k)\}$, by defining $\{\hat{g}(k)\}$ and $\{\hat{h}(k)\}$ as

$$\hat{g}(k) = \begin{cases} g(k), & 0 \le k \le N - 1 \\ 0, & N \le k \le 2N - 1 \end{cases} \qquad (9\text{-}38)$$

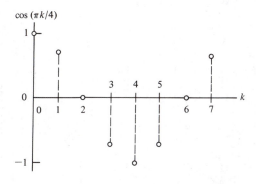

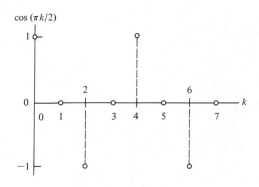

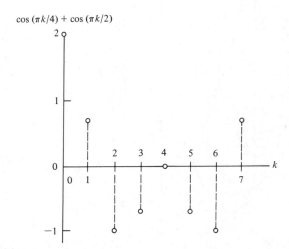

FIGURE 9-13. Signal sequence to illustrate filtering by means of the FFT.

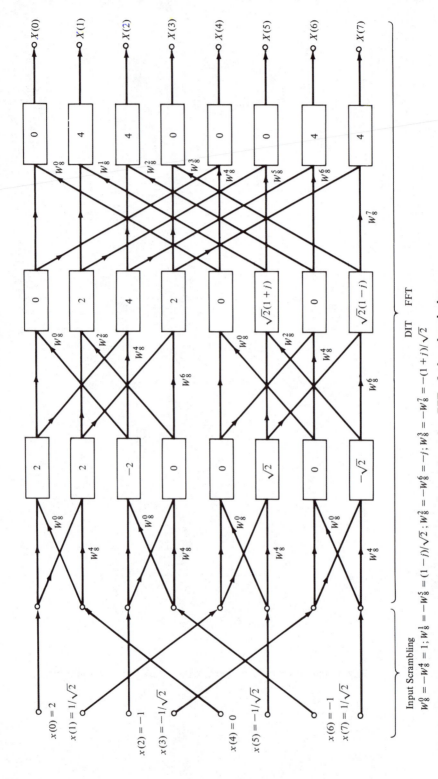

Input Scrambling

$w_8^0 = -w_8^4 = 1; w_8^1 = -w_8^5 = (1-j)/\sqrt{2}; w_8^2 = -w_8^6 = -j; w_8^3 = -w_8^7 = -(1+j)/\sqrt{2}$

DIT FFT

FIGURE 9-14. FFT flow graph calculation for eight-point FFT of the signal given by (9-37).

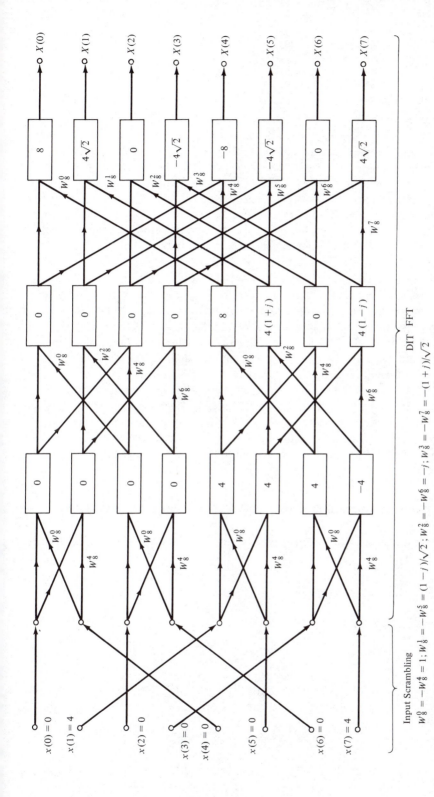

FIGURE 9-15. FFT flow graph computation of the inverse FFT of spectrum computed in Figure 9-14 with x(2) = x(6) = 0.

and

$$\hat{h}(k) = \begin{cases} h(k), & 0 \le k \le N - 1 \\ 0, & N \le k \le 2N - 1 \end{cases} \tag{9-39}$$

respectively.

If one of the sequences to be convolved is much longer than the other, transforming the entire sequence corresponding to both functions can be avoided by convolving the $N = 2M$-point zero-padded version of the shorter one, with M non-zero points and M points zero, with overlapping $2M$-point segments of the longer one. This is illustrated in Figure 9-16. The shorter sequence, designated $\{\hat{h}(k)\}$, can be thought of as the impulse response of a filter that is acting on the longer sequence, designated $\{g(k)\}$. The present output of the filter is simply a weighted sum of the last M samples it has seen. In the example of Figure 9-16, this means that the $2M$-term sequence $\{\hat{h}(k)\}$ convolved with one of the $2M$-term segments of $g(k)$ would result in M lags for which the correct M-term history is not available, and M lags for which the correct M-term history is available. Thus the first M lags computed are incorrect and the last M lags are correct. Only the right-most M points are kept, and the next convolution of $\{\hat{h}(k)\}$ with a segment of $\{g(k)\}$ is carried out with M new points and M old points. In each convolution, only the rightmost M points are saved. When all of these sets of M points are pieced together, the result is the desired convolution of $\{g(k)\}$ and $\{h(k)\}$. This procedure avoids having to pad out the long sequence with a string of zeros, and the length of the required FFT is determined by the length of the short sequence rather than the long one.† Because of the procedure used, this technique is called the *overlap-save* method of convolution. Another commonly used method is the *overlap-add* method, for which the student is referred to the references.

Energy Spectral Density and Autocorrelation Functions

If we consider a sequence $\{x(k)\}$ of length N whose DFT is $\{X(n)\}$, its energy is defined as

$$E = \sum_{k=0}^{N-1} |x(k)|^2 \tag{9-40}$$

In the frequency domain, the energy is

$$E = \sum_{n=0}^{N-1} \frac{|X(n)|^2}{N} \tag{9-41}$$

where Parseval's theorem, property 7 of the DFT, has been used. Equation (9-41) can be interpreted as the sum of an energy spectral density function

$$S_x(n) = \frac{1}{N} X(n)X^*(n) \tag{9-42}$$

---

†Actually, if an N-point FFT is used and the short sequence is of length M, zero padding with $M - 1$ zeros is sufficient with $N - (M - 1) \ge M$ "good" points obtained for each lag.

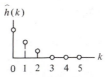

(a) Zero–padded impulse response for the desired convolution operation.

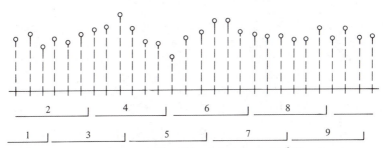

(b) Long sequence of samples to be convolved with $\hat{h}(k)$. Numbers correspond to successive segments of samples

(c) Periodic extension of time–reversed impulse response as implied by the DFT used to implement the circular convolution

FIGURE 9-16. Illustration of the overlap-save method of convolving a short sequence with a long sequence.

over all the frequency components to give total energy. Using appropriate theorems, it can be shown that

$$S_x(n) = \text{DFT}\ [R_x(m)] \tag{9-43}$$

where $R_x(m)$, defined as

$$R_x(m) = \frac{1}{N} \sum_{k=0}^{N-|m|-1} x(k)x^*(k + |m|) \tag{9-44}$$

is called the *autocorrelation function*. Equation (9-43) is known as the *Wiener–Khintchine theorem* for sample-data signals.

If $\{x(k)\}$ is a random sequence, care must be exercised in computing power spectra due to convergence properties of (9-42). One cannot blindly apply (9-42) or (9-43). A way to improve the convergence of the estimate of the energy spectral density is to employ window functions if (9-42) is used. A second way is to break $\{x(k)\}$ up into a number of shorter segments, compute the autocorrelation function of each segment according to (9-44), average these autocorrelation functions, and take the DFT to obtain an averaged approximation for the energy spectral density in accordance with (9-43).

System Identification†

We recall that the output of a linear system is given by the convolution of the input with the impulse response. Alternatively, as discussed in Chapter 3, we may obtain the Fourier transform of the output, $Y(f)$, by multiplying the transfer function, $H(f)$, by the Fourier transform of the input, $X(f)$:

$$Y(f) = H(f)X(f) \tag{9-45}$$

This frequency-domain multiplication of $H(f)$ and $X(f)$ can be approximated by employing the overlap-save method illustrated in Figure 9-16.

What if, instead of desiring the output, we wish to find the transfer function of the system? This problem is referred to as *system identification*. To see how we might accomplish this, consider the following series of operations. First, we multiply both sides of (9-45) by $X^*(f)$ to obtain

$$Y(f)X^*(f) = H(f)|X(f)|^2 \tag{9-46}$$

Dividing by the nonnegative, real function $|X(f)|^2$, we obtain

$$H(f) = \frac{Y(f)X^*(f)}{|X(f)|^2} \tag{9-47}$$

Which avoids division by a complex function. Comparing this with (9-42), we see a similarity; indeed, the quantity $(1/N)Y(f)X^*(f)$ is defined as the cross-spectrum of output with input and its inverse Fourier transform is the cross-correlation function of output with input. Thus, to ''identify'' the system, that is, determine $H(f)$, we perform the operations implied by (9-47). To obtain the impulse response [the inverse Fourier transform of (9-47)] we would, in effect, be cross-correlating the output with the normalized input $\mathcal{F}^{-1}[X^*(f)/|X(f)|^2]$. If $|X(f)|^2 = 1$, this corresponds to the operation

$$h(t) = \int_{-\infty}^{\infty} y(\lambda)x(t + \lambda) \, d\lambda \tag{9-48}$$

which is known as the *cross-correlation function* of $x(t)$ and $y(t)$.

What are the pitfalls of this operation when using the DFT? First, we must take care not to divide by zero. If $|X(f)|^2 = 0$ for any f, we must replace these values by a small quantity (perhaps the smallest number that the computer can represent). Second, if $x(t)$ and $y(t)$ are functions that are too long for the FFT operation, the segmenting procedure illustrated by Figure 9-16 must be used.

Signal Restoration or Deconvolution

Given the output of a fixed, linear system and its transfer function, we may determine the Fourier transform of its input by going through a derivation iden-

---

†All independent variables used here are represented as continuous-frequency and time variables for simplicity. Section 9-10 discusses appropriately choosing sampling rate and FFT size.

tical to the one which resulted in (9-47) except that $X(f)$ is replaced by $H(f)$. The result is

$$X(f) = \frac{Y(f)H^*(f)}{|H(f)|^2} \qquad (9\text{-}49)$$

In the time domain, this operation consists of cross-correlation of the output with a normalized impulse response of the filter given by

$$h_n(t) = \mathcal{F}^{-1}\left\{\frac{H^*(f)}{|H(f)|^2}\right\} \qquad (9\text{-}50)$$

In order to approximate this operation with the DFT, we may have to use the segmenting operation illustrated by Figure 9-16 since the output would, in general, be arbitrarily long.

9-9

WINDOWS AND THEIR PROPERTIES

The discussion given in relation to Figure 9-2 showed that sampling the spectrum of a windowed, time-domain signal implied a periodic expension of the signal. Unless the signal is periodic with an integer number of periods within the window or unless it smoothly approaches zero at each end of the interval, the resulting discontinuities generate additional spectral components. This is illustrated by Problem 9-3, part (b), wherein a noninteger frequency L for the rotating phasor generates spectral components at all DFT output frequencies. This is referred to as *spectral leakage*. The reason can be traced directly to the discontinuities implied at either end of the sampled signal. In order to minimize spectral leakage, the data samples can be multiplied by nonrectangular windows which approach zero smoothly at the beginning and end of the signal. Several window functions are available. A few common ones and their properties are listed in Table 9-4. Their effects on a cosine signal with a noninteger number of cycles within the DFT interval are shown in Figure 9-17. In general, Table 9-4 and Figure 9-17 show that it is not possible to have a narrow mainlobe and low sidelobes for the DFT of a window simultaneously. However, some windows provide a better compromise than others in trading off between these two parameters. For example, Table 9-4 shows that a Kaiser–Bessel window with $\alpha = 2.5$ has a highest sidelobe level of -57 dB and mainlobe 3-dB bandwidth of 1.57 frequency samples. A Hamming window, on the other hand, has a highest sidelobe of -43 dB and a 3-dB mainlobe width of only 1.3 frequency samples. Thus if one wishes to resolve closely spaced frequency components while minimizing leakage from one component to another, the Hamming window would be a logical choice. Alternatively, if suppression of leakage is of primary concern, a Kaiser–Bessel window would be a good choice.

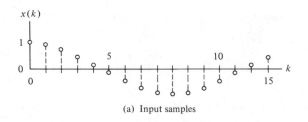

(a) Input samples

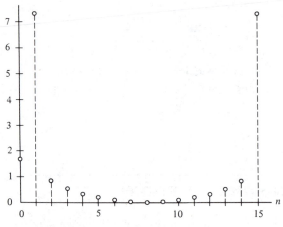

(b) Spectrum with rectangular windowing

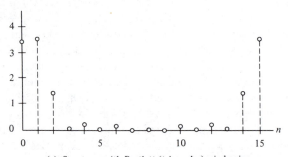

(c) Spectrum with Bartlett (triangular) windowing

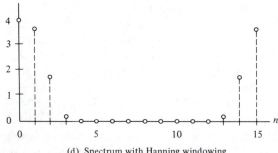

(d) Spectrum with Hanning windowing

FIGURE 9-17. **Effects of various windows on the spectrum of a sampled cosine waveform.**

TABLE 9-4. Various Windows and Their Characteristics

Window	Highest Sidelobe Level (dB)	3-dB Width Frequency (Bins)	Coherent Gain/N: $[\Sigma\ w(n)]^2/\Sigma\ w^2(n)$
1. Rectangular $w(n) = 1,\ n = 0, \ldots, N - 1$	-13	0.89	1.0
2.† Triangular $w(n) = 2(n + 0.5)/N$ $W(N - n - 1) = w(n),$ $n = 0, 1, 2, \ldots, N/2$	-27	1.28	0.75
3.† Hanning $w(n) = \frac{1}{2}\left[1 - \cos\frac{2(n + 0.5)\pi}{N}\right],$ $n = 0, 1, 2, \ldots, N - 1$	-32	1.44	0.67
4.† Hamming $w(n) = 0.54$ $-0.46\cos\frac{2\pi(n + 0.5)}{N},$ $n = 0, 1, 2, \ldots, N - 1$	-43	1.30	0.74
5. Kaiser–Bessel $w(n) = I_0\ (\pi\alpha\beta)/I_0(\pi\alpha)$ where $\beta = \sqrt{1 - \left(\frac{2n + 1}{N} - 1\right)^2}$ $n = 0, 1, 2, \ldots, N - 1$ $I_0(x) = $ modified Bessel function; order zero	$\alpha = 2;$ -46 $\alpha = 2.5;$ -57 $\alpha = 3.0;$ -69 $\alpha = 3.5$ -82	1.43 1.57 1.71 1.83	0.67 0.61 0.56 0.52

Source: F. J. Harris, "On the Use of Windows for Harmonic Analysis with the Discrete Fourier Transform," *Proc. IEEE*, Vol. 66, January 1978, pp. 51–83.

†The term $(n + 0.5)$ ensures that the window function is not zero at the first and last samples within the window, but is centered on the N samples that are windowed.

9-10

SELECTION OF PARAMETERS FOR SIGNAL PROCESSING WITH THE DFT

In processing a signal by means of the DFT or FFT, the sampling theorem requires a sampling rate of $f_s \geq 2W$ samples per second, where W is the bandwidth of the signal (assumed low-pass). Assuming a window width of T seconds and an N-point DFT, the sampling rate is $f_s = N/T$, which must satisfy

$$f_s = \frac{N}{T} \geq 2W \text{ Hz} \qquad (9\text{-}51)$$

The spacing between frequency samples is

$$\Delta f = \frac{f_s}{N} = T^{-1} \text{ Hz} \qquad (9\text{-}52)$$

which is the resolution in frequency.

Combining (9-51) and (9-52), we obtain

$$\Delta f = \frac{1}{T} \geq \frac{2W}{N} \qquad \text{Hz} \qquad (9\text{-}53)$$

Given the signal bandwidth, a desired resolution or frequency spacing dictates a required FFT size. To achieve a closer frequency spacing at an FFT output, the FFT size can be increased by adding zeros at the end of a transient input signal if necessary. This is called zero padding.

EXAMPLE 9-6

Consider a signal 1 s in duration whose bandwidth is 10 kHz. The spectrum is desired with a frequency resolution of 100 Hz or less. What should N be? Is zero padding necessary?

Solution: From (9-51) we have

$$f_s \geq 2W = 20 \text{ kHz}$$

From (9-52) we have

$$\Delta f = 100 \text{ Hz} = \frac{20{,}000}{N}$$

or

$$N = 200$$

Choosing the next largest power of 2, we let

$$N = 256$$

With $T = N/f_s = 256/20{,}000 = 1.28 \times 10^{-2}$, we see that the 1-s signal duration is more than sufficient to provide the desired resolution. Therefore, zero padding is unnecessary. ∎

EXAMPLE 9-7

The spectrum of a transient signal is to be determined by FFT analysis techniques. Its duration is 2 ms and it is determined that a sampling rate of 5kHz should be adequate to avoid significant aliasing. A spectral resolution of 100 Hz is desired. What number of FFT points must be employed? Is zero padding necessary?

Solution: The resolution desired, from (9-53), dictates that

$$T = \frac{1}{\Delta f} = 0.01 \text{ s} = 10 \text{ ms}$$

Therefore, zero padding is necessary to achieve the desired resolution since the transient signal is only 2 ms in duration. The required FFT size, from (9-51), is

$$N = f_s T$$

$$= (5 \times 10^3 \text{ Hz})(10 \times 10^{-3})$$

$$= 50$$

The next largest power of 2 is 64, so a 64-point FFT would be employed. ■

*9-11

FFT ALGORITHMS WITH ARBITRARY RADIXES†

So far we have considered FFT algorithms where the elementary computation, referred to as a butterfly, involved two input words and two output words. Such an algorithm is referred to as *radix 2*. Suppose that we desire an N-point DFT, where N is a composite number that can be factored into the product of integers

$$N = N_1 N_2 \cdots N_m \tag{9-54}$$

For example, if $N = 64$ and $m = 3$, we might factor N into the product $64 = 4 \times 4 \times 4$, and the 64-point transform can be viewed as a three-dimensional $4 \times 4 \times 4$ transform. Such an approach is sometimes taken for the following reasons. First, each elementary computation (i.e., each four-point transform in the above example) may involve fewer multiplications, additions, or memory cycles than the equivalent number of radix-2 operations. Second, this approach allows one to parallel operations using additional hardware and completely overlap memory cycles and computation cycles. Such a structure is referred to as "matched" in the sense that it is designed to be both memory limited and arithmetically limited simultaneously.

To proceed with the development of this approach, we begin by representing N by the product of two numbers:

$$N = N_1 N_2 \tag{9-55}$$

If N_1 and N_2 are both composite, they can be further broken down into the products of integers. The data to be transformed can then be arranged into a two-dimensional array and DFTs done on each row individually. Each element of the resulting array of numbers is then multiplied by a "twiddle factor," which is an appropriate power of $W_N = \exp(-j2\pi/N)$. The resultant matrix is then transformed column by column to obtain the final result.

If N is a prime number so that factorization of N is not possible, the original

---

†This section follows the theory developed in B. Gold and T. Bially, "Parallelism in Fast Fourier Transform Hardware," *IEEE Trans. Audio Electroacoust.*, Vol. AU-21, February 1973, pp. 5–16.

signal can be zero-padded and the resulting new composite number of points factored. If N can be expressed as $N = r^m$, and if the algorithm is carried out by means of a succession of r-point transforms, the resultant FFT is called a radix-r FFT. In a radix-r FFT, an "elementary computation," denoted EC, consists of an r-point DFT followed by multiplication of the r results by the appropriate twiddle factor. The number of ECs required is

$$C_r = \frac{N}{r} \log_r N \tag{9-56}$$

which decreases as r increases. Of course, the complexity of an EC increases with increasing r. For $r = 2$, the EC (the butterfly of our previous discussions) consists of a single complex multiplication and two complex additions; for $r = 4$, the EC requires three complex multiplications and several complex additions.

To see how the radix-r FFT algorithm is developed, we begin with the more general case where the composite number N factors as $N = LM$ and let n and k in the DFT, given by (9-4), be represented as

$$k = Ml + m, \qquad n = Lr + s \tag{9-57}$$

$$(s, l = 0, 1, \ldots, L - 1; r, m = 0, 1, \ldots, M - 1)$$

Substituting (9-57) into (9-4), we obtain

$$X(Lr + s) = \sum_{m=0}^{M-1} \sum_{l=0}^{L-1} x(Ml + m)W_N^{(Ml+m)(Lr+s)}$$
$$r = 0, 1, \ldots M - 1; \quad s = 0, 1, \ldots L - 1 \tag{9-58}$$

Now

$$(Ml + m)(Lr + s) = MLrl + Mls + Lrm + ms$$

and

$$W_N^{MLrl} = W_N^{Nlr} = 1$$

Hence (9-58) can be rearranged into the form

$$X(Lr + s) = \sum_{m=0}^{M-1} W_N^{Lmr} W_N^{ms} \sum_{l=0}^{L-1} x(Ml + m)W_N^{Msl} \tag{9-59}$$

Defining

$$q(s, m) = \sum_{l=0}^{L-1} x(Ml + m) \, (W_N^M)^{ls} \tag{9-60}$$

we can rewrite (9-59) as

$$X(L + s) = \sum_{m=0}^{M-1} q(m, s) \, W_N^{ms} (W_N^L)^{mr} \tag{9-61}$$

where it is important to note the following points:

1. For any fixed m, $q(s, m)$ can be interpreted as a DFT of each row of the array of $x(n)$'s arranged as

$$[x(n)] = \begin{bmatrix} x(0) & x(M) & \cdots & x(ML-M) \\ x(1) & x(M+1) & & x(ML-M+1) \\ \cdot & \cdot & & \cdot \\ \cdot & \cdot & & \cdot \\ \cdot & \cdot & & \cdot \\ x(M-1) & x(2M-1) & & x(ML-1) \end{bmatrix} \qquad (9\text{-}62)$$

Since each is an L-point transform, the proper W to be used is

$$W_L = \exp\left(\frac{-j2\pi}{L}\right)$$

$$= \exp\left(\frac{-j2\pi M}{ML}\right) = W_N^M \qquad (9\text{-}63)$$

which checks with (9-60).

2. Each element of the array $[q(s, m)]$ is multiplied by the twiddle factor $W_N^{sm} = W_N^{ms}$ (i.e., raised to the power of the product of the row and column indices) and an M-point DFT performed on each *column* [Note the reversal in the indices for $q(m, s)$ in (9-61) as compared with (9-60).] The proper W to be used is

$$W_M = \exp\left(\frac{-j2\pi}{M}\right)$$

$$= \exp\left(\frac{-j2\pi L}{ML}\right) = W_N^L \qquad (9\text{-}64)$$

which checks with (9-57).

In this fashion, an N-point DFT is broken down into an L-point DFT followed by an M-point DFT where $N = ML$.

To understand better how the DFT is computed by taking this approach, we consider several examples.

EXAMPLE 9-8

Consider $N = 4 = LM$, where $L = M = 2$. Assume that the signal for which the DFT is desired is

$$x(k) = \cos\frac{\pi k}{2}$$

Thus $x(0) = 1$, $x(1) = 0$, $x(2) = -1$, and $x(3) = 0$. From (9-60), we obtain the following results for $q(s, m)$:

$$q(0, 0) = \sum_{l=0}^{1} x(2l)(W_4^2)^0 = x(0) + x(2) = 0$$

$$q(1, 0) = \sum_{l=0}^{1} x(2l)(W_4^2)^l = x(0) + x(2)W_4^2$$

$$= x(0) - x(2) = 2$$

$$q(0, 1) = \sum_{l=0}^{1} x(2l + 1)(W_4^2)^0 = x(1) + x(3) = 0$$

$$q(1, 1) = \sum_{l=0}^{1} x(2l + 1)(W_4^2)^l = x(1) + x(3)W_4^2$$

$$= x(1) - x(3) = 0$$

where it is important to remember that the first index of q is s and the second one is m. We see that these calculations are precisely the same as the butterfly operations indicated in the first recursion of Figure 9-3.

Proceeding with the application of (9-61) and remembering the index reversal on $q(s, m)$, we obtain the following:

$$X(0) = \sum_{m=0}^{1} q(m, s = 0)W_4^0(W_4^2)^0$$

$$= q(m = 0, s = 0) + q(m = 1, s = 0)$$

$$= 0 + 0 = 0$$

$$X(1) = \sum_{m=0}^{1} q(m, s = 1)W_4^m(W_4^2)^0$$

$$= q(m = 0, s = 1) + q(m = 1, s = 1)W_4^1$$

$$= 2 + 0 \cdot W_4^1 = 2$$

$$X(2) = \sum_{m=0}^{1} q(m, s = 0)W_4^0(W_4^2)^m$$

$$= q(m = 0, s = 0) + q(m = 1, s = 0)W_4^2$$

$$= 0 + 0 = 0$$

$$X(3) = \sum_{m=0}^{1} q(m, s = 1)W_4^m(W_4^2)^m$$

$$= q(m = 0, s = 1) + q(m = 1, s = 1)W_4^1W_4^2$$

$$= 2 + 0 \cdot W_4^1W_4^2 = 2$$

These calculations are precisely those of the second recursion in Figure 9-3. ■

By proceeding with the computations in the matrix form alluded to in the discussion following (9-55), we may simplify considerably the procedure of taking the radix-r FFT. The next example illustrates the procedure for a radix-4 FFT.

EXAMPLE 9-9

Consider $N = 16$, $L = M = 4$, and $x(k) = \cos(\pi k/2)$. Arranging the data into a 4×4 array, we have

$$[x(k)] = \begin{bmatrix} 1 & 1 & 1 & 1 \\ 0 & 0 & 0 & 0 \\ -1 & -1 & -1 & -1 \\ 0 & 0 & 0 & 0 \end{bmatrix}$$

where $x(0)$, $x(1)$, $x(2)$, and $x(3)$ occupy the first column, etc. Applying (9-60) to each row, we obtain

$$q(s, m) = \sum_{l=0}^{3} x(4l + m)(e^{-j\pi/2})^{sl}$$

Thus the four-point EC is

$$q(0, m) = x(m) + x(4 + m) + x(8 + m) + x(12 + m) \qquad \text{(a)}$$

$$q(1, m) = x(m) - jx(4 + m) - x(8 + m) + jx(12 + m) \qquad \text{(b)}$$

$$q(2, m) = x(m) - x(4 + m) + x(8 + m) - x(12 + m) \qquad \text{(c)}$$

$$q(3, m) = x(m) + jx(4 + m) - x(8 + m) - jx(12 + m) \qquad \text{(d)}$$

In keeping with array notation, the second argument of $q(s, m)$ indicates the row. Applying this EC to the array $[x(k)]$, we obtain the array

$$[q(s, m)] = \begin{bmatrix} 4 & 0 & 0 & 0 \\ 0 & 0 & 0 & 0 \\ -4 & 0 & 0 & 0 \\ 0 & 0 & 0 & 0 \end{bmatrix}$$

The next step, according to (9-61), is to multiply by the twiddle factor W_{16}^{ms}. Since both nonzero factors in the $[q(s, m)]$ array are in the $s = 0$ row, there is no modification by the twiddle factor. The final step is to transform the array $[q(s, m)W_{16}^{ms}] = [q(s, m)]$ column by column in accordance with (9-61). [Note the index reversal in (9-61) on $q(m, s)$.] The EC is identical to that used to obtain $[q(s, m)]$. The results are as follows:

r	s	$X(Lr + s)$; $L = 4$ (output unscrambled)		
0	0	$X(0) = 0$	Eq. (a);	col. 0
0	1	$X(1) = 0$		col. 1
0	2	$X(2) = 0$		col. 2
0	3	$X(3) = 0$		col. 3
1	0	$X(4) = 8$	Eq. (b);	col. 0
1	1	$X(5) = 0$		col. 1
1	2	$X(6) = 0$		col. 2
1	3	$X(7) = 0$		col. 3
2	0	$X(8) = 0$	Eq. (c);	col. 0
2	1	$X(9) = 0$		col. 1
2	2	$X(10) = 0$		col. 2
2	3	$X(11) = 0$		col. 3
3	0	$X(12) = 8$	Eq. (d);	col. 0
3	1	$X(13) = 0$		col. 1
3	2	$X(14) = 0$		col. 2
3	3	$X(15) = 0$		col. 3

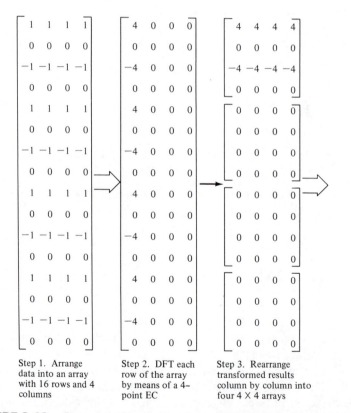

Step 1. Arrange data into an array with 16 rows and 4 columns

Step 2. DFT each row of the array by means of a 4-point EC

Step 3. Rearrange transformed results column by column into four 4 × 4 arrays

FIGURE 9-18. Steps used in performing a 64-point DFT by means of radix-4 ECs.

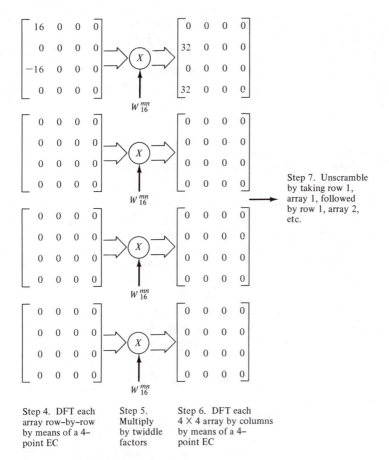

$$\begin{bmatrix} 16 & 0 & 0 & 0 \\ 0 & 0 & 0 & 0 \\ -16 & 0 & 0 & 0 \\ 0 & 0 & 0 & 0 \end{bmatrix} \Rightarrow X \Rightarrow \begin{bmatrix} 0 & 0 & 0 & 0 \\ 32 & 0 & 0 & 0 \\ 0 & 0 & 0 & 0 \\ 32 & 0 & 0 & 0 \end{bmatrix}$$

W'^{mn}_{16}

$$\begin{bmatrix} 0 & 0 & 0 & 0 \\ 0 & 0 & 0 & 0 \\ 0 & 0 & 0 & 0 \\ 0 & 0 & 0 & 0 \end{bmatrix} \Rightarrow X \Rightarrow \begin{bmatrix} 0 & 0 & 0 & 0 \\ 0 & 0 & 0 & 0 \\ 0 & 0 & 0 & 0 \\ 0 & 0 & 0 & 0 \end{bmatrix}$$

W'^{mn}_{16}

Step 7. Unscramble by taking row 1, array 1, followed by row 1, array 2, etc.

$$\begin{bmatrix} 0 & 0 & 0 & 0 \\ 0 & 0 & 0 & 0 \\ 0 & 0 & 0 & 0 \\ 0 & 0 & 0 & 0 \end{bmatrix} \Rightarrow X \Rightarrow \begin{bmatrix} 0 & 0 & 0 & 0 \\ 0 & 0 & 0 & 0 \\ 0 & 0 & 0 & 0 \\ 0 & 0 & 0 & 0 \end{bmatrix}$$

W'^{mn}_{16}

$$\begin{bmatrix} 0 & 0 & 0 & 0 \\ 0 & 0 & 0 & 0 \\ 0 & 0 & 0 & 0 \\ 0 & 0 & 0 & 0 \end{bmatrix} \Rightarrow X \Rightarrow \begin{bmatrix} 0 & 0 & 0 & 0 \\ 0 & 0 & 0 & 0 \\ 0 & 0 & 0 & 0 \\ 0 & 0 & 0 & 0 \end{bmatrix}$$

W'^{mn}_{16}

Step 4. DFT each array row–by–row by means of a 4–point EC

Step 5. Multiply by twiddle factors

Step 6. DFT each 4 × 4 array by columns by means of a 4–point EC

FIGURE 9-18. (continued). ∎

EXAMPLE 9-10

We now consider $N = 64$ and obtain the DFT of the signal $x(k) = \cos(\pi k/2)$ with a radix-4 algorithm. We employ (9-60) and (9-61) with $M = 16$ and $L = 4$. Thus we arrange the input data, consisting of the sequence $\{1, 0, -1, 0, \ldots\}$, into an array with four columns and 16 rows by filling column 0 first, column 1 next, etc. The result is shown as the left-hand array in Figure 9-18. Each row of the resulting array is then transformed by means of a radix-4 EC, which results in the second array from the left in Figure 9-18. However, rather than performing four 16-point column transforms, each column is next represented as a 4×4 array, shown as the third set of arrays from the left in Figure 9-18. These arrays are then transformed row by row using a radix-4 EC. The resulting arrays are multiplied element by element by the twiddle factors $W_{16}^{mn} = \exp(-j2\pi mn/16)$ since each array represents data for a 16-point DFT. For this example, there is no change in the arrays when this operation is performed, however, because the only nonzero array elements are in column 0, for which $W_{16}^0 = 1$. The resulting arrays are then transformed *column by column* by means of a radix-4 EC [recall the reversal of the index m

in $q(s, m)$ from (9-60) to (9-61)]. This completed DFT must be unscrambled by taking row 1 from array 1 followed by row 1 from array 2, etc.

Using the foregoing procedure allows one to obtain the DFT of a data sequence of length 4^n by succesively applying radix-4 ECs. A similar procedure would be followed to perform a DFT in any other radix, or by means of mixed radixes. ∎

*9-12

THE CHIRP z-TRANSFORM ALGORITHM

Several efficient algorithms for computing the DFT, given by (9-4), of a sequence of finite length, N, have been discussed in the preceding pages. In order to achieve this efficiency, N must be a highly composite number. The chirp z-transform provides a means for computing the DFT of a time series of length N for a sequence of M frequencies where $M \leq N$. Furthermore, the chirp z-transform is conveniently implemented as hardware by employing charge-coupled delay lines.

To derive the chirp z-transform, consider the z-transform of the sequence $x(n)$, $n = 0, 1, 2, \ldots, N - 1$, which for a discrete set of values of $z = z_k$, $k = 0, 1, 2, \ldots, M - 1$, can be expressed as

$$X(z_k) = \sum_{n=0}^{N-1} x(n)z_k^{-n}, \quad 0 \leq k \leq M - 1 \tag{9-65}$$

Let z_k be a point on a spiral centered at the origin in the z-plane. This can be accomplished by writing

$$z_k = AW^{-k}, \quad 0 \leq k \leq M - 1 \tag{9-66}$$

where

$$A = A_0 e^{j\theta_0} \tag{9-67}$$

is the point at which the spiral starts and

$$W = W_0 e^{-j\phi_0} \tag{9-68}$$

determines the rate of spiral and the angular separation of the points z_k. Figure 9-19 shows typical cases for $W_0 < 1$, for which the sequence of points $\{z_k\}$ spirals toward the origin, and for $W_0 > 1$, for which the sequence of points $\{z_k\}$ spirals away from the origin. If $|W_0| = 1$, the z_k-sequence lies of a circle of radius A_0.

Substitution of (9-66) into (9-65) yields

$$X(z_k) = \sum_{n=0}^{N-1} x(n)A^{-n}W^{nk}, \quad 0 \leq k \leq M - 1 \tag{9-69}$$

The next step in the derivation is to make the observation that

$$nk = \tfrac{1}{2}[n^2 + k^2 - (k - n)^2] \tag{9-70}$$

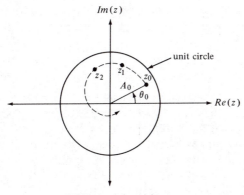

(a) Spiral toward the origin

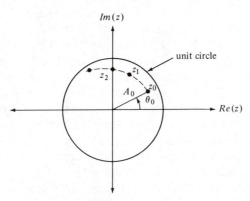

(b) Spiral away from the origin

FIGURE 9-19. *Z*-plane spirals for development of the chirp-*z* transform.

Thus (9-69) can be written as

$$X(z_k) = W^{k^2/2} \sum_{n=0}^{N-1} [x(n)A^{-n}W^{n^2/2}]W^{-(k-n)^2/2}, \quad 0 \le k \le M - 1 \quad (9\text{-}71)$$

With the aid of the definitions

$$h(n) \triangleq W^{-n^2/2} \quad (9\text{-}72)$$

and

$$g(n) \triangleq x(n)A^{-n}W^{n^2/2} \quad (9\text{-}73)$$

one may put (9-71) into the form

$$X(z_k) = \frac{1}{h(k)} \sum_{n=0}^{N-1} g(n)h(k-n), \quad 0 \le k \le M - 1 \quad (9\text{-}74)$$

The summation is recognized as a convolution of the sequences $\{g(n)\}$ and $\{h(n)\}$.

Thus (9-74) can be viewed as the series of operations represented by the block diagram of Figure 9-20. If A and W_0 are unity, the sequence $h(n)$ is given by

$$h(n) = (e^{-j\phi_0})^{-n^2/2}$$

$$= e^{jn^2\phi_0/2} \qquad (9\text{-}75)$$

which can be thought of as a complex exponential sequence with linearly increasing frequency. Such signals are known as chirp signals in radar system applications—hence the name *chirp-z transform* for (9-74).

All operations implied in Figure 9-20 could, of course, be carried out digitally. However, in recent years, the convolution operation for the chirp-z transform has been implemented by means of charge-transfer devices (CTDs), and CTD chirp-z transformers are available commercially.

Basically, a CTD can be viewed as a shift register for analog charge samples. As illustrated in Figure 9-21a, each stage of capacitance can weight the charge stored on it differently through employment of split electrodes and differential sensing of the charge. With the application of each clock pulse, the charge stored on each capacitor is transferred to the next capacitor in the chain and a new charge sample is stored on the first capacitor in the chain. The impulse response $h(n)$ of the convolution operation is implemented by varying the capacitance values of each stage of the CTD in accordance with the desired weighting for the convolution. Since $h(n)$ is complex, the convolution operation must be implemented to compute the real and imaginary parts of $h(n)*[(A^{-n}/h(n))x(n)]$.

The input and output chirping operations are implemented by means of multiplying analog-to-digital converters, with real and imaginary parts of the chirp coefficients, $h(n)$, being stored in read-only memories.

A schematic representation of a CTD-implemented chirp-z transformer is shown in Figure 9-21b. All complex operations are implemented as equivalent real operations. It is left for the student to show that the block diagram of Figure 9-21b implements the operations of the block diagram of Figure 9-20.

The major advantage of a CTD-implemented chirp-z transformer over the equivalent digital implementation appears to be cost. The inherent error sources associated with the CTD implementation are charge-transfer inefficiency and the inaccuracy associated with storing the chirping coefficients in the read-only memory as binary numbers. The former error source is due to the fact that the charge-transfer operation in the CTD is not 100% efficient. Typical values for transfer inefficiency ϵ are in the neighborhood of 10^{-3} for commercially available devices. That is, if $\epsilon = 10^{-3}$, then for each transfer, 0.1% of the charge

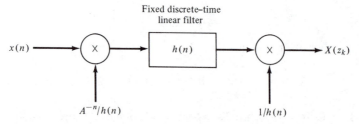

FIGURE 9-20. Block diagram representation of the chirp-z transform.

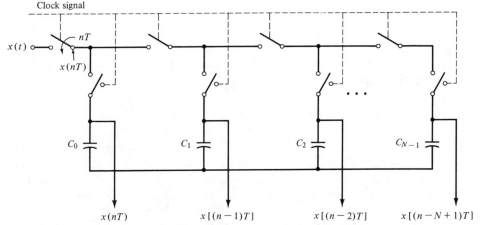

Clock signal

$x(t)$

nT

$x(nT)$

C_0 C_1 C_2 C_{N-1}

$x(nT)$ $x[(n-1)T]$ $x[(n-2)T]$ $x[(n-N+1)T]$

(a) Schematic representation of a charge transfer device
wherein each charge storage element can be weighted
independently of the others

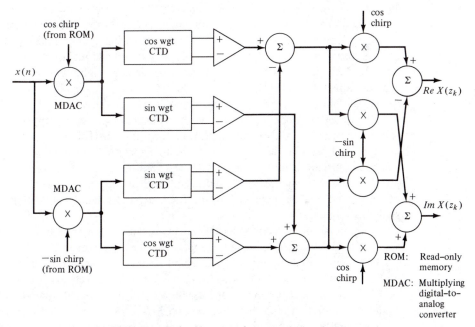

cos chirp
(from ROM)

cos
chirp

$x(n)$

×
MDAC

cos wgt
CTD

+
−

+

Σ

×

+

Σ

$Re\ X(z_k)$

sin wgt
CTD

+
−

×

−

−sin
chirp

MDAC

sin wgt
CTD

+
−

+

×

×

+

Σ

$Im\ X(z_k)$

cos wgt
CTD

+
−

+

Σ

×

+

−sin chirp
(from ROM)

cos
chirp

ROM: Read–only
memory

MDAC: Multiplying
digital–to–
analog
converter

(b) Block diagram of a chirp–z transformer employing charge
transfer devices

FIGURE 9-21. Illustration of the use of charge-transfer devices to implement a chirp-z transformer.

sample is left on a given electrode and 99.9% of it is transferred to the next one in the chain.

SUMMARY

In this chapter we have considered the discrete Fourier transform (DFT) as a tool for computing the spectra of sample-data signals. Several possible error sources inherent in approximating spectra of signals by means of the DFT were discussed and possible ways to minimize their effects were mentioned. These error sources were aliasing, leakage, and the picket-fence effect. Efficient algorithms for computing the DFT sum, known as fast Fourier transform (FFT) algorithms, were derived. Two basic types of FFT algorithms were presented—those based on decimation of time samples, and those based on decimation of frequency samples. Both types of algorithms required roughly the same number of multiplications and additions to compute an N-point FFT, and this number of computations is approximately $N \log_2 N$ versus N^2 for the DFT.

The DFT has several properties similar in nature to the Fourier transform including symmetry properties, superposition, time delay, and frequency shift. One important difference, however, is the convolution and multiplication properties. If the convolution of two sequences is obtained by multiplying their DFTs and obtaining the inverse DFT of the product, the result is a circular convolution or convolution of the periodic extensions of the sequences. This property has important consequences when the DFT is used in signal-processing applications; care must be taken to avoid aliasing in the time domain. This is accomplished through zero padding the sequences to be transformed or by zero padding in combination with selecting the good values from the result of segmented convolutions a long sequence with a short sequence. In addition to convolution, or filtering, the DFT has application to spectral analysis, system identification, and signal restoration.

To minimize the effects of leakage, windows are often employed. Several types of windows were presented together with their properties.

The selection of sampling rate and record length is dependent on the desired spectral resolution and the bandwidth of the signal being processed by means of a DFT. In general, sampling rate is proportional to the signal bandwidth and frequency resolution is inversely proportional to record length. Resolution can be increased by zero padding to increase record length, although caution must be exercised in doing so in the case of random signals.

Two final topics discussed were FFT algorithms for sample sets having lengths other than a power of 2 and the chirp-z transform. The former are referred to as mixed-radix FFT algorithms. The latter is a generalization of the DFT to computing frequency samples along spiral contours in the complex z-plane rather than along the unit circle as with the DFT.

FURTHER READING

Several good references are available on the FFT. In addition to the references cited in the body of the chapter, we give two that are well written.

E. O. BRIGHAM, *The Fast Fourier Transform*. Englewood Cliffs. N.J.: Prentice-Hall, 1974.

This book is concerned only with the DFT, FFT algorithms, their properties, and applications. It is written at a level which is comparable to that of this text.

A. V. OPPENHEIM and R. W. SCHAFER, *Digital Signal Processing*. New York: Wiley, 1974.

An excellent text that is written at a higher level than Brigham. It includes chapters on the DFT, FFT algorithms, the chirp-z transform, and errors in computing FFT algorithms digitally.

PROBLEMS

SECTION 9-2

9-1. Obtain the DFT of

$$x_k = 1, \qquad 0 \le k \le N - 1$$

by using (9-8) to sum the series (9-4). Why is this result reasonable?

9-2. **(a)** Obtain the DFT of

$$x_k = \begin{cases} 1, & 0 \le k \le K < N - 1 \\ 0, & K < k \le N - 1 \end{cases}$$

by using the sum for a geometric series to sum (9-4). Plot the periodically extended time series which corresponds to the inverse DFT of the resulting DFT (see Figure 9-2).

(b) Plot $|X(n)|$ versus n for $N = 8$ and $K = 2, 3$. Explain your plots in terms of the spectral analysis ideas developed in Chapter 3. What departures, if any, are there from the ideas of Chapter 3? (See Figure 9-2.)

SECTION 9-3

9-3. Let $\tilde{x}_k = A \exp [j(2\pi L k/N + \theta_0)]$ in (9-4), where $0 \le L \le N - 1$ and A and θ_0 are constant amplitude and phase, respectively.

(a) Use (9-8) to sum the series and show that

$$\tilde{X}(n) = \begin{cases} A \dfrac{\sin \pi(L - n)}{\sin [\pi(L - n)/N]} \\ \times \exp \left[j\left(\theta_0 + \pi(L - n)\left(1 - \dfrac{1}{N}\right)\right)\right], & L \ne n \\ NA \exp (j\theta_0), & L = n \end{cases}$$

(b) Plot $|X(n)|$ versus n for $A = 1$, $N = 8$, and (1) $L = 1$; (2) $L = 1.1$.
Answers: (1) For $L = 1$, $\{|X(n)|\} = \{0, 8, 0, 0, 0, 0, 0\}$; (2) for $L = 1.1$, $\{|X(n)|\} = \{0.74, 7.87, 0.89, 0.46, 0.34, 0.31, 0.33, 0.42\}$.
(*Note:* Part (b) illustrates the *leakage* phenomenon of the DFT.)

9-4. Using the result of Problem 9-3(a), obtain the DFT of

$$x_k = A \cos{(2\pi Lk/N + \theta_0)}$$

by writing it as $x_k = \dfrac{1}{2}\tilde{x}_k + \dfrac{1}{2}\tilde{x}_k^*$ and noting that the DFT is a linear transform.

9-5. Reproduce Figure 9-2 for the following signals. Plot magnitudes of transforms. Assume convenient values for the parameters involved.
(a) $x(t) = Ae^{-\alpha t}u(t)$
(b) $x(t) = A\Pi(t/\tau)$
(c) $x(t) = A \cos{\omega_0 t}$

SECTION 9-4

9-6. Show in detail that (9-18) and (9-19) result from (9-17).

9-7. Show that (9-20) and (9-21) result from (9-18) and (9-19).

9-8. Let $x_0(k) = 1$ for $k = 0, 1, 2, 3$ in the flow diagram of Figure 9-3a. Obtain the numerical values at all the nodes in the flow diagram.

9-9. Consider the pulse signal

$$x(000) = x(001) = 1$$

$$x(010) = x(011) = x(100) = x(101) = x(110) = 0$$

$$x(111) = 1$$

Using the flow diagram of Figure 9-5 obtain the DFT of this signal. Discuss the departure of this result from the result for the Fourier transform of the ideal pulse signal by comparing plots of their amplitude specta.

9-10. **(a)** Obtain the eight-point FFT of the following pulse signal using the flow diagram of Figure 9-5.

$$x(000) = x(001) = x(010) = x(011) = 1$$

$$x(100) = 0$$

$$x(101) = x(110) = x(111) = 1$$

Discuss the departure of this FFT from the result for the Fourier transform of the ideal pulse signal by comparing plots of their amplitude spectra.
(b) Obtain the inverse FFT of the result in part (a) by applying (9-10) and scaling by $N = 8$. Comment on how you have verified the identity (9-6).

9-11. **(a)** Obtain a matrix expression for the Sande–Tukey algorithm, expressed by

(9-23), similar to (9-20) and (9-21) for the Cooley–Tukey algorithm. Verify that the flow diagram of Figure 9-4 results.

(b) Extend the flow diagram for the Sande–Tukey algorithm to $N = 8$.

SECTION 9-5

9-12. Obtain an algorithm for determining the powers of W shown in Figure 9-9. Generalize to arbitrary N.

9-13. **(a)** Use the decimation-in-time approach to develop an FFT algorithm based on a four-point basic DFT computation for $N = 16$. Do this by grouping the input samples into four sets of four samples each.

(b) Repeat part (a) using decimation in frequency.

9-14. Write the Butterfly computation for the Cooley–Tukey algorithm, given by (9-22), entirely in terms of real computations. For this purpose, let $RX_m(p) = \text{Re}\ [X_m(p)]$, $IX_m(p) = \text{Im}\ [X_m(p)]$, etc. Also note that $W_N^s = \cos\ (2\pi s/N) - j \sin\ (2\pi s/N)$. Draw a block diagram of the hardware that would be used to compute the butterfly. Use the minimum possible number of multipliers and adders.

9-15. Repeat Problem 9-14 for the Sande–Tukey butterfly computation expressed by (9-24). Which butterfly requires the least number of real multiplications and additions?

SECTION 9-6

9-16. Prove the linearity property of the DFT sum.

9-17. **(a)** Prove the time shift property of the DFT by using the fact that $x(k - m)$ is a periodic sequence.

(b) Prove the frequency shift property by using the fact that $X(n - m)$ is a periodic sequence.

9-18. Prove the convolution property of the DFT by substituting for $y(k - m)$ in the convolution sum

$$x(k) \ \textcircled{N} \ y(k) = \sum_{m=0}^{N-1} x(m)y(k - m)$$

in terms of the inverse DFT of $Y(n)$, which represents the DFT of $y(k)$. Use the implied periodicity of $Y(n)$.

9-19. Show that

$$\sum_{k=0}^{N-1} |x(k)|^2 = \frac{1}{N} \sum_{n=0}^{N-1} |X(n)|^2$$

by writing the left-hand sum as $\sum_{k=0}^{N-1} x(k)x^*(k)$ and substituting for $x^*(k)$ in terms of the inverse DFT of $X(n)$.

9-20. Prove the multiplication property of the DFT by substituting for $Y(n-m)$ in terms of the DFT of $y(k)$.

9-21. Discuss how Property 10 in Section 9.6 could be used to obtain the DFT of a real sequence $2N$ points long by using an N-point DFT.

SECTIONS 9-8–9-10

9-22. Compute an appropriate sampling rate and DFT size N in order to analyze a signal spectrally with no significant frequency content above 10 kHz and with a resolution of 100 Hz. Make N a power of 2.

9-23. A 50-ms transient signal has no significant spectral content above 500 Hz. Choose an appropriate sampling rate and DFT size N to analyze this signal spectrally with a resolution of at least 10 Hz. Make N a power of 2. Is zero padding necessary?

9-24. Consider the convolution of a 50-point impulse response with a 1000-point random sequence. Choose an appropriate DFT size which is a power of 2 and state how many segments must be chosen for the 1000-point sequence. How many points are valid for each circular convolution operation when implemented by means of a DFT of the size you have chosen?

9-25. (This problem is meant to illustrate the steps in obtaining an energy spectral estimate as explained in Section 9-8 and does not contain numbers representative of a practical situation.) Consider the eight-point sequence $\{x(k)\}$ defined as

$$x(k) = \begin{cases} 1, & k = 0, 1 \\ 0, & k = 2, 3, 4, 5, 6, 7 \end{cases}$$

If $N = 8$, use (9-44) to compute the autocorrelation function $R_x(k)$. Use an eight-point FFT to compute $S_x(n)$ from (9-43). Next, compute $S_x(n)$ from (9-42) and compare with the result obtained from (9-43). For computing the FFTs, it will be helpful to use a flow graph such as that of Figure 9-14.

SECTION 9-11

9-26. Using the theory developed in Section 9-11, develop a 12-point FFT algorithm.

SECTION 9-12

9-27. Sketch the locus of points in the z-plane for the following parameter values for (9-66), (9-67), and (9-68):
(a) $M = 8$, $W_0 = 2$, $\phi_0 = \pi/16$ rad, $A_0 = 2$, and $\theta_0 = \pi/4$ rad.
(b) $M = 8$, $W_0 = \frac{1}{2}$, $\phi_0 = \pi/16$ rad, $A_0 = 2$, and $\theta_0 = \pi/4$ rad.

9-28. Show that the block diagram of Figure 9-21b provides an implementation of the operations for the chirp-z transform operation illustrated in Figure 9-20.

9-29. Compare the number of multiplications required for an FFT implementation of the chirp-z transform with $W_0 = 1$ for various values of N and M with the FFT assuming that the *frequency resolution* for both is the same.

Matrix Algebra

The purpose of this appendix is to summarize the basic concepts of matrix theory.

An $n \times m$ matrix $\mathbf{A}$ is an array of numbers a_{ij} given by

$$
\mathbf{A} = \begin{bmatrix}
a_{11} & a_{12} & \cdots & a_{1m} \\
a_{21} & a_{22} & \cdots & a_{2m} \\
\cdot & & & \\
\cdot & & & \\
\cdot & & & \\
a_{n1} & a_{n2} & \cdots & a_{nm}
\end{bmatrix}
$$

It has n rows and m columns. If $m = 1$, there is only one column and $\mathbf{A}$ becomes a column matrix or vector $\mathbf{a}$ given by

$$
\mathbf{a} = \begin{bmatrix}
a_1 \\
a_2 \\
\cdot \\
\cdot \\
\cdot \\
a_n
\end{bmatrix}
$$

The transpose of an $n \times m$ matrix $\mathbf{A}$, denoted $\mathbf{A}'$, is an $m \times n$ matrix obtained from A by interchanging rows and columns; that is,

$$
\mathbf{A}' = \begin{bmatrix} a_{11} & a_{21} & \cdots & a_{n1} \\ a_{12} & a_{22} & \cdots & a_{n2} \\ \cdot & & & \\ \cdot & & & \\ \cdot & & & \\ a_{1m} & a_{2m} & \cdots & a_{nm} \end{bmatrix}
$$

A square matrix $(n = m)$ is said to be *symmetric* if and only if $\mathbf{A}' = \mathbf{A}$. The transpose of a column vector is a row vector; that is,

$$
\mathbf{a}' = [a_1 \quad a_2 \quad \cdots \quad a_n]
$$

Two matrices (or vectors) which have the same dimensions may be added or subtracted. These operations are accomplished by adding or subtracting corresponding elements of the matrices. The operations may be defined as

$$
\mathbf{A} \pm \mathbf{B} = \mathbf{C}
$$

where the elements of $\mathbf{C}$ are obtained from those of $\mathbf{A}$ and $\mathbf{B}$ by the relation

$$
c_{ij} = a_{ij} \pm b_{ij}
$$

Matrix multiplication may be defined as follows. Let $\mathbf{A}$ be $n \times l$ with elements a_{ij} and $\mathbf{B}$ be $l \times m$ with elements b_{ij}. Then

$$
\mathbf{AB} = \mathbf{C}
$$

where $\mathbf{C}$ is $n \times m$ with elements c_{ij} given by

$$
c_{ij} = \sum_{k=1}^{l} a_{ik}b_{kj}
$$

Note that the number of columns of the first matrix must equal the number of rows of the second or multiplication cannot be accomplished. An an example, let

$$
\mathbf{A} = \begin{bmatrix} 1 & 2 & 1 \\ 3 & 0 & 1 \end{bmatrix}, \quad \mathbf{B} = \begin{bmatrix} 2 & 1 \\ 1 & 0 \\ 3 & 2 \end{bmatrix}
$$

Then

$$
\mathbf{AB} = \mathbf{C} = \begin{bmatrix} 7 & 3 \\ 9 & 5 \end{bmatrix}
$$

In general, matrix multiplication is not commutative; that is, in general

$$
\mathbf{AB} \neq \mathbf{BA}
$$

However, it is true that

$$A(BC) = (AB)C \quad \text{and} \quad A(B + C) = AB + AC$$

The identity matrix is a square matrix, I, having the property that for any square matrix A

$$AI = IA = A$$

The identity matrix is such that all principal diagonal elements are 1 and all other elements are zero. The identity matrix of dimension 2 is given by

$$I = \begin{bmatrix} 1 & 0 \\ 0 & 1 \end{bmatrix}$$

A square matrix A is said to be *nonsingular* if the determinant of A is nonzero. For a nonsingular matrix A there exists a matrix B with the property that

$$AB = BA = I$$

B is said to be the *inverse* of A, and is denoted as

$$B = A^{-1}$$

The inverse of a nonsingular matrix may be found from

$$A^{-1} = \frac{\text{adj } A}{|A|}$$

where adj A denotes the *adjoint* of A, which is the transpose of the matrix of cofactors of A.† $|A|$ is the determinant of A. As an example, let us find the inverse of

$$A = \begin{bmatrix} 0 & 1 & 0 \\ 0 & 0 & 1 \\ -2 & -3 & -1 \end{bmatrix}$$

We have

$$|A| = -2$$

$$\text{adj } A = \begin{bmatrix} 3 & -2 & 0 \\ 1 & 0 & -2 \\ 1 & 0 & 0 \end{bmatrix}' = \begin{bmatrix} 3 & 1 & 1 \\ -2 & 0 & 0 \\ 0 & -2 & 0 \end{bmatrix}$$

Then

$$A^{-1} = \begin{bmatrix} -\frac{3}{2} & -\frac{1}{2} & -\frac{1}{2} \\ 1 & 0 & 0 \\ 0 & 1 & 0 \end{bmatrix}$$

†The i–jth cofactor of a matrix is obtained by striking out the ith row and jth column, evaluating the resulting determinant, and multiplying by $(-1)^{i+j}$.

As a check, the student may verify that $\mathbf{AA}^{-1} = \mathbf{I}$.

The eigenvalues of an $n \times n$ matrix $\mathbf{A}$ are the roots λ_i, $i = 1, 2, \ldots, n$, of the characteristic equation

$$|\lambda\mathbf{I} - \mathbf{A}| = 0$$

which is an nth-degree polynomial in λ. Eigenvalues play an important role in system analysis.

As an example, we shall find the eigenvalues of the matrix

$$\mathbf{A} = \begin{bmatrix} 0 & 1 \\ -4 & -5 \end{bmatrix}$$

We have

$$|\lambda\mathbf{I} - \mathbf{A}| = \begin{vmatrix} \lambda & -1 \\ 4 & \lambda + 5 \end{vmatrix} = \lambda(\lambda + 5) + 4 = \lambda^2 + 5\lambda + 4 = 0$$

The polynomial may be factored and the roots found as

$$\lambda_1 = -1, \qquad \lambda_2 = -4$$

These are the eigenvalues of the matrix given. The procedure is exactly the same, but considerably more tedious, for larger matrices.

The characteristic equation is an important quantity associated with a matrix, having other uses than merely the determination of the eigenvalues. Let us denote the characteristic equation of an $n \times n$ matrix $\mathbf{A}$ as

$$\lambda^n + a_{n-1}\lambda^{n-1} + \cdots + a_1\lambda + a_0 = 0$$

The Cayley–Hamilton theorem states that a matrix satisfies its own characteristic equation. Therefore, $\mathbf{A}$ satisfies the *matrix* equation

$$\mathbf{A}^n + a_{n-1}\mathbf{A}^{n-1} + \cdots + a_1\mathbf{A} + a_0\mathbf{I} = 0,$$

from which

$$\mathbf{A}^n = -(a_{n-1}\mathbf{A}^{n-1} + a_{n-2}\mathbf{A}^{n-2} + \cdots + a_1\mathbf{A} + a_0\mathbf{I})$$

If we multiply both sides of this equation by $\mathbf{A}$, we obtain

$$\mathbf{A}^{n+1} = -(a_{n-1}\mathbf{A}^n + a_{n-2}\mathbf{A}^{n-1} + \cdots + a_1\mathbf{A}^2 + a_0\mathbf{A})$$

Clearly, we can substitute for $\mathbf{A}^n$ on the right to obtain an expression for $\mathbf{A}^{n+1}$ in terms of powers of $\mathbf{A}$ *no higher than* $n - 1$. The expression will be of the form

$$\mathbf{A}^{n+1} = b_0\mathbf{I} + b_1\mathbf{A} + \cdots + b_{n-1}\mathbf{A}^{n-1}$$

where the b_i can be determined from the a_i. By continuing this kind of operation and successively substituting, it is possible to write *any* power of $\mathbf{A}$ in terms of powers of $\mathbf{A}$ no higher than $n - 1$. This is the basis of one of the methods of finding the matrix exponential $e^{\mathbf{A}t}$ in Chapter 6.

Differentiation or integration of a matrix with respect to time is accomplished by performing the appropriate operation on each element of the matrix. For example, suppose that

$$\mathbf{A} = \begin{bmatrix} e^{-t} & te^{-t} \\ 0 & e^{-t} \end{bmatrix}$$

Then

$$\frac{d\mathbf{A}}{dt} = \begin{bmatrix} -e^{-t} & e^{-t} - te^{-t} \\ 0 & -e^{-t} \end{bmatrix}$$

Since the Laplace transform is an integral operation, the Laplace transform of a matrix is obtained by Laplace transforming each of the elements. The Laplace transform of $\mathbf{A}$ above is

$$\mathscr{L}[\mathbf{A}] = \begin{bmatrix} \dfrac{1}{s+1} & \dfrac{1}{(s+1)^2} \\ 0 & \dfrac{1}{s+1} \end{bmatrix}$$

Analog Filters

In Chapter 8 a number of methods were studied for the design of digital filters. Most of these design methods assumed knowledge of an analog prototype from which the digital filter could be derived. The design process for the digital filter was then carried out in order to obtain a digital filter whose time-domain or frequency-domain characteristics approximated the corresponding characteristics of the assumed analog prototype. For example, the impulse invariance design technique results in a digital filter whose unit pulse response is the sampled unit impulse response of the analog filter from which it was derived. The bilinear z-transform design technique results in a digital filter whose frequency response approximates the frequency response of the analog filter from which it was derived. The purpose of this appendix is to examine several analog prototypes in order to illustrate where those assumed analog filters used to derive digital filters originated.

The synthesis of analog filters is now a very mature subject area. Extensive sets of tables exist which give not only the frequency and phase responses of many analog prototypes, but also the element values necessary to realize those prototypes.† Many of the design procedures for digital filters have been developed in way that allows this wide body of analog filter knowledge to be utilized effectively.

Generally, filter tables contain the amplitude response, phase response, pole–zero

---

†See, for example, Zverev (1967). See "Further Readings."

locations, and element values for the *normalized filter*, which is defined to be a low-pass filter having a bandwidth of 1 rad/s. Then, using standard techniques, this filter can be transformed into a high-pass, bandpass, or notch filter having any desired bandwidth. We examine several of these transformation techniques later in this appendix.

GROUP DELAY AND PHASE DELAY

In Chapter 3 ideal filters were defined. Recall that the ideal filter resulted by investigating the conditions under which a linear system is distortionless. A distortionless system was defined as a system in which the system output is an amplitude scaled and time-delayed version of the system input. In other words, if the input $x(t)$ is applied to a distortionless system, the output $y(t)$ is

$$y(t) = Gx(t - \tau) \tag{B-1}$$

in which G is the system gain and τ is the system time delay between system input and output. By Fourier transforming (B-1) we were able to show that the transfer function of an ideal system is

$$H(f) = Ge^{-j2\pi f\tau} \tag{B-2}$$

The amplitude response is the constant

$$A(f) = G$$

and the phase response is the *linear* function of frequency

$$\phi(f) = -2\pi f\tau \tag{B-3}$$

Often, instead of specifying the phase response of a filter, the group time delay is specified. The group time delay is defined as

$$\tilde{T}_g(\omega) = -\frac{d}{d\omega}\tilde{\phi}(\omega) \tag{B-4}$$

or, equivalently,

$$T_g(f) = -\frac{1}{2\pi}\frac{d}{df}\phi(f) \tag{B-5}$$

where tilde indicates a function of ω. Since an ideal filter has a phase response that is a linear function of frequency, the group delay of an ideal filter is constant for all frequency. This is easily seen by substituting (B-3) into (B-5), which shows that the group delay of an ideal filter is

$$T_g(f) = -\frac{1}{2\pi}\frac{d}{df}(-2\pi f\tau) = \tau \tag{B-6}$$

Thus if an input to an ideal filter has a spectrum extending over a range of frequencies, the filter output will be distortionless if the filter gain and group delay are both constant over the nonzero range of the input spectrum.

Linear system theory tells us that if a sinusoidal input is applied to a linear time-invariant system, the output will also be sinusoidal, although the output may have a different amplitude and phase reference. Thus if a single spectral component is applied to a linear time-invariant system, the output is *always* distortionless. For this case we refer to another type of system time delay called *phase delay*, which is the actual time delay between input and output at the frequency of a single spectral component. Phase delay is easily illustrated by a simple example.

Assume that a system has input

$$x(t) = A \cos 2\pi f_0 t \tag{B-7}$$

and output

$$y(t) = B \cos (2\pi f_0 t + \theta)$$

The output can be written

$$y(t) = B \cos 2\pi f_0 \left(t + \frac{\theta}{2\pi f_0} \right) \tag{B-8}$$

Recognizing that since the output must be distortionless, $y(t)$ must be of the form

$$y(t) = B \cos 2\pi f_0 (t - t_0) \tag{B-9}$$

Thus, comparing (B-8) and (B-9), we see that the time delay between system input and system output is

$$t_0 = - \frac{\theta}{2\pi f_0} \tag{B-10}$$

which is the phase delay for this example. In general, the phase delay is written

$$T_p(f) = - \frac{\phi(f)}{2\pi f} \tag{B-11}$$

in which $\phi(f)$ is the system phase shift, in radians, at the frequency f (hertz). Ideal filters have equal group and phase delays.

B-2

APPROXIMATION OF IDEAL FILTERS
BY PRACTICAL FILTERS

Since ideal filters are not physically realizable (recall that the unit impulse response for an ideal filter is noncausal), the filtering problem is to select, from among the many available filter prototypes, that prototype which satisfactorily approximates the desired filter characteristic with a minimum of hardware. For an ideal filter, each value of frequency either falls within the filter passband, which is characterized by constant gain, or falls within the filter stop band, which is characterized by infinite attenuation. Practical filters have a region between each passband and stop band known as a transition band.

A typical set of filter requirements is illustrated in Figure B-1. Within the passband, the filter amplitude response is not required to be constant; some variation about the nominal gain G_0 is allowed. In Figure B-1a, this variation has a peak-to-peak value of Δ_R. Within the stop band the amplitude response is required to be below the value A. In order for the transition to be as small as possible, the amplitude response of the filter often passes through the two heavy dots shown in Figure B-1a. Typical applications also place requirements on the group delay of the filter, as illustrated in Figure B-1b. Some variation about the nominal value T_g is allowed. The peak-to-peak variation is dentoed by Δ_T in Figure B-1b. The group delay in the transition band and in the stop band is often unimportant. Although Figure B-1 is drawn for a low-pass filter, the extension to other types should be obvious.

Many different trade-offs are possible among the parameters depicted in Figure B-1, and each of the standard prototypes makes a different trade-off. We now briefly review some of the most popular filter prototypes in order to become familiar with the various trade-offs possible.

A sketch of the amplitude response of each prototype to be discussed is illustrated in Figure B-2. Each filter is fourth order, and the highest passband frequency is 1 rad/s. Thus these are the amplitude responses of the normalized filters.

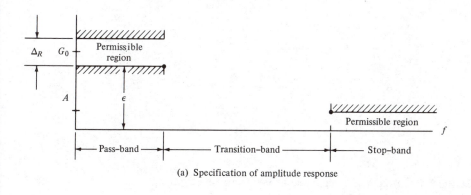

(a) Specification of amplitude response

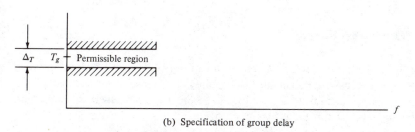

(b) Specification of group delay

FIGURE B-1. Typical set of filter requirements.

B-2 APPROXIMATION OF IDEAL FILTERS BY PRACTICAL FILTERS

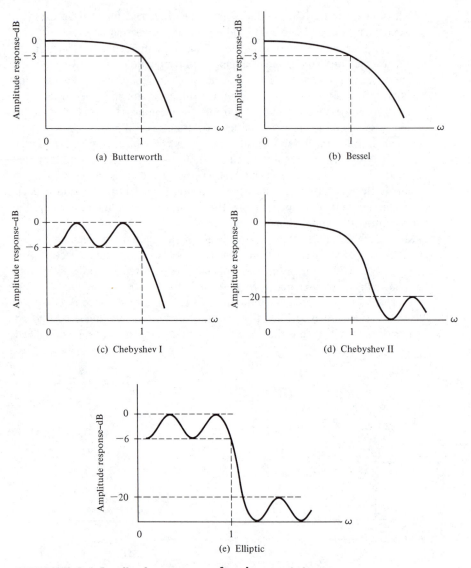

FIGURE B-2. **Amplitude response of various prototypes.**

Butterworth Filter

The Butterworth filter, which is perhaps the most popular of all filter prototypes, is designed to have a maximally flat amplitude response in the passband. The edge of the passband is usually defined to be the frequency where the amplitude response is 0.707 times the maximum value. Thus, ϵ in Figure B-1 is 0.707. The amplitude response of a Butterworth filter decreases monotonically in the transition band and in the stop band. Since the Butterworth filter is designed to have a maximally flat† passband amplitude response, no constraints are placed on the phase response or group delay. However, the group delay of a Butterworth filter is reasonably constant over most of the passband, as we shall observe later. The Butterworth filter is studied in more detail in the following section.

Bessel Filter

The Bessel filter is designed to exhibit a maximally linear phase response, which implies a maximally constant group delay. The Bessel filter has less stop band attenuation than a Butterworth filter of equal order and bandwidth. (See Problems B-3 and B-4 for more on Bessel filters.)

Chebyshev I Filter

The Chebyshev I filter allows the amplitude response to have ripple in the passband. This allows the amplitude response to have a steeper rolloff outside the passband for a fixed filter order. The ripples in the filter passband have equal amplitude and the amplitude response outside of the passband is monotonic. Where a given stopband attenuation is required, a Chebyshev I filter having lower order than a Butterworth filter can sometimes be used. This results in hardware simplifications assuming, of course, that the passband ripples can be tolerated in the given application. In Figure B-2c the passband ripple is 6 dB. This is an extremely large value but was done to illustrate this characteristic of the filter. We study the Chebyshev I filter in more detail in the following section.

Chebyshev II Filter

The Chebyshev II filter has a monotonic amplitude response and equal amplitude ripples in the stop band. For the filter illustrated in Figure B-2d, the minimum stopband attenuation is 20 dB.

Elliptic Filter

The elliptic filter is a combination of the Chebyshev I and Chebyshev II filters in that the elliptic filter has amplitude response ripples in both the passband and stop band. The advantage of the elliptic filter is that for a given allowable ripple in the passband and a minimum attenuation in the stop band, the width of the transition band is minimized. In Figure B-2e, the passband ripple is 6 dB and the minimum attenuation in the stop band is 20 dB.

---

†The nth order Butterworth filter amplitude reponse function has $(2n - 1)$ derivatives equal to zero at $\omega = 0$.

BUTTERWORTH AND CHEBYSHEV FILTERS

To gain a better understanding of filter characteristics, we now examine the Butterworth and Chebyshev I prototypes. Only normalized filters will be considered. In the following section we shall see how to convert these normalized filters to other types of filters.

Butterworth Filters

The Butterworth filter having a 3-dB bandwidth of 1 rad/s has the steady-state amplitude response

$$|H_B(\omega)| = \frac{1}{\sqrt{1 + \omega^{2n}}} \tag{B-12}$$

where $\omega = 2\pi f$ and n is the order of the filter. As stated in the preceding section, the Butterworth filter is designed to have a maximally flat passband amplitude response. The system function corresponding to (B-12) is

$$H_B(s) = \frac{(-1)^n s_1 s_2 \cdots s_n}{(s - s_1)(s - s_2) \cdots (s - s_n)} \tag{B-13}$$

The poles of the Butterworth filter $(s_1, s_2, \ldots, s_n)$ all lie on a circle having a radius of 1 rad/s and centered on the origin of the s-plane. The poles are symmetrical with respect to the real axis. Figure B-3 illustrates the pole locations for $n = 2, 3, 4$ and 5. Note that the angle between the poles is $180°/n$ and that odd-order filters have one real pole at $s = -1$. Even-order Butterworth filters have no real poles.

The transfer function for an nth-order Butterworth filter is given by

$$H_B(s) = \frac{1}{B_n(s)} \tag{B-14}$$

where $B_n(s)$ represents the Butterworth polynomial of degree n. These polynomials are easily derived from knowledge of the pole locations (see Problem B-2). The first six Butterworth polynomials are tabulated in Table B-1. The amplitude responses, in decibels, of several Butterworth filters are illustrated in Figure B-4. All Butterworth filters have a gain at $\omega_c = 2\pi f_c = 1$ of

$$|H(\omega_c)| = \frac{1}{\sqrt{1 + 1^{2n}}} = 0.707 \tag{B-15}$$

which expressed in decibels is

$$20 \log |H(\omega_c)| = 20 \log 0.707 = -3 \tag{B-16}$$

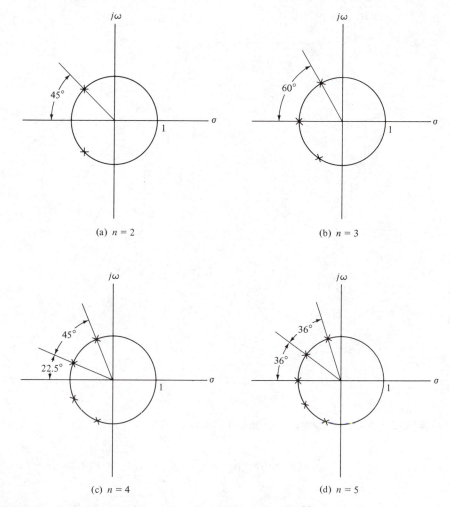

FIGURE B-3. Pole locations for normalized Butterworth filters.

TABLE B-1. Butterworth Polynomials $1 \leq n \leq 6$[a]

$B_1(s) = s + 1$

$B_2(s) = s^2 + 1.4142136s + 1$

$B_3(s) = s^3 + 2.0000000s^2 + 2.0000000s + 1$

$B_4(s) = s^4 + 2.6131259s^3 + 3.4142136s^2 + 2.6131259s + 1$

$B_5(s) = s^5 + 3.2360680s^4 + 5.2360680s^3 + 5.2360680s^2 + 3.2360680s + 1$

$B_6(s) = s^6 + 6.8637033s^5 + 7.4641016s^4 + 9.1416202s^3 + 7.4641016s^2 + 6.8637033s + 1$

[a] $H_n(S) = 1/B_n(S)$.

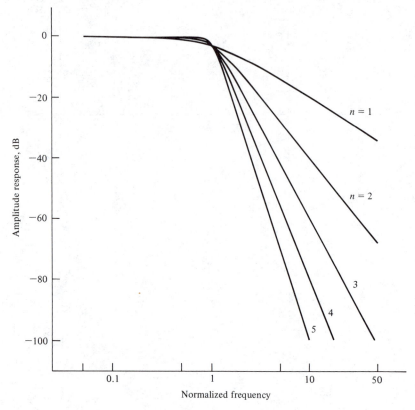

FIGURE B-4. Amplitude response of Butterworth filters.

The group delay of the nth-order Butterworth filter is given by[†]

$$\tilde{T}_g(\omega) = \frac{1}{1 + \omega^{2n}} \sum_{m=0}^{n-1} \frac{\omega^{2m}}{\sin (2m + 1)(\pi/2n)} \qquad \text{(B-17)}$$

and is illustrated in Figure B-5 for several values of n. It can be seen that the group delay is reasonably constant over most of the filter passband.

Chebyshev I Filter

The amplitude response of an nth-order Chebyshev I filter is defined by the equation

$$H_c(\omega) = \frac{1}{\sqrt{1 + \epsilon^2 C_n^2(\omega)}} \qquad \text{(B-18)}$$

in which $C_n(\omega)$ is the Chebyshev polynomial and ϵ is a parameter to be discussed later. The Chebyshev polynomial is defined by

$$C_n(\omega) = \cos (n \cos^{-1} \omega) \qquad \text{(B-19)}$$

[†]See Weinberg (1962), p. 497.

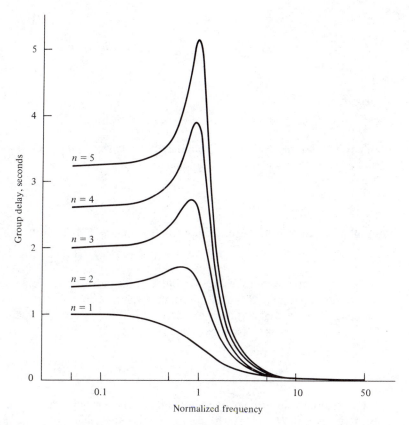

FIGURE B-5. Group delay of Butterworth filters.

Since the argument of the cosine function is only a real angle for ω in the range $-1 \le \omega \le 1$, it is convenient to define $C_n(\omega)$ in two ranges as

$$C_n(\omega) = \cos{(n \cos^{-1} \omega)}, \qquad |\omega| \le 1 \qquad \text{(B-20)}$$

and

$$C_n(\omega) = \cosh{(n \cosh^{-1} \omega)}, \qquad |\omega| > 1 \qquad \text{(B-21)}$$

Equation (B-20) describes the filter in the passband and (B-21) describes the filter outside of the passband.

From (B-20), we see that at zero frequency

$$C_n^2(0) = \cos^2 \frac{n\pi}{2} = \begin{cases} 0, & n \text{ odd} \\ 1, & n \text{ even} \end{cases} \qquad \text{(B-22)}$$

so that

$$|H_c(0)| = 1, \qquad n \text{ odd} \qquad \text{(B-23)}$$

and

$$|H_c(0)| = \frac{1}{\sqrt{1 + \epsilon^2}}, \qquad n \text{ even} \qquad (B-24)$$

Since

$$C_n^2(1) = \cos^2 0 = 1, \qquad \text{all } n \qquad (B-25)$$

both even-order and odd-order Chebyshev filters have an amplitude response at band edge, $\omega = \omega_c = 2\pi f_c = 1$, equal to

$$|H(\omega_c)| = \frac{1}{\sqrt{1 + \epsilon^2}} \qquad (B-26)$$

for all n. Thus, at band edge, the amplitude response of a Chebyshev filter is

$$20 \log |H(f_c)| = -10 \log (1 + \epsilon^2) \qquad (B-27)$$

The parameter ϵ establishes the amplitude of the passband ripple discussed in the preceding section. Since the maximum value of $|H_c(\omega)|$ within the filter passband is 1, and the minimum value of $|H_c(\omega)|$ within the passband is $1/\sqrt{1 + \epsilon^2}$, the peak-to-peak passband ripple is

$$\Delta_R = 1 - \frac{1}{\sqrt{1 + \epsilon^2}}$$

The total number of points within the passband (including $\omega = 0$) at which the derivative of $|H_c(\omega)|$ with respect to ω is zero is equal to the order of the filter. This observation makes it relatively easy to sketch the frequency response of a Chebyshev filter.

It is relatively simple to compute the additional stop band attenuation given by the Chebyshev filter over the Butterworth filter by comparing the amplitude responses of both filters for $\omega \gg 1$. For the Butterworth filter

$$|H_B(\omega)| = \frac{1}{\sqrt{1 + \omega^{2n}}} \qquad (B-28)$$

so that for $\omega \gg 1$,

$$|H_B(\omega)| \simeq \frac{1}{\omega^n} \qquad (B-29)$$

which expressed in decibels is

$$20 \log |H_B(\omega)| = -20n \log \omega \qquad (B-30)$$

For the Chebyshev filter with $\omega \gg 1$, we have

$$|H_c(\omega)| = \frac{1}{\epsilon C_n(\omega)} \qquad (B-31)$$

even for small ϵ (see Problem B-7).

Placing (B-31) in a form that allows comparison with (B-30) requires some effort. Since $\omega \gg 1$, we use (B-21)

$$C_n(\omega) = \cosh (n \cosh^{-1} \omega)$$

For large x

$$\cosh^{-1} x \simeq \ln 2x \qquad \text{(B-32)}$$

so that

$$C_n(\omega) \simeq \cosh(n \ln 2\omega) = \cosh [\ln (2\omega)^n] \qquad \text{(B-33)}$$

Equation (B-32) results from the fact that for large x

$$\cosh x \simeq \tfrac{1}{2} e^x$$

Also, for $\omega \gg 1$, we have

$$C_n(\omega) \simeq \tfrac{1}{2} \exp [\ln (2\omega)^n] = \tfrac{1}{2} (2\omega)^n \qquad \text{(B-34)}$$

Substituting (B-34) into (B-31) yields

$$|H_c(\omega)| \simeq \frac{2}{\epsilon} (2\omega)^{-n}, \qquad \omega \gg 1 \qquad \text{(B-35)}$$

which, expressed in decibles, is

$$20 \log |H_c(\omega)| \simeq 20 \log 2 - 20 \log \epsilon - 20n \log 2\omega$$

or

$$20 \log |H_c(\omega)| \simeq -20(n - 1) \log 2 - 20 \log \epsilon - 20n \log \omega \qquad \text{(B-36)}$$

The additional *attenuation* of the Chebyshev filter over the Butterworth filter, expressed in decibels, is

$$\Delta_A = 20 \log |H_B(\omega)| - 20 \log |H_c(\omega)| \qquad \text{(B-37)}$$
$$\simeq 20(n - 1) \log 2 + 20 \log \epsilon$$

For $\epsilon \ll 1$, the value of Δ_A can be negative, indicating less stop band attenuation for the Chebyshev filter. However, if $\epsilon < 1$, the bandwidth of the Chebyshev filter is effectively larger than the bandwidth of the Butterworth to which it is being compared. Letting $\epsilon = 1$, which yields 3-dB attenuation at $\omega = 1$, results in

$$\Delta_A \simeq 20(n - 1) \log 2 = 6(n - 1) \text{ dB} \qquad \text{(B-38)}$$

indicating equal stop band attenuation for $n = 1$. The value of Δ_A increases as n and ϵ increase above these values.

It is worth noting that for $n = 1$

$$C_1(\omega) = \omega$$

so that

$$|H_c(\omega)| = \frac{1}{\sqrt{1 + (\omega\epsilon)^2}} \qquad \text{(B-39)}$$

We shall show in the following section that this is simply a first-order Butterworth filter with a scaled 3-dB frequency of $1/\epsilon$ radians per second.

It is also worth noting that since Δ_A is not a function of ω, the slopes of the stop-band amplitude response of both the Butterworth and Chebyshev filters for large ω are equal.

The amplitude responses of several Chebyshev I filters are illustrated in Figure B-6 for 1-dB passband ripple and in Figure B-7 for 3-dB passband ripple. A comparison of Figure B-6 and B-7 illustrates that, as the passband ripple increases, the stop-band attenuation also increases for fixed frequency and filter order. A comparison of Figure B-6 and B-7 with Figure B-4 illustrates the increased stop-band attenuation of the Chevyshev I filter compared to the Butterworth filter.

The pole locations of a Chebyshev filter are considerably more difficult to specify than the pole locations of a Butterworth filter having the same order. The difficulties involved can be appreciated by considering, as an example, a fourth-order Chebyshev filter. This is illustrated in Figure B-8. In order to determine the pole locations of a normalized Chebyshev filter, the first step is to draw two circles having radii a and b, respectively. The values of a and b

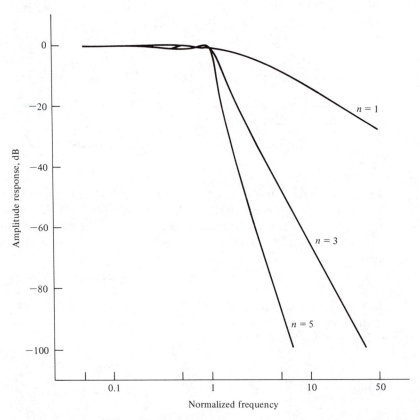

FIGURE B-6. Amplitude response of Chebyshev I filter with 1-dB passband ripple.

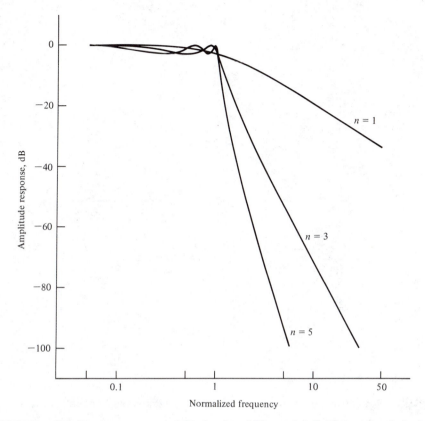

FIGURE B-7. **Amplitude response of Chebyshev I filter with 3-dB passband ripple.**

are determined by the parameter ϵ according to

$$b, a = \tfrac{1}{2} [(\sqrt{1 + 1/\epsilon^2} + 1/\epsilon)^{1/n} \pm (\sqrt{1 + 1/\epsilon^2} + 1/\epsilon)^{-1/n}]$$

in which n is the order of the filter. These two circles define the major and minor axes of an ellipse, which is shown by the dashed line in Figure B-8. In the circle of radius b, the pole positions of a Butterworth filter of order n are located as shown by the heavy dots. Horizontally projecting these points onto the ellipse determine the pole locations of the Chebyshev filter.

The angles illustrated in Figure B-8 are the same angles between the poles as defined for a Butterworth filter of equal order. This can be seen by comparing Figure B-8 with Figure B-3c. The values of a and b in the preceding expression were defined for $\omega_c = 1$. For different values of ω_c the values of a and b are replaced by $a\omega_c$ and $b\omega_c$, respectively.

B-4

FILTER TRANSFORMATIONS

In the previous sections emphasis has been placed on the normalized filter prototypes, that is, low-pass filters having a bandwidth of 1 rad/s. In this section

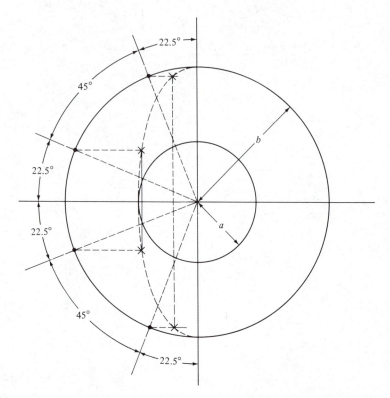

FIGURE B-8.   Pole locations for a fourth-order low-pass Chebyshev filter.

we see how to convert the normalized filter to low-pass filters having arbitrary bandwidth and also how to convert the normalized filter to bandpass and notch (band reject) filters.

Frequency Scaling

The 1-rad/s bandwidth of a normalized prototype is converted to a bandwidth of ω_c radians per second by substituting s/ω_c for s in the normalized prototype system function $H(s)$. This is equivalent to substituting ω/ω_c or f/f_c in the sinusoidal steady-state transfer function.

EXAMPLE B-1

The amplitude response of an nth-order Butterworth filter having a 3-dB bandwidth of 150 Hz is, from (B-28),

$$|H_B(\omega)| = \frac{1}{\sqrt{1 + [\omega/2\pi(150)]^{2n}}} = \frac{1}{\sqrt{1 + (f/150)^{2n}}} \qquad \text{(B-40)}$$

■

EXAMPLE B-2

The transfer function of a second-order Butterworth filter with a 3-dB bandwidth of ω_c radians per second is found by using $B_2(s)$ in Table B-1 with s

replaced by s/ω_c. The result is

$$H(s) = \frac{1}{(s/\omega_c)^2 + \sqrt{2}\,(s/\omega_c) + 1}$$

which yields

$$H(s) = \frac{\omega_c^2}{s^2 + \sqrt{2}\,\omega_c s + \omega_c^2} \tag{B-41}$$

EXAMPLE B-3

The system function $H(s)$ for a third-order Butterworth filter having a 3-dB bandwidth of 150 Hz is determined by first substituting

$$\frac{s}{2\pi(150)}$$

for s in $B_3(s)$ in Table B-1. The required system function is then

$$H(s) = 1/B_3\left(\frac{s}{2\pi(150)}\right)$$

Since

$$B_3\left(\frac{s}{2\pi(150)}\right) = \frac{s^3 + 2[2\pi(150)]s^2 + 2\,[2\pi(150)]^2\,s + [2\pi(150)]^3}{[2\pi(150)]^3}$$

which is

$$B_3\left(\frac{s}{2\pi(150)}\right) = \frac{s^3 + 1.88496(10^3)s^2 + 1.77653(10^6)s + 8.37169(10^8)}{8.37169(10^8)}$$

the required system function is

$$H(s) = \frac{8.37169(10^8)}{s^3 + 1.88496(10^3)s^2 + 1.77653(10^6)s + 8.37169(10^8)} \tag{B-42}$$

Low-pass-to-Bandpass Transformation

Assume that $H(s)$ represents the system function of a low-pass filter having a bandwidth of 1 rad/s (i.e., a normalized prototype). This filter is easily converted into a bandpass filter having a *geometric center frequency* of ω_c radians per second and a bandwidth of ω_b radians per second by substituting for s in $H(s)$ the quantity

$$\frac{s^2 + \omega_c^2}{s\omega_b}$$

In other words, the bandpass transfer function is defined by

$$H_{bp}(s) = H\left(\frac{s^2 + \omega_c^2}{s\omega_b}\right)$$

(B-43)

This transformation illustrates that the order of the bandpass filter is twice the order of the low-pass filter from which it is derived.

The bandwidth of the low-pass prototype is $\omega = 1$. Thus the corresponding critical frequencies† of the bandpass filter are given by the two solutions of the quadratic equation

$$\left.\frac{s + \omega_c}{s\omega_b}\right|_{s=j\omega} = \frac{-\omega^2 + \omega_c^2}{j\omega\omega_b} = j1$$

(B-44)

This yields

$$\omega^2 - \omega\omega_b - \omega_c^2 = 0$$

(B-45)

whose solutions are

$$\omega = \tfrac{1}{2}\omega_b \pm \tfrac{1}{2}\sqrt{\omega_b^2 + 4\omega_c^2}$$

(B-46)

The upper critical frequency ω_u is obtained by using the plus sign. Thus

$$\omega_u = \tfrac{1}{2}\omega_b + \tfrac{1}{2}\sqrt{\omega_b^2 + 4\omega_c^2}$$

(B-47)

Using the minus sign in (B-46) clearly results in a negative quantity. This is reconciled by realizing that, since the amplitude response is an even function of frequency, the lower critical frequency ω_l can be written

$$\omega_l = -\tfrac{1}{2}\omega_b + \tfrac{1}{2}\sqrt{\omega_b^2 + 4\omega_c^2}$$

(B-48)

Taking the difference between (B-47) and (B-48) illustrates that

$$\omega_u - \omega_l = \omega_b$$

(B-49)

which is the standard definition of bandwidth.

The relationship between the two critical frequencies and ω_c is also easily derived. The product of the two critical frequencies is, from (B-47) and (B-48),

$$\omega_u\omega_l = \tfrac{1}{4}\left(\omega_b + \sqrt{\omega_b^2 + 4\omega_c^2}\right)\left(-\omega_b + \sqrt{\omega_b^2 + 4\omega_c^2}\right)$$

which yields

$$\omega_u\omega_l = \omega_c^2$$

Thus

$$\omega_c = \sqrt{\omega_u\omega_l}$$

(B-50)

which states that the center frequency of a bandpass filter, derived by use of the

---

†Usually, we speak in terms of 3-dB break frequencies. However, for some filter prototypes the attenuation at $\omega = 1$ is not 3 dB (consider, for example, the Chebyshev filter). Thus we must be careful in our terminology.

transformation (B-43), is the geometric mean of the upper and lower critical frequencies. Thus, we refer to ω_c as the geometric center frequency.

It should be noted that if $\omega_c \gg \omega_b$, which is typically the case, (B-47) and (B-48) become

$$\omega_u \simeq \omega_c + \tfrac{1}{2}\omega_b \qquad \text{(B-51)}$$

and

$$\omega_l \simeq \omega_c - \tfrac{1}{2}\omega_b \qquad \text{(B-52)}$$

respectively. Adding these two equations yields

$$\omega_c \simeq \tfrac{1}{2}(\omega_u + \omega_l) \qquad \text{(B-53)}$$

An example will illustrate the application of the low pass-to-bandpass transformation.

EXAMPLE B-4

In this example the transfer function of a fourth-order Butterworth filter having a geometric center frequency of 1000 Hz and a bandwidth of 100 Hz will be determined. Equation (B-43) illustrates that a second-order low-pass filter will yield a fourth-order bandpass filter upon application of the low pass-to-bandpass transformation. Thus the required transfer function is

$$H(s) = \frac{1}{B_2 \left[(s^2 + \omega_c^2)/s\omega_b \right]} \qquad \text{(B-54)}$$

Using $B_2(s)$ from Table B-1 yields

$$H(s) = \frac{1}{\left[(s^2 + \omega_c^2)/s\omega_b \right]^2 + 1.41421 \left[(s^2 + \omega_c^2)/s\omega_b \right] + 1} \qquad \text{(B-55)}$$

which can be written as

$$H(s) = \frac{\omega_b^2 s^2}{s^4 + 1.41421\omega_b s^3 + (2\omega_c^2 + \omega_b^2)s^2 + 1.41421\omega_c^2\omega_b s + \omega_c^4} \qquad \text{(B-56)}$$

The required specifications are realized by letting

$$\omega_c = 2\pi(1000) = 2000\pi$$

and

$$\omega_b = 2\pi(100) = 200\pi$$

Substituting these values into (B-56) yields

$$H(s) =$$

$$\frac{3.94784(10^5)s^2}{s^4 + 8.88577(10^2)s^3 + 7.93516(10^7)s^2 + 3.50796(10^{10})s + 1.55855(10^{15})} \qquad \text{(B-57)}$$

■

Low-pass-to-Band Reject (Notch) Transformation

A band-reject filter can be obtained from a standard low-pass prototype by using the transformation

$$H_{br}(s) = H\left(\frac{s\omega_b}{s^2 + \omega_c^2}\right) \tag{B-58}$$

which is the reciprocal of the low-pass-to-bandpass transformation. The analysis of the resulting parameters (bandwidth and center frequency) is exactly parallel to the analysis of the bandpass filter; thus it will not be pursued here. However, it can be easily shown that the bandwidth of the notch filter is ω_b radians per second and that ω_c is the geometric center frequency of the notch in radians per second.

FURTHER READING

L. WEINBERG, *Network Analysis and Synthesis*, New York: McGraw-Hill, 1962.
    This is the classic text on filter synthesis. It is written at the senior or beginning graduate level and is extremely complete. The transfer functions of all the filter prototypes discussed in this chapter are derived in detail. Implementations using lumped-parameter, passive circuit elements are discussed.

A. I. ZVEREV, *Handbook of Filter Synthesis*, New York: Wiley, 1967.
    This volume contains extensive collections of amplitude response and group delay characteristics that are valuable in learning the characteristics of the various prototypes. It also contains extensive sets of tables giving the required element values for various implementations.

L. P. HUELSMAN and P. E. ALLEN, *Introduction to the Theory and Design of Active Filters*, New York: McGraw-Hill, 1980.
    This book is a standard text on active filter theory, which is an important area not covered in this book. The student who has an interest in the implementation of analog filters should be aware of the concept of the active filter and should appreciate the techniques used in implementing active filters.

M. E. VAN VALKENBURG, *Analog Filter Design*. New York: Holt, Rinehart and Winston, 1982.
    This very recent book contains much information on classical filter theory as well as information on the synthesis of active analog filters based on a large number of prototypes.

PROBLEMS

SECTION B-3

B-1. It can be shown that the coefficients of the nth-order Butterworth polynomial

$$B_n(s) = \sum_{k=0}^{n} a_k s^k$$

are given by the recursion formula

$$a_{k+1} = \frac{\cos(k\pi/2n)}{\sin(k+1)(\pi/2n)} a_k$$

with $a_0 = 1$.† By using this recursion formula, derive $B_n(s)$ in Table B-1 for $n = 2, 3,$ and 4.

B-2. Use the fact that the poles of the Butterworth filter are distributed on the unit circle as shown in Figure B-3 to derive the Butterworth polynomials $B_2(s)$ and $B_3(s)$.

B-3. It can be shown that the system function for a low-pass Bessel filter is given by

$$H(s) = \frac{k}{Y_n(s)}$$

where $Y_n(s)$ is the Bessel polynomial of degree n and k is a suitably chosen constant. Assume that $Y_2(s)$ is given by‡

$$Y_2(s) = s^2 + 3s + 3$$

(a) Determine k so that the dc gain is 1.
(b) Plot the amplitude response of both the second-order Bessel and Butterworth filters together using a single coordinate system. Comment on your results.
(c) Plot the phase response of both the Bessel and Butterworth filters together using a single coordinate system and linear scales. Comment on your results.
(d) The bandwidth of a low-pass Bessel filter is defined to be

$$f_c = \frac{\omega_c}{2\pi} = \frac{1}{2\pi t_0}$$

where t_0 is the nominal passband group delay of the filter. Using this definition, compare the bandwidths of the Bessel and Butterworth filters.

B-4. Repeat Problem B-3 for third-order Bessel and Butterworth filters assuming that

$$Y_3(s) = s^3 + 6s^2 + 15s + 15$$

B-5. Determine the values of ϵ required if a Chebyshev filter is to have a passband ripple of (a) 0.2 dB, (b) 1.0 dB, and (c) 6 dB.

B-6. Sketch, approximately to scale, amplitude response of a Chebyshev filter having 3-dB passband ripple for $n = 2, 3, 4,$ and 5. Pay particular attention to the response at $\omega = 0$ and $\omega = 1$. Also pay particular attention to the number of passband ripples.

B-7. Show by plotting $C_n(\omega)$ as a function of ω in the range $0 \leq \omega \leq \omega_x$, where $\omega_x > 1$, that $C_n(\omega)$ oscillates in the filter passband and increases rapidly for $\omega > 1$. Let $n = 2, 3,$ and 4.

†Weinberg (1962), p. 494.
‡Weinberg (1962), p. 500.

SECTION B-4

B-8. Derive the transfer function for a second-order Butterworth low-pass filter having a 3-dB bandwidth of 1000 Hz. Plot accurately the amplitude and phase response of the lowpass filter.

B-9. Derive the transfer function for a second-order Butterworth bandpass filter having a geometric center frequency of 1000 Hz and a bandwidth of 600 Hz. Verify your result by plotting the amplitude and phase responses.

B-10. Determine the transfer function for a fourth-order Butterworth bandpass filter with a geometric center frequency of 100 Hz and a bandwidth of 80 Hz. Determine the upper and lower 3-dB frequencies.

B-11. Repeat Problem B-10 assuming that the bandpass filter is to be sixth order.

B-12. Determine the transfer function of a fourth-order Butterworth notch filter with a notch geometric center frequency of 1000 Hz and a notch bandwidth frequency of 600 Hz. Plot accurately the amplitude and phase response of the notch filter.

Functions of a Complex Variable—Summary of Important Definitions and Theorems

To cover complex variable theory in any detail in an appendix of the length of this one is impossible. Therefore, only the definitions and theorems important to the material in this text are summarized with few proofs given.

Let $s = \sigma + j\omega$ denote a complex variable. Another complex variable $W = U + jV$ is said to be a *function of the complex variable s* if, to each value of s in some set, there corresponds a value, or a set of values, of W. We denote the rule of correspondence between s and W by writing $W = F(s)$. If for each value of s there is only one value of W, $F(s)$ is said to be a *single-valued* function of s. If more than one value of W corresponds to some value or set of values of s, $F(s)$ is *multivalued*.

EXAMPLE C-1
(a) The function $W = F(s) = s^2$ is single-valued. We can express U and V in terms of σ and ω as

$$W = U + jV = (\sigma + j\omega)^2$$
$$= \sigma^2 - \omega^2 + j2\sigma\omega$$

or

$$U = \sigma^2 - \omega^2$$

and

$$V = 2\sigma\omega$$

For each value of $s = \sigma + j\omega$ there corresponds one, and only one, value for $W = U + jV$.

(b) The function $W = F(s) = \sqrt{s}$ is double-valued. To show this, we write s in polar form as

$$s = re^{j\phi}$$

so that

$$W = (re^{j\phi})^{1/2}$$

Now $e^{j\phi} = e^{j(\phi + 2k\pi)}$, where $k = 0, 1$. Therefore,

$$W = r^{1/2}e^{j(\phi/2 + k\pi)}, \qquad k = 0, 1$$

and for each value of s there corresponds two values of W, which are

$$W_1 = r^{1/2}e^{j\phi/2}$$

and

$$W_2 = r^{1/2}e^{j(\phi/2 + \pi)}$$

where $r^{1/2}$ is the positive square root of r. For example, with $s = -4$, these two values of W are

$$W_1 = 2e^{j\pi/2} = 2j \quad \text{and} \quad W_2 = 2e^{j(\pi/2 + \pi)} = -2j$$

(c) $W = s^*$, where the asterisk denotes the complex conjugate, is a single-valued function of s. If $s = \sigma + j\omega$, then $W = F(s) = \sigma - j\omega$. ∎

We now state a few definitions pertaining to functions of a complex variable.

Definiton 1: The *neighborhood* of a point s_0 in the complex plane is the open circular disk $|s - s_0| < \rho$ where $\rho > 0$ is a real arbitrary constant.

Definition 2: A function $F(s)$ is said to have the *limit L* as s approaches s_0 if $F(s)$ is defined in a neighborhood of s_0 (except perhaps at s_0) and if, for every positive real number $\epsilon > 0$, we can find a real number δ such that

$$|F(s) - L| < \epsilon \tag{C-1}$$

for all values of $s \neq s_0$ in the disk $|s - s_0| < \delta$ $(\delta > 0)$. Note that s may approach s_0 from *any direction* in the complex plane.

EXERCISE

Show that the limit of $F(s) = s^2$ as $s \to 0$ is zero. Note that the value of $F(s)$ at $s = 0$ is immaterial.

Definition 3: The function $F(s)$ is said to be *continuous* at $s = s_0$ if $F(s_0)$ is defined and

$$\lim_{s \to s_0} F(s) = F(s_0) \tag{C-2}$$

EXERCISE

Show that $F(s) = s^2$, all s, is continuous at $s = 0$, and in fact for all s.

Definition 4: A function $F(s)$ is said to be *differentiable* at a point $s = s_0$ if the limit

$$F'(s_0) = \lim_{\Delta s \to 0} \frac{F(s_0 + \Delta s) - F(s_0)}{\Delta s} \tag{C-3}$$

exists. Letting $s = s_0 + \Delta s$, we may also write this as

$$F'(s_0) = \lim_{s \to s_0} \frac{F(s) - F(s_0)}{s - s_0} \tag{C-4}$$

Note that s may approach s_0 from any direction and the resulting limiting values of (C-4) must be the same regardless of the direction of approach.

EXAMPLE C-2

Consider the derivative of the complex function $F(s) = s^2$. We find that

$$F(s) = \lim_{\Delta s \to 0} \frac{(s + \Delta s)^2 - s^2}{\Delta s}$$

$$= \lim_{\Delta s \to 0} \frac{2s\Delta s + (\Delta s)^2}{\Delta s}$$

$$= 2s$$

at any point s. ∎

At this point, we simply state that all the familiar rules for differentiation of functions of a real variable continue to hold for functions of a complex variable. For example, the derivative of $\sin s$ is $\cos s$, etc. The following example gives a function that does not have a derivative anywhere.

EXAMPLE C-3

Consider $F(s) = s^* = \sigma - j\omega$. For this function,

$$\frac{F(s + \Delta s) - F(s)}{\Delta s} = \frac{(\sigma + \Delta \sigma) - j(\omega + \Delta \omega) - (\sigma - j\omega)}{\Delta \sigma + j\Delta \omega}$$

$$= \frac{\Delta \sigma - j\Delta \omega}{\Delta \sigma + j\Delta \omega}$$

where $\Delta s = \Delta \sigma + j\Delta \omega$. The limit of this ratio depends on the way Δs approaches zero. For example, if $\Delta \sigma \to 0$ first and then $\Delta \omega \to 0$, we obtain as the limit -1. On the other hand, if $\Delta \omega \to 0$ first and then $\Delta \sigma \to 0$ the limit is $+1$. Hence this function is not differentiable anyplace since s was arbitrarily chosen. ∎

Definition 5: A function $F(s)$ is said to be *analytic* (regular) at a point $s = s_0$ if it is defined and has a derivative at every point in some neighborhood of s_0.

Note that this definition is stronger than simply saying that a function is differentiable at a point, since the derivative must exist everywhere in a neighborhood of the point.

Definition 6: A function $F(s)$ that is analytic at every point in a domain D is said to be *analytic* in D. (*Note:* A domain is simply an open, connected set of points.)

Definition 7: A function $F(s)$ that has at least one analytic point is called an *analytic function*.

Definition 8: A point s_0 at which the analytic function $F(s)$ is not analytic is called a *singular point*.

It can be shown that an analytic function $F(s)$ of a complex variable has derivatives of all orders, except at singular points, and can be represented as a Taylor series in a neighborhood of each point s_0 where $F(s)$ is analytic:

$$F(s) = \sum_{n=0}^{\infty} \frac{F^{(n)}(s_0)}{n!} (s - s_0)^n \tag{C-5}$$

where $F^{(n)}(s_0)$ denotes the nth derivative of $F(s)$ evaluated at $s = s_0$. For example,

$$e^s = \sum_{n=0}^{\infty} e^{s_0} \frac{(s - s_0)^n}{n!}$$

about any point s_0 in the complex plane. In fact, this is how the transcendental functions e^s, $\sin s$, $\cos s$, etc., of a complex variable are defined.

The function $F(s) = G(s)/(s - s_0)^m$, where $G(s)$ is analytic at $s = s_0$, is singular at $s = s_0$ because of the factor $(s - s_0)^{-m}$. Such a singularity is called a *pole* of order m. This type of singularity is said to be isolated and, in the neighborhood of an isolated singularity, the representation

$$F(s) = \sum_{n=-\infty}^{\infty} C_n(s - s_0)^n \tag{C-6}$$

can be used. More specifically:

Definition 9: If there is some neighborhood of a singular point s_0 of a function $F(s)$ throughout which F is analytic, except at the point itself, then s_0 is called an *isolated singular point* of F.

The series (C-6) is known as the *Laurent series expansion* for the function at $s = s_0$. If $C_n = 0$, $n < -m$, $F(s)$ is said to have an mth-order pole at $s = s_0$ ($m > 0$); if $m = 1$, $F(s)$ is said to have a simple pole at $s = s_0$; if $C_n \neq 0$ for infinitely many negative values of n, $F(s)$ has an *essential singularity* at $s = s_0$. Examples of functions with isolated singular points are $1/s$ and $e^s/(s - 1)^2$. The function $1/\sin(\pi/s)$ has isolated singular points at $s = \pm 1$, $\pm 1/2$, $\pm 1/3$, ... At $s = 0$ this function has a nonisolated singular point since every neighborhood of $s = 0$ contains other singular points.

Theorem 1: A necessary condition that a function $F(s) = U + jV$ of a complex variable $s = \sigma + j\omega$ be analytic at a point s_0 is that its real and imaginary parts satisfy the *Cauchy–Riemann equations*

$$\frac{\partial U}{\partial \sigma} = \frac{\partial V}{\partial \omega} \quad \text{and} \quad \frac{\partial U}{\partial \omega} = -\frac{\partial V}{\partial \sigma} \tag{C-7}$$

Theorem 2: Sufficient conditions for a function $F(s)$ of a complex variable s to be differentiable at a point $s = s_0$ is that $U(\sigma, \omega)$ and $V(\sigma, \omega)$ have continuous first partial derivatives that satisfy the Cauchy–Riemann equations at the point $s = s_0$.

Theorem 3 (Cauchy's Integral Theorem): Let a function of a complex variable $F(s)$ be analytic everywhere on and within a simple closed curve C (by simple, we mean that the curve does not cross itself). Then

$$\oint_C F(s)\ ds = 0 \tag{C-8}$$

where $\oint_C (\cdot)\ ds$ denotes the line integral in a counterclockwise direction along C.

Figure C-1 illustrates the circumstances of the theorem. Suppose that $F(s)$ has an isolated singularity at $s = s_0$ which is inside C as shown in Figure C-2. Is $\oint_C F(s)\ ds = 0$ still? No, because $F(s)$ is not analytic everywhere within and on C. However, consider the simple closed curve shown in Figure C-3. Since $F(s)$ is analytic within and on the simple closed curve $C + C_1 + C' + C_2$ it follows by Cauchy's integral theorem that

$$\oint_{C+C_1+C'+C_2} F(s)\ ds = 0 \tag{C-9}$$

Breaking the line integral up into the sum of four integrals over the separate curves C, C_1, C_2, and C', we obtain

$$\oint_C F(s)\ ds + \oint_{C_1} F(s)\ ds + \oint_{C_2} F(s)\ ds + \oint_{C'} F(s)\ ds = 0 \tag{C-10}$$

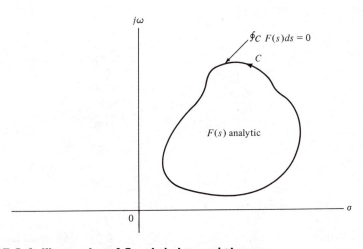

FIGURE C-1. Illustration of Cauchy's integral theorem.

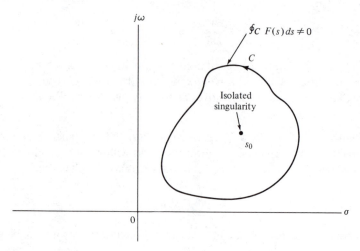

FIGURE C-2. Line integral around a closed contour that is not zero because of a singularity within the contour.

Now, as the gap between the curves C_1 and C_2 approaches zero, the second and third integrals will approach equal magnitude, opposite in sign values (both have the same integrand, are integrated over the same length, but are taken in opposite directions). It follows, therefore, that

$$\oint_C F(s)\ ds\ =\ -\oint_{C'} F(s)\ ds, \qquad \epsilon \to 0 \qquad \text{(C-11)}$$

We now focus our attention on the integral over C' as $\epsilon \to 0$. For any point s on C', we may write

$$s\ =\ s_0\ +\ \delta e^{j\theta} \qquad (s \text{ on } C') \qquad \text{(C-12)}$$

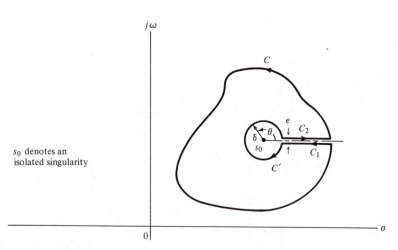

FIGURE C-3. Composite closed curve, $C + C_1 + C_2 + C'$, chosen so as to satisfy Cauchy's integral theorem.

Therefore, in the integral over C' of (C-11) we may substitute

$$ds = j\delta e^{j\theta}\, d\theta \qquad\qquad\text{(C-13)}$$

which results by taking the derivative of (C-12) (note that for any point on C', both s_0 and δ are fixed). Also, we may express $F(s)$ in the neighborhood of s_0 as a Laurent series as given by (C-6). In particular, on the curve C', s is given by (C-12), so that

$$s - s_0 = \delta e^{j\theta} \qquad\qquad\text{(C-14)}$$

and (C-6) may be expressed as

$$F(s) = \sum_{n=-\infty}^{\infty} C_n (\delta e^{j\theta})^n$$

$$= \sum_{n=-\infty}^{\infty} C_n \delta^n e^{jn\theta} \qquad (s \text{ on } C') \qquad\text{(C-15)}$$

Substituting (C-13) and (C-15) into the integral over C', we obtain

$$\oint_{C'} F(s)\, ds = \int_{\theta=2\pi}^{0} \left[\sum_{n=-\infty}^{\infty} C_n \delta^n e^{jn\theta} \right] j\delta e^{j\theta}\, d\theta \qquad\text{(C-16)}$$

where the line integral can be written as an integral over the angle θ because of the circular path. Note that the integral over θ goes from $\theta = 2\pi$ to $\theta = 0$ because of the clockwise orientation of C'. Because the Laurent series for $F(s)$ is uniformly convergent to $F(s)$ in the neighborhood of s_0, the integral and sum in (C-16) can be interchanged to yield

$$\oint_{C'} F(s)\, ds = -\sum_{n=-\infty}^{\infty} jC_n \delta^{n+1} \int_0^{2\pi} e^{j(n+1)\theta}\, d\theta \qquad\text{(C-17)}$$

where the integral is now taken from $\theta = 0$ to $\theta = 2\pi$ with a minus sign inserted to account for the reversal of limits. Now, for $n + 1 \neq 0$, the integral of $\exp[j(n+1)\theta]$ from 0 to 2π is zero (the student may show this by integration and substituting the limits). For $n + 1 = 0$ the integral in (C-17) evaluates to 2π. Thus, all terms in the sum are zero except the term for $n = -1$, which gives

$$\oint_{C'} F(s)\, ds = -2\pi j C_{-1} \qquad\qquad\text{(C-18)}$$

Therefore, as δ and ϵ approach zero, (C-11) evalautes to

$$\oint_C F(s)\, ds = 2\pi j C_{-1} \qquad\qquad\text{(C-19)}$$

where C_{-1} is the residue of $F(s)$ at $s = s_0$. If $F(s)$ has more than one isolated singular point within C, say at the points $s_1, s_2, \ldots, s_N$ with residues $K_1, K_2, \ldots, K_N$, then we may follow the procedure outlined above for each singular point and arrive at the *residue theorem:*

Theorem 4 (Residue Theorem): Let $F(s)$ be analytic everywhere within and on the simple closed curve C except at the isolated singular points $s_1, s_2, \ldots, s_N$. Then

$$\oint_C F(s)\, ds = 2\pi j \sum_{n=1}^{N} K_n \tag{C-20}$$

where $\oint_C (\cdot)\, ds$ is the line integral in a counterclockwise sense of $F(s)$ around C and $K_1, K_2, \ldots, K_N$ are the residues of $F(s)$ at $s_2, s_2, \ldots, s_N$.

We complete this appendix with several theorems pertaining to the single-sided Laplace transform.

Theorem 5 (Absolute Convergence of the Single-Sided Laplace Transform): The integral

$$X(s) = \int_0^\infty x(t)e^{-st}\, dt \tag{C-21}$$

converges absolutely for all Re $(s) > c$ if

$$\int_0^L |x(t)|\, dt \le K < \infty \tag{C-22}$$

for $0 < L < \infty$ and $0 < K < \infty$, and

$$|x(t)| \le Ae^{ct}, \qquad t > L \tag{C-23}$$

for some real A and c.

Proof: According to the definition on absolute convergence, we must show that

$$I = \int_0^\infty |x(t)e^{-st}|\, dt < \infty \qquad \text{for } \sigma = \text{Re } (s) > c$$

We divide I up into two integrals as

$$I = \int_0^L |x(t)|e^{-\sigma t}\, dt + \int_L^\infty |x(t)|e^{-\sigma t}\, dt$$

$$= I_1 + I_2$$

Now

$$I_1 \le e^{|\sigma|L} \int_0^L |x(t)|\, dt \le Ke^{|\sigma|L}$$

and

$$I_2 \le A \int_L^\infty e^{-(\sigma - c)t}\, dt = \frac{Ae^{-(\sigma - c)L}}{\sigma - c} \qquad \text{since } \sigma > c$$

Hence,

$$I = I_1 + I_2 \le Ke^{|\sigma|L} + \frac{Ae^{-(\sigma - c)L}}{\sigma - c} < \infty \qquad \text{for } \sigma > c$$

Definition 10: (Uniform convergence of an integral): The integral

$$I(s) = \int_0^\infty g(t, s)\, dt$$

is said to *converge uniformly* for s in a specific range (say 0 to ∞) if, for each $\epsilon > 0$, we can find a d such that

$$\left| \int_0^b g(t, s)\, dt - I(s) \right| < \epsilon \qquad \text{for } b > d$$

Theorem 6: If the integral

$$X(s) = \int_0^\infty x(t)e^{-st}\, dt$$

converges absolutely for $s_0 = \sigma_0 + j\omega_0$, it also converges absolutely and uniformly in the half plane Re $(s) \geq \sigma_0$.

Theorem 7: If the single-sided Laplace transform integral converges absolutely for $\sigma > \sigma_a$, then $X(s)$ is analytic for $\sigma > \sigma_a$.

This theorem tells us that the singularities of $X(s)$ must lie to the left or on the line Re $(s) = \sigma_a$. By a process known as *analytic continuation*, $X(s)$ may be defined and shown to be analytic on the entire s-plane except, of course, at its singularities.

Theorem 8: If $x_1(t)$ and $x_2(t)$ both have the same single-sided Laplace transform, then

$$x_1(t) = x_2(t)$$

except, possibly, at a set of points with "measure zero"; that is, the integral of $|x_1(t) - x_2(t)|$ over the range of integration is zero.

FURTHER READING

The books by Churchill and Kreyzig cited as references in Chapter 4 are recommended if the student wishes to have a more complete treatment of complex variable theory than that given here.

Mathematical Tables

The Sinc Function

z	sinc z	sinc$^2 z$	z	sinc z	sinc$^2 z$
0.0	1.0	1.0	1.6	-0.18921	0.03580
0.1	0.98363	0.96753	1.7	-015148	0.02295
0.2	0.93549	0.87514	1.8	-0.10394	0.01080
0.3	0.85839	0.73684	1.9	-0.05177	0.00268
0.4	0.75683	0.57279	2.0	0	0
0.5	0.63662	0.40528	2.1	0.04684	0.00219
0.6	0.50455	0.25457	2.2	0.08504	0.00723
0.7	0.36788	0.13534	2.3	0.11196	0.01254
0.8	0.23387	0.05470	2.4	0.12614	0.01591
0.9	0.10929	0.01194	2.5	0.12732	0.01621
1.0	0	0	2.6	0.11643	0.01356
1.1	-0.08942	0.00800	2.7	0.09538	0.00910
1.2	-0.15591	0.02431	2.8	0.06682	0.00447
1.3	-0.19809	0.03924	2.9	0.03392	0.00115
1.4	-0.21624	0.04676	3.0	0	0
1.5	-0.21221	0.04503			

Trigonometric Identities

Euler's theorem: $e^{\pm ju} = \cos u \pm j \sin u$

$\cos u = \frac{1}{2}(e^{ju} + e^{-ju})$

$\sin u = (e^{ju} - e^{-ju})/2j$

$\sin^2 u + \cos^2 u = 1$

$\cos^2 u - \sin^2 u = \cos 2u$

$2 \sin u \cos u = \sin 2u$

$\cos^2 u = \frac{1}{2}(1 + \cos 2u)$

$\sin^2 u = \frac{1}{2}(1 - \cos 2u)$

$\sin(u \pm v) = \sin u \cos v \pm \cos u \sin v$

$\cos (u \pm v) = \cos u \cos v \mp \sin u \sin v$

$\sin u \sin v = \frac{1}{2}[\cos(u - v) - \cos(u + v)]$

$\cos u \cos v = \frac{1}{2}[\cos(u - v) + \cos (u + v)]$

$\sin u \cos v = \frac{1}{2}[\sin(u - v) + \sin(u + v)]$

Indefinite Integrals

$\int \sin (ax) \, dx = -\frac{1}{a} \cos (ax)$

$\int \cos (ax) \, dx = \frac{1}{a} \sin (ax)$

$\int \sin^2 (ax) \, dx = x/2 - \sin (2ax)/4a$

$\int \cos^2 (ax) \, dx = x/2 + \sin (2ax)/4a$

$\int x \sin (ax) \, dx = [\sin(ax) - ax \cos (ax)]/a^2$

$\int x \cos (ax) \, dx = [\cos(ax) + ax \sin (ax)]/a^2$

$\int x^m \sin (x) \, dx = -x^m \cos (x) + m \int x^{m-1} \cos (x) \, dx$

$\int x^m \cos (x) \, dx = x^m \sin (x) - m \int x^{m-1} \sin (x) \, dx$

$\int \sin (ax) \sin (bx) \, dx = \frac{\sin(a - b)x}{2(a - b)} - \frac{\sin(a + b)x}{2(a + b)}, \ a^2 \neq b^2$

$\int \sin (ax) \cos (bx) \, dx = -\left[\frac{\cos(a - b)x}{2(a - b)} + \frac{\cos(a + b)x}{2(a + b)}\right], \ a^2 \neq b^2$

$\int \cos (ax) \cos (bx) \, dx = \frac{\sin(a - b)x}{2(a - b)} + \frac{\sin (a + b)x}{2(a + b)}, \ a^2 \neq b^2$

$\int e^{ax} \, dx = e^{ax}/a$

$\int x^m e^{ax} \, dx = \frac{x^m e^{ax}}{a} - \frac{m}{a} \int x^{m-1} e^{ax} \, dx$

$$\int e^{ax} \sin (bx) \, dx = \frac{e^{ax}}{a^2 + b^2} [a \sin (bx) - b \cos (bx)]$$

$$\int e^{ax} \cos (bx) \, dx = \frac{e^{ax}}{a^2 + b^2} [a \cos (bx) + b \sin (bx)]$$

Definite Integrals

$$\int_0^\infty \frac{a \, dx}{a^2 + x^2} = \pi/2, \, a > 0$$

$$\int_0^{\pi/2} \sin^n (x) \, dx = \int_0^{\pi/2} \cos^n (x) \, dx = \begin{cases} \dfrac{1 \cdot 3 \cdot 5 \cdots (n - 1)}{2 \cdot 4 \cdot 6 \cdots (n)} \dfrac{\pi}{2}, \, n \text{ even}, \, n \text{ an integer} \\[4mm] \dfrac{2 \cdot 4 \cdot 6 \cdots (n - 1)}{1 \cdot 3 \cdot 5 \cdots (n)}, \, n \text{ odd} \end{cases}$$

$$\int_0^\pi \sin^2 (nx) \, dx = \int_0^\pi \cos^2 (mx) \, dx = \pi/2, \, n \text{ an integer}$$

$$\int_0^\pi \sin (mx) \sin (nx) \, dx = \int_0^\pi \cos (mx) \cos (nx) \, dx = 0, \, m \neq n, \, m \text{ and } n \text{ integer}$$

$$\int_0^\pi \sin (mx) \cos (nx) \, dx = \begin{cases} 2m/(m^2 - n^2), \, m + n \text{ odd} \\[2mm] 0, \, m + n \text{ even} \end{cases}$$

$$\int_0^\infty \frac{\sin(ax)}{x} \, dx = \frac{\pi}{2}, \, a > 0$$

$$\int_0^\infty \frac{\sin^2 x}{x^2} \, dx = \frac{\pi}{2}$$

$$\int_0^\infty e^{-a^2 x^2} \, dx = \sqrt{\pi}/2a, \, a > 0$$

$$\int_0^\infty x^n e^{-ax} \, dx = n!/a^{n+1}, \, n \text{ an integer and } a > 0$$

$$\int_0^\infty x^{2n} e^{-ax^2} \, dx = \frac{1 \cdot 3 \cdot 5 \cdots (2n - 1)}{2^{n+1} a^n} \sqrt{\frac{\pi}{a}}$$

$$\int_0^\infty e^{-ax} \cos (bx) \, dx = \frac{a}{a^2 + b^2}, \, a > 0$$

$$\int_0^\infty e^{-ax} \sin (bx) \, dx = \frac{b}{a^2 + b^2}, \, a > 0$$

$$\int_0^\infty e^{-a^2 x^2} \cos (bx) \, dx = \frac{\sqrt{\pi}}{2a} e^{-b^2/4a^2}$$

Answers to Selected Problems

1-6.   (a) $1/50$ s;   (c) $1/50$ s;   (d) 0.1 s

1-8.   (a) $\mathrm{Re}\{4 \exp [j(100\pi t - \pi/2)]\}$
    (d) $\mathrm{Re}\{3 \exp [j(100\pi t - \pi/2)] + \exp [j120 \pi t_1$

1-9.   (a) $2 \exp [j(100\pi t - \pi/2)] + 2 \exp [-j(100\pi t - \pi/2)]$
    (c) $\frac{1}{2} \exp [j(100\pi t - \pi/2)] + \frac{1}{2} \exp [-j(100\pi t - \pi/2)]$
        $+ \exp [j(200\pi t - \pi/2)] + \exp [-j(100\pi t - \pi/2)]$

1-10.   (a) amplitude: line of height 4 at $f = 50$ Hz
        phase: line of height $- \pi/2$ at $f = 50$ Hz

1-11.   (a) amplitude: lines of height 2 at $f = \pm 50$ Hz
        phase: line of height $-\pi/2$ at $f = 50$ Hz
            line of height $\pi/2$ at $f = -50$ Hz

1-17.   (a) $u(t) + r(t-1) - 2 r(t-2) + r(t-3) - u(t-4)$
    (c) $r(t) - r(t-1) - r(t-3) + r(t-4)$

1-19.   (a) $4 e^{-2}$;   (c) $5 - 10\pi$

1-20.   (a) $C_1 = 5, C_2 = 4, C_3 = 7$

1-22.   (a) $20/3$ J;   (c) $8/3$ J

1-24.   (a) 8 w;   (c) 2.5 w

1-26.   (a) power;   (c) energy;   (e) power

1-29. $2u(t) - u(t-2) + u(t-4) - r(t-6) + r(t-8)$

1-31. (a) $A[2u(t) - 2u(t-T) + u(t-2T) - u(t-3T)]$
(c) $r(t-1) + \frac{1}{2} u(t-1.5) - 2r(t-2) - \frac{1}{2}u(t-2.5) + r(t-3)$

1-34. (a) one; (c) zero; (d) two

2-1. (a) $y(t) = 0, t \le -1; y(t) = [1 - \exp(-10(t+1))]/5, -1 < t < 1;$
$y(t) = [\exp(-10(t-1)) - \exp(-10(t+1))]/5, t > 1$
(c) $y(t) = 0, t \le 2; y(t) = [1 - \exp(-10(t-2))]/5, t > 2$

2-5. $h(t) = (R/L) \exp(-Rt/L) u(t)$

2-7. $h(t) = (R/L) \exp(-Rt/L) u(t) - (R/L)^2 t \exp(-Rt/L) u(t)$

2-10. $y(t) = 0, t < 0; y(t) = [1 - \exp(-Rt/L)], 0 \le t \le 1;$
$y(t) = \exp(-R(t-1)/L) - \exp(-Rt/L), t > 1$

2-12. (a) $a(t) = [1 - \exp(-Rt/L)] u(t)$

2-14. (a) $a_s(t) = \exp(-Rt/L) u(t)$
(b) $a_r(t) = (L/R)[1 - \exp(-Rt/L)] u(t)$

3-4. $a_0 = 0; a_{2m+1} = 4(-1)^m A/(2m+1)\pi, m = 0, 1, 2, \ldots$
$b_n = 0$, all n

3-7. $Y_n = jn\omega_0 X_n$

3-9. (a) $P[|nf_0| \le \tau^{-1}] = 0.9053$ of total

3-11. Double sided amplitude spectrum: lines of $2A/3\pi$ at $\pm 2f_0$; lines of $2A/15\pi$ at $\pm 4f_0$; lines of $2A/35\pi$ at $\pm 6f_0$.
Double sided phase spectrum: lines of $-\pi$ rad. at $2f_0, 4f_0, 6f_0, \ldots$; lines of π at $-2f_0, -4f_0, -6f_0, \ldots$

3-16. (a) Amplitude distortion; (c) no distortion.

3-17. (a) $|X_a(f)| = A/\sqrt{\alpha^2 + (2\pi f)^2}$
$\underline{/X_a(f)} - \tan^{-1}(2\pi f/\alpha)$
(c) $|X_c(f)| = 2A\alpha/[\alpha^2 + (2\pi f)^2]$; phase $\equiv 0$

3-23. $X(f) = \frac{1}{4} A\tau \operatorname{sinc}(f\tau - 1) + \frac{1}{4} A\tau \operatorname{sinc}(f\tau + 1) + \frac{1}{2} A\tau \operatorname{sinc}(f\tau)$

3-25. (a) $E(|f| \le \alpha/2\pi) = A^2\pi/2\alpha$

3-28. (a) $X_a(f) = 2 \operatorname{sinc}(f) \cos(3\pi f)$
$X_b(f) = -2j \operatorname{sinc}(f) \sin(3\pi f)$

3-30. $x(t) = 4 \operatorname{sinc}(4t) + 2 \operatorname{sinc}(2t)$

3-33. (a) $X(f) = \exp(-j2\pi f)/[1 + j2\pi(f-1)]$

3-35. (b) $x(t) = 45 \operatorname{sinc}^2(3t)$

3-40. (b) Phase or delay distortion only
(d) Both amplitude and phase distortion

3-43.   (a) $X_a(f) = \frac{1}{2} \sum\limits_{n=-\infty}^{\infty} \text{sinc}^2(n/2)\delta(f - n/6)$

(c) $X_c(f) = (3/8) \sum\limits_{n=-\infty}^{\infty} \text{sinc}^2(3n/8)\delta(f - n/8)$

4-1.   (a) $X_a(s) = 2/s(s+2)$
(c) $X_c(s) = [1 - \exp(-10\ s)]/s$

4-3. $X(s) = [1 - \exp(-s)]/s^2$

4-5. $i(t) = 2\ \exp(-t/50)u(t)$

4-7. $x_a(t) = [\exp(-4t)\cos(2t) + 3\exp(-4t)\sin(2t)]u(t)$
$x_c(t) = [\exp(-3t)\cos(3t) - \exp(-3t)\sin(3t)]u(t)$

4-10.   (a) $x(0+) = 1; x(\infty) = 0$
(c) $x(0+) = $ doesn't exist; $x(\infty) = 0$

4-12.   (a) $y(t) = [(20/3)\exp(-4t) - (5/3)\exp(-t)]u(t)$
(c) $y(t) = 4\ \exp(-\sqrt{2}\ t)\cos(\sqrt{2}\ t + \pi/4)\ u(t)$

4-13.   (b) $x(t) = [\exp(-t) + \exp(-3t) + \frac{1}{2}\exp(-t)\sin(2t)]u(t)$

4-15. $x(t) = [-(27/2)t^2\exp(-3t) + \exp(-t)]u(t)$

4-16.   (b) $x(t) = \{(1/4)\exp(-t) - (1/4)\exp(-t)[\cos(2t) + \frac{1}{2}\sin(2t) - t\cos(2t)]\}u(t)$

4-21.   (a) $X_a(s) = (4A/Ts^2)[1 - \exp(-sT/4)]^2$

5-5. $v_0(t) = (5/6)[\exp(-t/3) - \exp(-4t/3)]u(t)$

5-6. $i(t) = \dfrac{1/R}{1 + (\omega L/R)^2}\ [\exp(-Rt/L) + \cos(\omega t) + (\omega L/R)\sin(\omega t)]u(t)$

5-10.   (a) $|H(j\omega)| = \sqrt{\omega^2 + 100}\ /\ \sqrt{(\omega - 19.6)^2 + 16}\ \sqrt{(\omega + 19.6)^2 + 16}$
(c) $|H(j\omega)| = |\omega|/\sqrt{(\omega - 3)^2 + 9}\ \sqrt{(\omega + 3)^2 + 9}$

5-12.   No. of RHP roots: (a) none;   (c) 1

5-13.   (b) $0 < K < 3/4$

5-23. $\dfrac{Y}{R} = \dfrac{(G_1 + G_4)G_2G_3}{1 + G_1G_2H_1 + G_1G_2G_3H_2}$

6-4. $e^{At} = \begin{bmatrix} e^{-t} & 0 \\ 0 & e^{-4t} \end{bmatrix}$

6-7. $\mathbf{x}(t) = \begin{bmatrix} 1 + e^{-t} \\ \dfrac{3}{4} + \dfrac{1}{4}e^{-4t} \end{bmatrix},\ t \geq 0.$

6-12. $y(t) = (1/2 + e^{-t} - 1/2\, e^{-4t})\, u(t)$

6-17. **(b)** $A = \begin{bmatrix} 0 & 1 \\ -25 & -14 \end{bmatrix}, \quad b = \begin{bmatrix} 0 \\ 25 \end{bmatrix}$

7-6. 63.7 dB

7-12. **(a)** $x(0) = 1$, $x(T) = 1$, $x(7T) = 0.5$, $x(nT) = 0$ otherwise

    (b) $x(nT) = 5$ for $n = 0,1,3,4$, $x(nT) = 0$ otherwise

        $x(nT) = n = 0, 3;\quad x(nT) = -5, n = 1, 4;\quad x(nT) = 0$, otherwise

    (c) $x(0) = 6$, $x(T) = \dfrac{-9}{2}$, $x(2T) = \dfrac{9}{8}$, $x(3T) = \dfrac{-3}{32}$, $x(nT) = 0$ otherwise

7-14. **(a)** $x(0) = 5$, $x(T) = -5.3$, $x(2T) = -7.4$, $x(3T) = 10.7$

    (b) $x(0) = 2$, $x(T) = 3.6$, $x(2T) = -1.72$, $x(3T) = 3.344$

7-16. **(a)** $x(nT) = 2 - \left(\dfrac{1}{2}\right)^n$, $n \geq 0$

    (b) $x(nT) = \dfrac{2}{3} + \dfrac{1}{3}\left(-\dfrac{1}{2}\right)^n$, $n \geq 0$

    (c) $x(nT) = 3(0.3)^n - 2(0.2)^n$, $n \geq 0$

    (d) $x(nT) = -\dfrac{25}{21}(-1)^n + \dfrac{6}{7}(-0.3)^n + \dfrac{1}{3}(0.2)^n$, $n \geq 0$

7-19. **(a)** $x(nT) = 1.1547 \cos 60°n$, $n \geq 0$

    (b) $x(nT) = \sin 60°n + 1.7321 \sin 60°n$, $n \geq 0$

    (c) $x(nT) = \cos 90°n$, $n \geq 0$

7-21. **(a)** $H(z) = \dfrac{1}{1 - z^{-1}}$

    $h(nT) = 1$, $n \geq 0$

    (b) $H(z) = \dfrac{1}{(1 - z^{-1})^2}$

    $h(nT) = n + 1$, $n \geq 0$

    · (e) $H(z) = \dfrac{1}{1 - 0.707z^{-1} + 0.25z^{-2}}$

    $h(nT) = \delta(n) + 0.707\left(\dfrac{1}{2}\right)^{n-1} \cos\,[45°(n-1)]u(n-1)$

8-11. $H(z) = 30\,\dfrac{1 + z^{-1}}{1 - 0.5z^{-1} - 0.5z^{-2}}$

8-21. **(a)** $H(z) = \dfrac{5}{4}\,T\left[\dfrac{1}{1 - e^{-Tz^{-1}}} - \dfrac{1}{1 - e^{-5Tz^{-1}}}\right]$

(b) $H(z) = T\left[\dfrac{1}{5}\dfrac{1}{1-z^{-1}} - \dfrac{1}{4}\dfrac{1}{1-e^{-T}z^{-1}} + \dfrac{1}{20}\dfrac{1}{1-e^{-5T}z^{-1}}\right]$

(c) $H(z) = T\left[\dfrac{25}{18}\dfrac{1}{1-e^{-T}z^{-1}} - \dfrac{5}{2}\dfrac{1}{1-e^{-5T}z^{-1}} + \dfrac{10}{9}\dfrac{1}{1-e^{-10T}z^{-1}}\right]$

8-24. $H(z) = \dfrac{3.27492z^{-1}}{1-1.637462z^{-1}+0.67032z^{-2}}$

8-28. $H(z) = \dfrac{10+20z^{-1}+10z^{-2}}{430.22883-578.31190z^{-1}+188.08307z^{-2}}$

9-8. $X(0) = 4, X(1) = X(2) = X(3) = 0$

9-10. $X(0) = 7; X(1) = X(3) = X(5) = X(7) = 1$
$X(2) = X(4) = X(6) = -1$

9-15. 8 real multiplies and 8 real adds.

9-22. Choose $N = 256$. Gives actual resolution = 78.1 Hz

B-5. For $x_{dB} = 0.2$, $\epsilon = 0.21709$
For $x_{dB} = 1.0$, $\epsilon = 0.50885$
For $x_{dB} = 6.0$, $\epsilon = 1.72658$

B-8. $H(s) = \dfrac{3.94784(10^7)}{s^2+8.88577(10^3)s+3.94784(10^7)}$

B-10. $H(s) = \dfrac{2.52662(10^5)s^2}{s^4+7.10861(10^2)s^3+1.04223(10^6)s^2+2.80637(10^8)s+1.55855(10^{11})}$

$f_u = 147.7$ Hz
$f_l = 67.7$ Hz

Index